VOLUME ONE

A New Form of Energy:

Its Description
Its Generation
&
Its Applications

Stanley North,
Colin A. Slate,
Peter M. South

Cataloguing in Publication on file with Library and Archives Canada

ISBN 978-1-926582-21-4

DEDICATION

This book is dedicated to all those brave, noble, and intelligent researchers throughout all of man's history who have so diligently sought for the secrets of the universe and the solution to man's sorry and degraded state. This includes one of the greatest geniuses of all time, - *Nicola Tesla,* the nearly forgotten and abused genius who did much to bring us into the modern technological era.

Three cheers should also be given to *Wilhelm Reich,* an American who did much to discover the scientific principles and characteristics of the so-called 'Life - Force', - despite great adversity. His government returned the favour by imprisoning him, burning his books, and causing his death by heart failure.

Mention should also be made of *Baron Von Reichenbach* who in the nineteenth century also contributed much to our esoteric knowledge.

Mr. John Hutchison of British Columbia, Canada also deserves a rousing "Bravo". He has done many years of research in a new field of endeavour which can only be described as 'quantum-electronics'. John, who has been featured on television shows - may well be called a modern-day Nicola Tesla for his work and bizarre results, - including many instances of levitation. His contribution to new knowledge was also rewarded by the Gestapo-like confiscation of all his electronic equipment (several million dollars worth) - without compensation.

The book is written for the *elite*. This elite is not the group we think of as the super-rich or the intelligencia, the politicos, the highly educated, or those professionals in the arts and sciences. The *elite* I am referring to are those who occupy the upper echelons of the spirit, mind, and thought.

O

IMPORTANT NOTE: It is of paramount importance that the Reader begin this book at the first page and work his way incrementally forward until finally he gets to the last page. The Reader should understand the material as it is presented *sequentially*. If he does not adhere to this rule, he will not understand the material, the reasoning and logic, or how all the parts are inter-related to form a coherent whole. He will suffer the loss of new viewpoints, gained perception, insight, and the understanding of the universe (and indeed his own mind) and the way it works.

Some articles - if taken out of the context of the entirety, may seem irrelevant. They are not irrelevant however, and the Reader will come to understand this upon his completion of the book. The book contains a great deal of information and it is highly condensed, therefore the Reader will probably enjoy reading the book several times, each time gaining new, additional knowledge, insight, and understanding.

There is an Index at the back, but in addition, it is highly recommended that he keep notes all the while since this will make it much easier to refer to previous material which is of interest to him. If he uses the downloaded version of the book, so much the better.

O

DISCLAIMER: It may be imprudent of me to omit a *Disclaimer*. I am not a physician and any statements made regarding medical conditions must be understood in the context of this book. It is not intended to provide a cure or suggested cure or treatment for any physical, psychosomatic, or psychological ailment or sickness. If you have an ailment requiring treatment, go to a qualified practitioner or physician. Any statements or opinions made throughout the contents of the text are those of the author.

After reading the book in its entirety however, I believe the Reader will be able to intelligently conclude for himself the tremendous potential of the technology described herein. Likewise, any suggested experiments using electricity should be done with the supervision of an expert.

> "there is hope for the future - and when the world is ready for a new and better life, all these things will some day come to pass - in God's good time" (*Jules Verne*: 'Twenty Thousand Leagues Under The Sea').

> 'There is one thing stronger than all the armies in the world, and that is an idea whose time has come'… Victor Hugo.

Δ

My book is the result of nearly thirty years of research and study. I believe it represents a genuine breakthrough in the sciences and indeed, proposes a completely new physics and technology. These researches have been in the various fields of physics (including particle dynamics, astrophysics, quantum mechanics, electronics, nuclear fusion, and so on). These fields also include electrodynamics, *Einsteinian* physics, Relativity, energy and space-time peculiarities, Black Holes, - as well as more esoteric studies in 'para-physics', such as with Poltergeist and Ball-lightning phenomena.

Herein, the concept of 'Relativity' is shown to be *flawed*, as well as are the current theory and efforts of Nuclear *Fusion*; - (which is why it doesn't work!). Inertia and gravity are also key subjects herein.

The book describes a *truly* Unified Field philosophy which embraces *all* physical phenomena as any such professed Theory should.

Inevitably - as physicists will tell you - when one researches *quantum physics* and related phenomena, one is confronted with philosophical questions such as; "what are the true natures of reality, illusion, subjectivity, objectivity, perception, viewpoint, the physical universe", and so on. Hence, this book deals not only with a completely new type of physics and technology, but also with a relevant philosophy and, inevitably, even includes pertinent *religious* fundamentals. In other words, this book brings together and explains the intimate relationships between physical technology and the cosmic, mind, philosophy, and religion.

(The inclusion of religion is found to be inevitable. For example when we look up into the clear night sky, we see the fundamental expression of the physical universe, and everyone - almost without exception, will wonder about its beauty and its origins. 'Where does the atom originate, why is it such a marvel of logic and construction. Why should not forces yet unrecognized still persist? Does life exist on other planets? These are the types of questions which religion is based upon; - indeed this *is* religion).

Consequently, it is a study into the deepest and the most fundamental realities of matter, energy, space, and time, - their origins and interrelationships. It will be shown herein that *new ways* of manipulating matter, energy, space, and time do exist, - as well as new and practical principles, - based on the technology herein of 'Quantum-Electronics'.

Also revealed are the true natures of gravity and inertia, and how they may be *controlled* through this technology. It also reveals a very practical and unique method of creating a controlled, sustained, and contained *Nuclear Fusion* process, - the energy which powers the Sun and stars.

The great enigmas still remaining in philosophy and the physical sciences are the questions of gravity and inertia, nuclear fusion, time, space, and the existence of natural phenomena whose presences are unexplained and annoying to physicists, - such as Poltergeist activity, ball-lightning, Doppelgangers, apparitions, 'worm-holes', 'time-tunnels', and so on. It is of more than passing interest to realize that these paradoxical mysteries are in fact *all related* - and furthermore, are related through a new form of energy! This Energy is described with a simple equation as are gravity and inertia.

It will be seen herein that indeed, *truth really is stranger than fiction!*

> Electricity - created in the nineteenth century - has transformed the world completely, both laterally and vertically in a social sense. Nothing has remained untouched by this very simple energy form. The same will again be true many times over for 'Quantum Energy' and quantum technology.

> "I can believe anything provided it is incredible"... Oscar Wilde.

> (Regarding science); "it will free Man from the remaining chains, the chains of gravity which still tie him to this planet. It will open to him the gates of heaven": (Werner Von Braun).

> "There are more things in heaven and earth, Horatio, than are dreamt of in your philosophy"... Shakespeare ('Hamlet').

Δ

THE EIGHT DYNAMICS OF EXISTENCE AND SURVIVAL

There are eight dynamics of physical existence and survival for all life-forms [please see note] and they may be stated as follows; - the *first* dynamic is survival through and for the self as an individual. The *second* is through sex and the family group and the *third* is through and for groups such as companies and organizations, etc.. The *fourth* is through Mankind as a global group and the *fifth* is through all life-forms.

The *sixth* is through the <u>physical universe</u>. The *seventh* is through the non-physical and immaterial agencies as they interact with the physical. The *eighth* is survival through and for the non-physical universe.

This book is an effort to explain and describe completely the <u>sixth</u> dynamic at the most fundamental levels. These eight dynamics may be considered to represent a spectrum such as the electromagnetic spectrum - so that each of these dynamics inevitably merge one with another.

Just as energy and all physical phenomena cannot exist or survive without the full and complete electromagnetic spectrum, so also an individual or group, etc., cannot survive without the full complement of these dynamics. Hence we find that any thorough examination of one dynamic must reach into and include its neighboring dynamics. Therefore a book such as this – devoted as it is to the sixth dynamic, must likewise also reach into the fifth and seventh dynamics. This is inevitable and may not be avoided. The emphasis however is on the Sixth. (These dynamics of survival have been discovered by *L. Ron. Hubbard*, the originator and founder of the modern mental sciences of Scientology and Dianetics).

The present work then - while ostensibly being a thorough examination of the physical universe, may also be considered to be a philosophical and religious work in the traditions of Buddhism and Hinduism. Indeed this book may be considered to be the knowledge sought for by Man through the ages - since it apparently succeeds in uniting the physical and the non-physical.

PREFACE - Part 2

> "when we finally discover the means for control, it will very likely… be something which we have encountered frequently and failed to recognize". (Charles Tilger, Jr., of Grumman Aircraft writing in a paper on gravity).

Here is a thought provoking question. If someone were to offer you a book which promised to reveal the 'secrets of the universe', what kind of information would you expect to find within? Suppose the concept of 'mind over matter' were not only a reality but were carried to its ultimate conclusion. In what ways would this fact manifest itself? What potential effects would this signify for you, for your mind, for your body, for your personal life, for your environment, and for society? What are the relationships between the mind, illusion, reality, objectivity, subjectivity, perception, Einsteinian physics, quantum mechanics, matter, energy, space, and time? You will find the answers to these questions within this book.

Δ

Have we discovered all the physical laws of nature?

This book is an instruction manual for a completely new physics and technology. This is not idle chatter; it is a promise to my Readers. This book has the potential to change your world, the entire world, and the history of Mankind. Read this book and you will never be the same.

It may be said by some that the new technology and material presented herein is too advanced for this age and too far ahead of its time. This may well be true, but consider the world situation as it is today…

Could it be said that the introduction of conducted and controlled electricity was ahead of its time, - or that the introduction of steam engines and mass transportation were ahead of their times? – Or that automobiles and computers were ahead of their times? These are all matters of viewpoint and subjective consideration.

Vacancies existed in society wherein these technologies were able to find a niche - finally burgeoning into wide markets, - and they may well be said to be ahead of their times. Furthermore, they were intrusive technologies onto a previously settled order, - made so by their very popularity; - some say for better and some for worse. New technology however, whether considered good or bad, will inevitably find a way to change our lives. This new technology will germinate either from obscure independent members of society or already established private organizations, quasi-government agencies, or perhaps even by the government itself. Unfortunately the only new technology which interests the government is the military kind.

It is the older generations who have settled down to an established and pre-dictable lifestyle who lack interest in the new and the revolutionary. It is the young and adventurous who are enthralled by new and dramatic knowledge, technology, and procedures. It is the young who change the world through ideology, whether this ideology is good or bad.

> "All is a mystery, but he is a slave who will not struggle to penetrate the dark veil"… Benjamin Disraeli.

Electronics has, by and large - proved itself to be primarily a technology of peace. In this context, we may truly say it is a 'passive' technology ('passive', in fact means 'peaceful').

Quantum technology is in a class parallel to electronics and I therefore believe and hope it will also be dedicated to peaceful purposes. Like any technology how-ever, it will be a two-edged sword. You will be the judge.

This technology is indeed far in advance of what we have today; - so then should we wait another fifty or a hundred - two hundred - three hundred years for it to ap-pear when it is available here at this moment.

It may be a truism that if the time is right and the people are ready for it, a new social or technological idea will flourish and blossom. The marvelous visionary and writer *Jules Verne* once wrote, "there is hope for the future - and when the world is ready for a new and better life, all these things will some day come to pass - in God's good time": ('Twenty Thousand Leagues Under the Sea').

If this book be said to have one central theme, that theme would be as an in-struction manual for the construction and operation of an aerial ship, powered by nuclear fusion, lifted by quantum Energy, moved through an altered space-time, and guided by an operator's will.

Δ

> "Any technology sufficiently advanced is indistinguishable from magic", (Arthur C. Clarke).

> "the existence of a specific life force seems highly probable to me, — "it is present certainly in the formative process of crystals". (Biologist Kammerer). (This was also a belief of Nicola Tesla and H.g. Wells).

> "The discovery of the energies associated with psychic events will be as im-portant, if not more important, than the discovery of atomic energy", (*Dr. Leonard L. Vasiliev*, psychologist and winner of the Lenin Prize).

> "At this point we are, then, it seems, faced with the need of another order of energy, not radiant", (*Prof. J.B. Rhine, Ph.D.*, Duke University, USA).

Δ

Forget everything you have been taught by your education system and your peers about physics and science - for today dawns a new technological age for Mankind. Much of present day knowledge will be rendered _obsolete_.

> Please Note: many articles in this book deal with a philosophy pertinent to this technology. Read these articles without omission for your mind is about to undergo a transformation, and must therefore be prepared with adequate groundwork.

< why I wrote this book >

In the first place, this book exists because I am interested in physics and borderline research, i.e. physics beyond the permitted reach of orthodox science. Such studies are often referred to as 'Para-physics' or the 'Para-normal'. (The prefix 'Para' simply means 'parallel to' so that para-physics means 'parallel to known physics').

I wrote this book partially because I truly believe that Mankind can and must be elevated above the spiritual squalor and degradation that he now experiences. This is not man's fault alone. Man is capable of much higher aspirations, as has been demonstrated in past civilizations, but he is being degraded by those who would control him and own him.

For the most part, those who actually fight wars are noble and honorable. Those who plot, plan, and create those wars however are not warriors, nor soldiers, nor even politicians; - for the most part they are the truly degraded and often super-wealthy who work stealthily and covertly behind the spotlight. These armchair instigators of war and degredation are not the ones who experience mind-bending fright, loss of loved ones, or get their arms and legs blown off. The technology revealed herein has the potential to raise Man to the noble position which is rightly his. [Please see note].

Another primary reason for this book is to open the Reader's mind and to introduce him to possibilities which he may never have before considered. In order to do this however the Reader must leave the door of his mind unlocked*!*

NOTE: The ancient Egyptian civilization for example, elevated all men to the level of spiritual Beings. In the Egyptian's perception, the physical world was merely a dull reflection of a higher and more ethereal existence. It may be said that Man reached his highest pinnacle of civilization here.

Δ

> "Most of the things worth doing in the world had been declared impossible before they were done": (Louis D. Brandeis).

It seems appropriate at the outset of our adventure of discovery to imagine if you will that you are a newly born entity or viewpoint. Now in this context, "newly born" does not mean that you have a body, biological or otherwise. You are simply an awareness or a point of view which exists in a Void of nothingness. You know nothing, for indeed you have as yet created nothing to know. There is no energy and no motion, and so there can be no time, - and hence no space nor matter either.

The subject of quantum physics is not a very popular one in the general society. There are several reasons for this. First of all, it is a relatively new area of science and is not yet adequately understood by the physicist in the field, so-to-speak. The Reader likely feels that if the scientists do not understand it, then he won't either. This however, is far from the truth.

Secondly, the subject of mainstream physics generally involves much esoteric mathematics and long, dry explanations which of course have little interest for the general reading public. Thirdly, it involves the study of particles and their energies, which again do not present the drama which appeals to the average Reader, since they occupy a field beyond his experience, visual impact, and reality.

The Reader is more interested in the visual results that he can relate to on some level. For example, the data surrounding the science of atomic fission is of little interest for the above given reasons, - but the results, i.e. the explosions of atomic bombs, have the appeal of visual drama. I have therefore endeavoured to write a book which is at once informative, entertaining, and hopefully - "mind-blowing".

> "Nothing is impossible. There are ways that lead to everything, and if we had sufficient will we should always have sufficient means": (Francois De La Rochefoucauld).

Quantum physics can be an extremely fascinating subject if one excludes the more mundane elements such as mathematics and the lengthy dry, monotonous intonements. He is more interested in the popular applications of science. Other examples of dramatic effects are those associated with Poltergeist phenomena or Ball-lightning, etc.. To see an object levitating off the table or passing through a wall unscathed is truly a breathtaking adventure.

Quantum physics is the ultimate physical science, and as such deals with the very fundamental beginnings and behaviours of Matter, Energy, Space, and Time: [please see note 1]. As we shall see, the mind itself is not entirely excluded from its effects. There is in fact a mutual interaction between the mind and quantum energies - and indeed, as we shall see - it is quantum Energies which form the bridge between the mind and the physical. Therefore, if we are to hope for the solutions to enigmas such as Poltergeist phenomena and levitation, etc., then we should look inevitably to the answers in quantum physics. [See note 2 below].

The biological body is of course composed ultimately of atoms, electrons, protons, and molecules, etc.. Since quantum physics is the science which deals primarily and exclusively with these, should we therefore not expect to find solutions to ailments in quantum technology, - especially ones which even today cannot be explained - such as cancer?

Conditions such as tuberculosis, AIDS, blindness, deafness, arthritis, withering, the common cold, and so and on, - should find their ultimate causes to be connected with the atomic and molecular structures of biological tissues, their quantum energies, and the link with the mind.

If there are to be any ultimate solutions applied to fundamental problems at all, they must therefore lie within the realm of quantum mechanics. We can assume this for indeed, the body itself is a source of quantum energies such as radio-waves, microwaves, heat radiations, infra-red light, - and even visible and ultra-violet light photons: [note 3].

Likewise one can continue along the spectrum through ultra-violet light and even X-rays and electrons. These, as may be expected - are emitted in very low intensity but it does show that even biological organisms are composed of "quantum-active" matter. It will be seen throughout the book that verily, the body is associated with quantum energies and will therefore respond to them on a quantum level: [note 3].

NOTE 1: These four 'elements' which compose the physical universe do not have the apparent stability one normally associates with them. They are all 'flexible' and subject to alteration, manipulation, re-structuring, and de-structuring. Our so-called Constants (K), such as the velocity of light, the flow rate of time, specific gravity, and so on, are likewise unstable and fickle, - like the direction of the wind or density of the air.

NOTE 2: such subjects, although well documented and recorded - are simply ignored and 'forgotten' by those who call themselves dedicated scientists. Yet true science is the investigation of all natural phenomena, however bizarre they may seem to be - and however much they may seem to flaunt the thus-far discovered laws of physics.

NOTE 3: Light photons generated and emitted by the bio-organism cannot normally be seen by the unaided eye simply because they are emitted with such low intensity. However there are cases on record where sick people have become not only electrically charged but have also actually *emitted light* (a glow) visible to other observers. Furthermore, there are recorded cases of individuals such as *Sir Wm. Crookes* who was able to emit visible light at will. These are of course explored and explained in the text.

PREFACE – part 3

Now as particle physicists widely acknowledge, the quantum behavior of particles is inevitably influenced by particle dynamics and the presence of the observer - and by inference, the observer's mind: [see note 3].

Since particles are ultimately composed of organized energy, i.e. "Quantum Energy", and are in constant relative motion, - we have then three factors composing the quantum technology trilogy. They are Particles, Quantum Energy, and Mind. (Matter of course is composed of particle groups and associated electronic fields). When particles are discussed it is almost always in a dynamic context. This is the way the universe works.

Throughout the ages, Man has sought long and in vain for the knowledge which would bring together the principles embodied in spirit, religion, mind, and science. Hence, this new technology bridges the final gap between the physical world of matter, energy, space, and time, - and that of the mind. This means that a grand new opera of adventure and experience is now Mankind's if he will accept this promise of a new world.

Δ

One subject in particular which is of major importance is that of Nuclear Fusion. Now nuclear fusion as it is presently conceived of has not, cannot, and *will not* succeed as a viable energy source since it is based on *false assumptions and misconceptions*: (explained in section 07). The NF process - when properly engineered, will convert hydrogen into Helium with the *direct release of electrical energy*.

Hence it is plain to see that this new system - described herein - by-passes the usual configurations of heat exchangers, steam turbines, and mechanical generators. Obviously therefore the costs of equipment maintenance are virtually eliminated - not to mentioned the tremendously gained efficiency.

The importance of this book simply cannot be overstressed or exaggerated, - it has the power to transform the entire world into something Man has only dreamed of throughout all of history. The Reader who uses this practical manual has it in his power to do this. Hence in a very real sense this book may be thought of as a "Pandora's Box"…

(However it should not allude directly to the Pandora of Greek mythology whose Ark, when opened - unleashed chaos and misery. The results of the knowledge contained within these pages will only bring true order, reason, beauty, and happiness): [see note 2].

Either the technology presented in these pages is a valid technology or it is not, the Reader will decide. Either it is the ultimate physical technology or it is not. If for some reason the Reader decides that it is not, then he must put aside this book. If however he is a diligent seeker of truth and the sincere promise of a new world, he will find himself obliged to commit himself to the knowledge contained herein.

Δ

Charles Tilger Jr. of the Grumman Aircraft Engineering Corporation once wrote a paper entitled," Why I think the Force of Gravity Is Controllable". Reporting to the 'Gravity Day Meeting' he said, "*when we finally discover the means for control, it will very likely… be something which we have encountered frequently and failed to recognize*". (my Italics).

Some well known (American) companies who are or have been engaged in anti-gravity research are, - Glenn L Martin, General Electric, Hughes Aircraft, Lear Inc., Gluharoff Helicopter Airplane Corp, Clarke Electronics, Sperry Rand, NASA, and Boeing Aircraft.

One researcher into this field is *Leonard G. Cramp, M.S.I.A.*, who is or was a member of the British Interplanetary Society. In his book 'Space, Gravity, and the Flying Saucers' he wrote, "… it may well be, even in our time *that men will discover the key with which to unlock a force that is all about us and so enable us to use it for all our industries, transport, and finally, space travel*". (Italics mine).

In Interavia, published by the Swiss Review Of World Aviation, it is reported that *some materials have been reduced in weight by 'energizing' them.* (Italics mine).

NOTE: Quantum Energy is essential to the fundamental structure and function of the physical universe. The physical universe is comprised of surprisingly simple 'elements' regardless of its apparent complexity. Any complex structure is composed essentially of very simple components and so it is with the physical universe.

In its simplest and most elementary fundamentals, the universe is composed of two essential ingredients. They are the Electric field and Motion. These two are the sole criteria upon which all else is constructed and predicated. The primary result of these two is Quantum Energy - and the sources, nature, and applications of quantum Energy are the topics which will be discussed at great length herein.

NOTE 2: I like to remind myself of *Prometheus* of Greek mythology: "The Bringer of Fire and Knowledge to Mankind".

NOTE 3: in this book, whenever the word mind occurs, it implies the mind/will , otherwise called the "Being" or 'non-physical agency'.

PREFACE - part 4

My original purpose as a writer and as a physicist was to create a book dealing exclusively with Nuclear Fusion, but as I researched and wrote, it became clear to me that NF was a subject which inevitably reached beyond mere atomic energy. The realization gradually dawned on me that NF and its inherent function had far reaching implications – especially with its intimate association with quantum Energy and electro-dynamics.

I further realized that I was dealing with the very structure and function of stars and fundamental particles, and indeed - even creation itself. This was a great revelation to me and as I proceeded, I was to have other such important revelations.

NF is intimately associated with both quantum Energy and electro-dynamics. Further, q. Energy is in fact the very same Energy involved in Poltergeist phenomena. Since levitation is a phenomenon also associated with Poltergeist activity it thus became evident and apparent that I was researching an energy form related also to gravity and hence inertia. Similarly, since Poltergeist activity also included the bizarre phenomenon of objects passing through walls, etc., I was obliged to consider other similar quantum phenomena both known and unknown to physicists.

It was perceived also that Poltergeist phenomena are likewise associated with dynamic anomalies of matter, energy, space, and time. I was therefore obliged to consider electro-dynamics and q. Energy as they relate to these other constituents of the physical universe.

Finally this book should "prove" to both layman and physicist alike that a comprehensive and well rounded knowledge of physics (including electronics) allied with philosophical reasoning, religious circumspection, and intuitive insight will certainly uncover the 'Secrets of the Universe'.

Δ

DIAGRAMS AND LEGENDS

The following is a list of the diagrams and their descriptions. Please notice that the numbers are not all in direct sequence although they are sequential. This is due to the way in which the book was compiled and will not inconvenience the Reader.

01: the electro-magnetic spectrum showing the known gap between X-rays and the electron at rest.

03: illustrates the magnetic field surrounding a bar magnet. It is important to notice that this type of field is toroidal (i.e. like a torus) in structure and forms a "torosphere" with the bar at its center. (please see Glossary).

06: shows the charge accumulating in an electric storm cloud and the dynamic phenomena associated with the discharge.

08: Shows an atom with its electron shells (S) and nucleus (N) impinged upon by a wave-train pulse of quantum Energy. (V) Shows the vector of the wave-train.

10: shows the internal behavior of a conductor under the influence of an electron flow.

11: describes the dynamic phenomenon of 'population inversion' in a conductor.

16: a section view of the hydrogen atom showing its energy band levels.

17: illustrates a section view of the iron atom and the reconsidered dynamic structure and function of all atoms generally together with their electron flow patterns. It also shows the flange generated by its magneto-electro-dynamic function.

25: Shows an electrical discharge between two terminals (T) generating an ora-spherical Em pulse of wave energy (W). (C) Shows the geometric center of the energy radiation.

26: Shows the annular/toroid magnetic field surrounding a moving electron or other electrically charged particle in motion.

29: a section view of a conductor showing the dynamic behavior and flow patterns of electric particles as they flow within it.

30: illustrates the toroid flow structure of a "universal particle".

31: the simple helical electromagnetic coil, its electron flow vectors, and surrounding generated magnetic field.

32: Shows an annular ring (R) surrounded by a wire (W) having helical windings. An electric current passes through the wire via the terminals (T). This generates a toroid magnetic field corresponding to the geometry of the ring. This is the essential structure of a transformer coil and core.

35 A: Shows the geometry of MED (Magneto-Electro-Dynamic) lines of force surrounding the planet Jupiter, (Saturn), the sun, and other stars. (F) Represents the MED condensed electronic flow surrounding the planet.

35 B: a section view of the planet Jupiter showing detail of the structure of the MED fields and the gaseous and semi-plasma envelop surrounding the central star engine (S).

42: the "overhead" view of a rotating vortex of particles in a fluid medium such as liquids, gases, plasmas, or electronic fields, showing the particulate and electronic flow patterns.

44: illustrates a toroid sphere or "torosphere", i.e. the "universal particle". The vectors illustrate the flow patterns of energy and shows the fundamental dynamic structure of discreet self-enclosed energy systems upon which the physical universe apparently operates.

45 B: the initial torus formed by an emitted particle of energy before it "collapses" and adopts the orispherical (spheroid) form [diag. 44], as in the case of a photon of violet light.

47: Shows a hollow metallic sphere (S) conducting an electronic flow between the terminals (T). (F) Shows the flow under the influence of a self-generated magnetic field. (E) shows the equatorial flow path of a high density current flow which in effect behaves like a circular energized coil.

48: describes the internal and outer structure and function of an EGV.

51: illustrates the helical electrodynamic flow patterns and fields within a high energy wire conductor.

Shows an electrical conductor (C) supporting a current flow (E). This flow generates a magnetic field (M) which in turn interacts with the flow thereby causing it to describe a helical path.

53: the upper and lower energy levels within the electron shell of an atom, and the formation of an emitted photon by an electron moving from one energy band to another.

Shows the generation of a photon (P) by an electron jumping from an energized electron shell (S) to its ground state (G).

54: the behavior of electric particles within a magnetic field under the influence of an electro-motive force perpendicular to the field. Shows how an electrically charged particle behaves in a magnetic field. Flowing from A to B it follows a helical path. The vector 1: Indicates the direction of the particle's e.m.f. 2: Indicates the direction of the magnetic field lines of force and vector 3: Shows the resulting direction of these two forces.

55: the section view of a functioning nuclear fusion plasma spheroid. Shows a torotating spheroid of plasma with its dynamic particle flow and resulting helical vectors of the 'shell' and the central column engine.

56: illustrates a randomized plasma cloud and its subsequent organization under the influence of electro-dynamic forces. Shows the difference between an unstructured randomly dispersed gas or plasma and dynamically structured plasma.

57: section view of a toroid plasma structure showing its structure, flow patterns, and associated magnetic fields.

58: Shows a section view of a proposed EGV with its central fusion plasma engine (C). Dashed lines represent vectors of electronic flow (E) within the cylinder walls and hull (H). (F) the flange surrounding the hull.

60 B & C: the structure and function of lightning and the formation of ball-lightning. Shows the activity within a lightning strike and the formation of conditions which give birth to a lightning spheroid plasma. The central core (C) of the electronic flow is surrounded by a plasma sheath or "sleeve" (S) and a simultaneously self-generated magnetic field (M) which surrounds and "squeezes" the plasma core due to magnetic-striction.

60 B: Shows the detail of a lightning strike. (M) represents the main core of the current flow path. (R) is the return flow path or plasma sheath induced by the main flow path. (A) shows the resulting induced annular vortex of plasma created by these two interacting flow paths. (B) shows the "pinched" flow path due to magneto-striction with (C) its attendant surrounding induced toroidal magnetic field. The annular magnetic field is strongest surrounding the "pinched" portion of the main flow path.

60 C: shows section views of the initial "bagel-like" structure of a ball-lightning before it assumes the conventional spheroid form.

The electronic core flow is surrounded by an opposite flow induced in the sheath. These two flows in juxtaposition interact to create an annular vortex (V) which is the embryonic ball lightning.

61: a hollow metal sphere and internally contained magnet. Shows a hollow metal sphere (S) having two terminals (T). Within the sphere is a bar-magnet with its surrounding field lines of force (F). This field permeates the non-magnetic metal (e.g. copper, aluminium, etc.) and influences the flow vectors of the current passing between the two terminals through the sphere's shell.

64: a section view of the galaxy, its MED fields, and the central generating Black Hole. Shows the galaxy, its central feature (C) (i.e. the Black Hole), and the surrounding MED field (F) generated by the Black Hole. (D) Shows the dynamic vortexial field and electronic flows as they emanate from this source.

66: the cylinder engine/generator of the EGV. Shows the cylindrical fusion engine and its optional fluted helix design. Note the right-hand helical geometry of the cylinder conforming to the right hand rule for electronic flow.

67 A & B: section views of the EGV. Shows two alternate section views of the EGV craft. The hull (H) of the craft incorporates a circular flange buss-bar (F) and the internal central cylinder plasma (P) engine (C).

70: illustrates the structure of the nuclear fusion cage or "chamber". Shows the plasma cage engine with its bars (B). The terminals (T) and the spark gaps (S) are functional elements in creating the fusion process at the cage center (C).

101: shows a section view of a hypothetical hollow planet and various gravitational vectors. These vectors demonstrate that anyone standing at a point (P) on the inner surface of the hollow cavity would experience an overall pull 'downwards' towards that surface. These vectors are valid regardless of the relative size of the internal cavity.

103: shows the graph form of the universal energy inflow/outflow and threshold behavior. The knowledge of this behavior is of fundamental importance to understanding the dynamic physical universe.

105: the transmutation of hydrogen to Helium in nuclear fusion.

106: the behavior of electrons and protons in a dynamic structured electronic plasma.

107: the pair bonding of electrons.

Δ

The following diagrams do not have legends and are therefore explained here.

08: the atom;   N = nucleus
            S = electron shell
            W = quantum Energy wave-train
            V = vector of wave-train

25: spark gap;   T = terminals of spark discharge
            C = discharge
            W = electromagnetic (radio) wave fronts

26: the magnetic field surrounding a charged particle in motion;
            V = vectors of particle motion, the arrows show
            the polarization of the magnetic field.

31: helical coil;   This is the simplest form of coil and is found
            in electrical transformers

32: toroid coil;   T = terminals of coil
            W = coil wire
            R = toroid core

35 A: Magneto-electro-dynamic fields of Jupiter;
            C = planet
            M = dynamic fields surrounding Jupiter
            F = the equatorial flange of the surrounding field

35 B: section view of Jupiter showing structure and dynamic fields;
            E = outer cloud 'shell' of Jupiter
            S = inner fusion plasma or 'proto-star'

47: electrodynamic flows in a conducting hollow metal sphere;
            T = electrical terminal
            S = spherical shell
            F = vectors of electronic flow
            E = equator of the sphere and the path of
            condensed electronic flow.

51: linear conductor illustrating structure of dynamic flow patterns.
            C = conductor
            E = main path of flow and its helical structure

M = resulting magnetic field lines of force

53: emission of a photon by an electron moving between energy bands.

E = electron

S = energized level of electron shell

G = ground level of electron shell

P = emitted photon

54: the spiral vectors of charged particles in a magnetic field;

A = originating terminal of flow

B = output terminal

1 = direction of electric field potential

2 = direction of incident magnetic field

3 = resulting overall vector of electronic flow

48 & 58: section view of EGV;

H = hull

F = equatorial flange

C = central column/engine showing flow patterns

E = electronic flow path patterns

60 B: lightning bolt, section view;

M = central core flow

R = return flow by induction in the plasma sheath

A = toroid vortex formed in the sheath

B = pinched core flow due to magneto-striction

C = magnetic field surrounding pinched core flow

61: hollow conducting metal sphere containing magnet;

T = electric terminal

S = metal sphere

F = magnetic field lines of force

64: section view of galaxy with the central Black Hole and associated surrounding magneto-electro-dynamic fields;

C = Black Hole

D = dynamic helical vortex associated with B.H.

F = electrodynamic 'lines of force' and flow patterns

67 B: alternate section view of EGV;

C = central cylinder engine

H = hull

F = circular flange surrounding the craft

P = nuclear fusion plasma

70: nuclear fusion plasma generator and container;

T = spark-gap terminals

S = spark-gap

B = 'cage bars'

C = central focus area supporting plasma

Δ

INTRODUCTION – PART 1

> "There are more things in heaven and earth, Horatio, than are dreamt of in your philosophy", (Wm. Shakespeare).

> "there is hope for the future, and when the world is ready for a new and better life, all these things will some day come to pass - in God's good time" (*Jules Verne*: 'Twenty Thousand Leagues Under The Sea').

You may perceive that the simple discovery of electricity has totally and completely transformed every aspect of human life. From manufacturing, trade, transportation, computers, machinery, finance, construction of buildings and cities, medicine, arts and entertainment, the media, education, and so on and on. Not the slightest corner or crevice of society has been untouched by this simple force. This is an everyday reality which we comfortably accept and which everyone uses.

Now imagine that another new and simple, but equally powerful and far reaching potential form of energy were to be introduced to the world. Is it not possible then that we would see another new technological revolution which would likewise transform our lives for the better?

There is an old saying that truth is stranger than fiction. This adage is often presented to us on television and late-night radio programs that deal with unusual subjects such as UFO's, ghosts, anti-gravity, angels, strange healing powers, Poltergeist activity, hollow Earth and Moon theories, "time tunnels", "space warps", "parallel universes", and bottomless pits.

This book is an attempt to bring together all data concerning such bizarre phenomena and merge them into a coherent and understandable body of practical scientific and workable knowledge.

> "You can recognize the truth by its beauty and simplicity", (Richard Feynman, Nobel Prize winner in quantum electro-dynamics).

The *common denominator* of these various fields of research and knowledge is a new form of Energy - and a new method of application of an already existing energy form. The former is actually a very ancient form of energy, having been known under many names throughout the ages. It is currently known and used under the names of 'Kundalini' in India, '*Chi*' in China, '*Ki*' in Japan, and by many other names in other times and places, - including the West.

What is new now – and in the context of this book – is that first of all, this energy form has been determined to be fundamentally *electronic* in nature and more specifically, *quantum-electronic*, - (this term will be adequately explained). Secondly, with the new technology presented herein, this Energy may be generated in phenomenally large quantities using presently available technological means. Furthermore, it may then be applied in many fields of endeavor to perform truly "magical" feats.

The new science and physics presented herein may be alluded to as a "new age" technology and includes some principles and pronouncements which will not be comfortably accepted by some. The scientist who reads this will be confronted with some startling notions, - some quite bizarre - and some just plain unbelievable.

> "I can believe anything provided it is incredible": (Oscar Wilde).

Δ

This book is for the most part an examination into long held mysteries of our world and an effort to provide explanations for each of them in the light of modern physics and science, borderline researches, - and new knowledge concerning the now widely acknowledged reality of quantum Energy (*Chi*). This is an effort also to precisely describe from a physicist's point of view the role of this Energy in various phenomena, - such as atomic structure and behavior, levitation, Poltergeist activity, and so on. Furthermore, the Reader will be introduced to a description of the precise quantum natures of gravity and inertia, and how they may be manipulated and controlled.

Ultimately this is the Energy of <u>creation</u>, - the quasi-physical link connecting matter, electromagnetic energy, gravity, and the mind itself.   Some subjects such as Ball Lightning have been researched, - and in spite of all the strong evidence arguing for their existence, are still relegated by some scientists to the realm of the paranormal, (i.e. the "unbelievable").

(To be fair, these scientists are only refusing to accept second hand reports and demanding only first-hand evidence, which is simply good science. Too much circumspection however becomes dogmatism and paranoia, and closes the doors of the mind).

However, the material contained herein has been carefully and well researched, and is taken from respected and reputable reports. My own function has been merely to take selected and readily available information and fit it all together in an intelligible manner to yield heretofore unguessed at solutions to certain problems.

This research has allowed me a rare insight into, and an understanding of the nature of energy and matter - and space and time - and how they may interact in unusual ways. Readers of this book will encounter very startling concepts. It is therefore strongly advised that he read the book in its entirety <u>starting at the beginning</u> and working his way through it sequentially toward the end.

In this way, he will gain a better insight and grasp on the concepts presented. In fact, this is standard practice for the composition of any technical document dealing with for example, mathematics, mechanics, electronics, language, history, and so on. The idea of this is so that basic and fundamental data may be absorbed easily to form a base upon which further ideas may be constructed. [A very good example of this is the series of Propositions found in a high school text on geometry. Every Proposition presented is founded upon the one previous and cannot be understood unless the previous one has also been understood].

He will eventually recognize the underlying current of truth which connects and pervades all the phenomena explained and described herein. He will also make other very startling fundamental realizations about the universe, life, and existence.

Now it must be understood by the Reader that the field of physics *per se* is simply a major branch of *Philosophy*. The book therefore is as much a book of philosophy as it is of pure physics. Since the book deals with the very fundamentals of the physical universe, it is then inevitable that philosophy should be - and *must* be included. (Indeed, the first scientists in our traditions were called 'natural philosophers').

This book, like any other good work of science – is written so that the information contained is presented gradiently. This is done to permit the easy assimilation of the data presented. It is *especially* true and necessary in this case since the gradient is rather steep. Furthermore, the Readers' mind – like a well-inflated balloon – will be stretched in its capacity and in its ability to accept further input. (So look to your mental latex*!*).

The book may be likened to a Mexican stepped pyramid whose sides rise in an incremental fashion to the apex. Upon completion of the book and the attainment of the apex, one may possibly experience a kind of "vertigo" or "headiness", similar to a feeling of unsure-ness or ethereal surrealism. His footing will be unsure perhaps as he gazes tremulously over the precipice to view the material covered previously.

This is where the stability and strength of mind, - and good perspective are essential. The final result of course will be the ability to see a vast panorama never before suspected to exist. Moreover, he will now have in his possession a knowledge which has the power to change the world.

Δ

The book - of necessity - covers several parallel themes and develops them simultaneously, - again like the multiple sides of a pyramid. These include: quantum Energy and electro-dynamics with its laws and principles as newly interpreted and explained herein. The relationships of q. Energy to electronics and atomic physics are explored along with other subjects, - not the least of which is the mind and its relationship to all of the above mentioned subjects.

Also introduced gradually will be other parallel subjects such as the quantum-dynamic behavior of matter and energy, - and even space and time. The inter-relationships of all the above mentioned subjects will be discussed and presented in the following pages.

Guidelines have been set forth to ease the negotiation of the "pyramid". Eventually you will discover that a practical new way of extracting energy from matter is possible, - and how it may be done. The riddles of gravity and inertia have been solved with q. physics, along with many other useful discoveries, including superconductivity, merging, and yes - even space-travel. Further revealed will be the ultimate source of biological health, healing, and regeneration at a level heretofore considered incredible, - even impossible.

Δ

It is inevitable that such a unique book as this which explores the very fundamentals and origins of the physical universe should include a good share of philosophy. These two subjects (physics and philosophy) are intimately related and ultimately inseparable, - like the two elements of a dichotomy pair - or like the two elements oxygen and hydrogen which, when combined - yield an entirely new and 'unexpected' substance (i.e. water). One may therefore call the data herein an "applied philosophy".

The book is necessarily also a study of the <u>mind</u> since it must be perceived that any book which includes philosophy as a practical engine must also include the mind and the ways in which it operates.

We may therefore construct a triune composed of these three inter-related subjects, - i.e. physics, philosophy, and mind. This triad itself then represents an experience which we may call "the study of being and existence". The title of "Electro-atomic Field Wave Emission Energy" (or E-energy, or indeed - "*quantum Energy*") which I have chosen, may give the startling impression of involved and obscure concepts in physics.

Do not be alarmed for this is not the case however as the concepts and materials presented are easy to digest and understand for any intelligent and reasonably educated layman. The title is apt for it describes accurately a certain phenomenon, otherwise known as "Life-energy" (*Wilhelm Reich*) or 'Chi' (and 'Ki') – as it is known in the Orient.

'Quantum Energy' is the fundamental force underlying these phenomena and is essentially <u>electronic</u> in nature. The concepts herein outlined are easy to assimilate for anyone having a rudimentary (i.e. High school) grasp of basic electronics, atomic structure and behavior, electromagnetic phenomena, and /or 'Chi', etc..

It is also suggested that the Reader familiarize himself with at least some of the materials listed in the Bibliography section, since this will give him a more indepth understanding of this work. Diagrams are included which will serve to elucidate the descriptions given.

Δ

Ultimately all fascinating mysteries of nature and physics can be described and explained as fundamentally electronic phenomena, - whether it be a melting snowflake, the growth of a mushroom, a dripping tap, the blink of an eye, the wag of a dog's tail, or even a sneeze, etc.. It is recognized that these - as is everything composing the physical universe - are the result of quantum-electronic phenomena.

In order to truly understand physics and the powers of nature, one must approach each subject with an holistic view and a truly open mind – like that of a child - that is, in a truly *scientific spirit* - questioning <u>everything</u>. The true scientist must be willing to accept and examine diligently any and <u>*all*</u> data, no matter how bizarre, no matter how jarring they may be to our sensibilities, - or how ludicrous

— 24 —

or incredible they may appear to be.

(It is interesting to note that although children have a healthy curiosity about everything, apparently nothing seems to be bizarre or ludicrous to them, - no doubt because they have not yet been indoctrinated by society and given a "real" version of life the way they "should" see it. They quite simply accept something as it exists and as they perceive it. Animals too are very lucky in this respect, and luckier still in that they presumably and evidently do not have a written or spoken language nor any grasp of esoteric or abstract topics. This of course is an assumption!

If scientists and physicists are to strive to discover the true and fundamental nature of the material universe, then they must not refuse to investigate unusual and "bizarre" phenomena. These phenomena (like Ball-lightning, Poltergeist activity, and levitation, etc.) – whether we like it or not, whether we choose to ignore them or not, are nonetheless real and integral parts of our universe, - and hence must be confronted, examined, and ultimately *explained*.

Most physicists are at a great disadvantage in their researches and studies. They are restricted in their scope, they are overburdened with data and mathematical formulae. They have been coerced by the education system, peer pressure, and other systems to which they owe allegiance, - and to focus on extremely narrow confines of study. They do not have the freedom or scope for investigation and research that the "natural philosophers" of the nineteenth century had. They have "indoctrinated tunnel vision" (ITV).

(It is also interesting to note here that as our knowledge increases in scope and detail, our individual investigations become ever more narrow and confined. We can see here then a kind of dichotomy where quantity is rapidly increasing while quality is just as rapidly diminishing. One could also take a reverse viewpoint and say that the quantity of scope is diminishing while the quality of precision and focus is increasing. It all depends upon your *perception* and point of view!).

An analogy in nature is the vortex - like the whirlpool or the tornado, and its cousin the water-spout. As one delves deeper into the complexities which our technology creates, our views and our fields of activity become ever narrower and limited, - and indeed more hectic as we go round and round ever faster before the final plunge into oblivion!

My own scope of research on the other hand, being unhampered – has included a wide variety of subjects, many of which seemed at times hardly to be connected with physics at all.

> I am a searcher of truth and have always been keen to better understand those subjects seemingly obscured to a degree by a veil of mystery - (veils placed perhaps intentionally) - and to perchance, "lift the curtain and gaze upon the forbidden truth, - revealed at last"! (In this context, I am happy to call myself truly a "natural philosopher").

I am by intrinsic nature a physicist – self-taught, having spent nearly a quarter of a century researching all manner of subjects and many fields of physics. As such, I have had the tremendous advantage and freedom of being able to research and study any subject and material I chose, pertaining, however remotely, to my ultimate goal. Such subjects include philosophy, religion, the mind, sociology, ancient lore, even economics, - and of course physics as a general field.

Assembling this body of knowledge was like putting together a huge jigsaw puzzle where all of the pieces were indiscriminately multi-colored, and where some of the pieces didn't even belong. But every piece had to be tested somewhere in the process of constructing the overall matrix and the "big picture".

Further, on occasion I discovered that some pieces were missing - or false pieces added. I moved them this way and that, reversing them and inverting them, placing them here and placing them there, trying to complete the picture, - and all the while not even sure if indeed there were missing or additional pieces at all, - or even a picture to be completed!

And furthermore unknown to me, the key pieces might be up on a shelf hidden from view. Would it occur to me to look for them there? Why should I look there or anywhere else at all since I knew not of their existence? [see note 1 below]. These are typical of the traps and pitfalls which face any such researcher or explorer into unknown territory. The upside of this however is that if you are dedicated and have the required confidence to succeed, it really is a lot of fun! [Please see note].

These newly discovered pieces of potential, once gained, can then be weighed and balanced against observation and other known quantities. Also like a jigsaw puzzle, pieces of data can be set aside and notes taken until they may be comfortably placed in their appropriate location in the matrix. Finally, the datum or data may become a new element(s) which will withstand the rigors of analysis and practical application.

From two new pieces of confirmed data, one can then go on to find, deduce, or extrapolate a third, and so on. This is reminiscent of the method used by *Champollion* in his deciphering of the Rosetta Stone found in Egypt during *Napoleon's* explorations. (or by *Sir A. Conan Doyle's Sherlock Holmes* in his deciphering "The Dancing Men").

There are many who claim that there is no shortcut to knowledge. This bit of seemingly sensible and prosaic reasoning has consequently become a dogmatism and is never questioned by the "socially well-adjusted".

I personally suspect that this is not necessarily so however since I am not "socially well-adjusted". (In any event, to which social structure are we referring?). For the vast majority however - living as they do in the demanding and practical physical world, there is only one fundamental road to knowledge, and that is through - "trial and error". This of course implies 'not knowing' something in the first place.

Upon this philosophy is constructed the major part of twentieth-century technology: [see note 2]. Without trial and without error, it is perceived and believed by modern Man that there can be little gain of knowledge. It is easy for one to create ideas and speculation, - it is even easier for another to *ridicule* any possibility of advancement.

Like a computer, mathematics will only yield answers according to the way in which the guiding mind programs the machine and poses the questions. ("We have to remember that what we observe is not nature herself, but nature exposed *to our methods of questioning*", <*Werner Heisenberg* >). (My italics).

> I recall that when I was a young child in primary school, I would often tell the other children that one day I would design a space-ship. At home, I would often take pencil and paper and happily draw them. When I grew older it became the decade of the '70's. This period saw the introduction of a new genre of popular reading and a plethora of material on the subjects of 'UFOs', 'Pyramid power', Stonehenge, Poltergeist phenomena, psycho-kinesis, strange energies, (including 'Orgone', the 'Odic Force', 'Bio-plasma', 'Earth energy', Yoga, meditation, - and subjects written by authors such as *Colin Wilson, Francis Hitching, Joseph Goodavage, Baron Von Reichenbach, Wilhelm Reich, Vincent Gaddis*, and many other marvelous writers and researchers of the borderline and the mysterious).

I was deeply intrigued and fascinated. This fascination drew me into a quest to find out whatever I could towards achieving, - what I believed at that time – to be the ultimate goal and the 'Holy Grail' of physics, - namely the discovery of the mysterious force of levitation or "anti-gravity". (Which by then I had become convinced was a very real form of energy).

NOTE: So with all these possibilities and unknowns, why did I 'soldier on'? The answer can only be ascribed to *Faith*. That is, - faith in the conviction that somewhere about there were answers to the questions I was asking. I can tell you now that <u>faith</u> - in yourself and in your goals - while seemingly insignificant in practical life, - can be a very powerful tool. It has taught me that literally nothing is impossible, - nothing*!*

INTRODUCTION - PART 2

If you have the heart of a true scientist, - and I don't mean simply a scientist who has become a professional, - I mean one who has the true *Scientific Spirit* and *Scientific Soul* - then you have no choice but to accept at least the possibility that such phenomena as "table turning", "psycho-kinesis", "spoon-bending", and Poltergeist activity, etc., do indeed exist.

Furthermore, in the knowledge that these phenomena have been reported for hundreds, if not thousands of years in all parts of the globe, a true scientist would have to admit that such occurrences are worthy of investigation. They are worthy - not only for the sake of adding to scientific knowledge, but also for our own richer and fuller understanding of the universe in which we live. Further, any avenue which proposes to serve and to improve and advance the lot of Mankind should be seriously investigated in any event, - and most certainly not ignored!

When all is said and done however, one should at last question what science and physics is all about. What is the purpose of all this research, the gathering of data, and investigation into the secrets of nature, matter, energy, and so on? What else can it be but a grand quest for the "Holy Grail", the "Rosetta Stone", the "Ultimate Discovery", - the one single, all encompassing answer of what the universe is all about?

> "…There must be one overall truth. One truth that blends all into a unified whole. It does not have any contradictions, any falsehoods, or any inconsistencies. Every branch of knowledge must blend into it, harmoniously". (*John Strong* in his book, 'The Doomsday Globe'). He is of course referring to a truly 'Unified Field Theory'. [Please see note 3].

I personally believe that when this 'Holy Grail' is at last uncovered, science and physics as we know it now will cease to exist. In their places we should expect to find a more philosophical and gentler approach to life and those around us. Included also will be a greater understanding of our place in the great scheme, and a release from our isolation on Earth.

I believe the Reader will discover with delight that this is a unique book, the like of which he has never before encountered. I truly believe also that he will find it not just interesting, but fascinating, - and I believe I can promise – even <u>exciting</u> – in a very real adrenaline sense.

Even further than that, I can say from my heart that this may be <u>the most important book of the twenty-first century.</u> I don't say this out of conceit or braggadocio, - I say it because I earnestly believe it!

> What I have attempted to do (as I have already said) is bring together all the important mysteries of the physical world, science, electronics, "meta-physics", "para-physics", astrophysics, Poltergeist, ball-lightning, etc., and knit them all together into a coherent whole, - using the legendary fiber of a fine elusive Energy which exists all around us.

This truly ubiquitous Energy has been continually found, researched, - then the results discarded, lost, or hidden. It has been found and forgotten again and again, simply because it was never generally acknowledged (or admitted) to be a true part of the physical world. It still exists however in the Oriental martial and healing arts.

This "new" Energy is ultimately electronic in nature and properly belongs in the field of quantum physics; - for indeed it is a sub-atomic energy and its behavior, characteristics, and wavelengths belong to this level.

As may be surmised, with the introduction of a new form of energy - new laws of physics will inevitably be forthcoming. These new laws will be explored herein and their potentials analyzed and described.

This of course is not intended to be a fictional work. It is intended to be a factual and scientific one, but its subject matter deals with a new kind of science and physics that has only been dreamed of by Man and written about by science-fiction authors. Its efficacy and practicality therefore must be judged in the final analysis by the Reader, whatever his background.

INTRODUCTION - PART 3

It is becoming increasingly evident in the world of physics that the universe can no longer be divided into the separate categories of Mind and Matter in the old Newtonian or nineteenth century style traditions. The two are apparently – in some "mysterious" and "obscure way", inter-dependant and "inter-functional". This has become decidedly and overtly apparent in the field of quantum mechanics.

In fact there are presently fields of research in physics, science, and 'Para-physics' in which this fact is made abundantly clear, i.e. such as *The Thomson Experiment*. There are many unanswered questions in physics, although physicists generally don't like to admit their failings.

For example, there are the mysteries of electricity and even common clay, which even up till today are still not completely understood nor explained. Gravity and inertia are unexplained enigmas *without definitions*. Neither is simple fire completely understood, - not to mention Poltergeist activity, ball-lightning, and other Para-physical phenomena.

(Interestingly, we may say that when we can define behaviour, this implies that we are able to predict and control it. Likewise if we can control behaviour we should logically then define or describe it. It follows then that if we cannot define or describe behaviour, we cannot control it, and conversely if we cannot control it, we cannot define it! The control of behaviour then lies in our ability to define it, and vice-versa).

The orderly array of atoms in a crystal is not understood, and even the ability of 'polarized' lenses to pass 'polarized' light, - none of which are understood. The list of such unexplained matters is long indeed and in many if not in all cases, the answers lie with the new quantum Energy.

INTRODUCTION - PART 4

I consider this book to be the culmination of my life's work and purpose – the accomplishment of my life's mission. On the last day of my bodily life here I will be able to say that my purpose here is complete. In this I consider myself very fortunate.

I am giving this knowledge - not to a privileged few who would without doubt allow it stagnate and be forgotten, - or worse - to be hidden and locked away. Likewise - not to industrialists who would likely consider it either too valuable or too dangerous to their interests; - but to the real people who are the backbone of society - the people who do the valuable work, who are the creators, and who make positive things happen; and to those who are motivated by more than mere profit.

My desire is for this hard-won knowledge to be disseminated broadly so that it cannot be cached, hidden, or otherwise suppressed by vested interests. To any intelligent and politically aware individual, it will be evident that the avenue I have chosen, i.e. widespread publication, is ultimately perhaps the only one available which may bear real lasting fruit and benefit to Mankind.

The subject matter within deals with the fundamental physics of electronics and atomic structure. It is explained simply and precisely without the use of obscure and confusing mathematics or Latin terminology. There are many simple diagrams and analogies which will help the Reader to relate more readily to the descriptions and explanations given.

The book is necessarily divided into sections – each one dealing with a general topic, although *inevitably* one or more of these topics will overlap, like the colors in the light spectrum. These topics include subjects such as: Quantum Energy (the central theme), Quantum-electro-dynamics and how this broad subject relates to the theme, the nature of Gravity and inertia and countermeasures. Nuclear-fusion is a major theme explaining how such a process may be initiated, sustained, contained, controlled, and applied. Sections are also included which explain the curious interrelationships of matter, energy, space, time, - and the mind - on a quantum Energy level. ('Quantum-electro-dynamics' as a subject can be found in some already existing Physics books. However, in this book this term refers to a different subject and describes different phenomena).

Finally you will discover how all these elements may be brought together in the construction of a flying craft powered by nuclear fusion, lifted by quantum Energy, and controlled by an operator within it.

> Very little has yet been said herein about Nuclear Fusion, a new form of atomic energy. This happens to be the second of the three main and central themes discussed in this book. This powerful new source of energy, its true nature, its initiation, containment, sustainment, and control may all be realized through the correct application of quantum-electro-dynamics.

The information hereby presented to you is hard won and of a most fundamental nature - regarding the very underlying secrets of our material universe, - many of which will be revealed herein perhaps for the first time in man's history. If you have the courage and temerity to read this book - to its end, - it will change forever your perspectives, your attitudes, your viewpoints, and your outlook on life and the world in which we live.

"The reasonable Man adapts himself to the world, but the unreasonable Man tries to adapt the world to himself - therefore all progress depends upon the unreasonable man", (Samuel Butler).

> WARNING: The technology revealed in this book is as important to Man as are the inventions of fire, the wheel, and electricity, - and all of them combined. This book is the springboard for a gigantic leap forward for Mankind. It is the technology that will bring Man truly into a New 'Golden' Age.

It is said that great ideas make their appearance when the proper time warrants their arrival. It is now time for planet Earth to really and truly enter the galactic space-age - for great things await us. This is not a belief, nor a threat, nor a promise, - it is a statement of fact and reality. The new 'Golden Age' of Man is actively upon us, even as you read this.

As with all new technology, there are those who will inevitably try to exploit its sinister side. Since this new technology deals with the very fundamentals of the physical universe, the potentials for evil use are far beyond any which have yet been conceived. It can be used in ghastly and frightening ways, - almost beyond imagination.

It is therefore absolutely essential that the intelligences of the world create a legal system of ethics to safeguard the use of such technology.

The responsibility for this will not lie with governments nor government agencies, for these institutions have shown themselves to be inept and often irresponsible. The responsibility will be yours, the Reader - the dynamic flesh and bones of society.

If you, the Reader have a mind to organize with others, to take the knowledge contained herein, to use it, and promote it, - then you will be assuming a great responsibility. It will then be your duty to ensure that a legal system of ethics is put into place to safeguard – not only this technology, - but *to safeguard against what may be done with it*.

This technology is for everyone to use in an ethical, beneficial, and happy way, so that such a system of ethics may be entered into the curriculum of the United Nations world government and every land. *Furthermore, if you wish for this technology to succeed and create a better world where we can all prosper and be free, you must be dedicated to exclusion of all else, and have the will to succeed.*

THE FUNCTION OF THIS BOOK – 1

> "I can believe anything provided it is incredible"…*Oscar Wilde*

The purpose and function of this book is ultimately very simple. It is to change the world and create a better one wherein Man can live closer to the heavens and stars, both figuratively and literally*!*

It would not be overstating the case here to say that "secrets of the universe" are herein revealed. Three of these are fundamental to understanding the physical aspect of the universe and man's place and function in it, including the mind he possesses.

The first of these "secrets" is the knowledge of the existence of a new form of energy, which in fact is not really new since it has been known about and used throughout man's history. What is new however is the understanding that this energy is essentially *electronic* – operating at the quantum (i.e. subatomic) level. When produced in large quantities as herein explained, this Energy is capable of unbelievable marvels and true miracles in the Biblical and Mythological senses.

The second "secret" revealed and explained is the fact that all aspects of the physical universe (i.e. matter, energy, space, and time) are ultimately quantum-dynamic in nature and may therefore be manipulated and modified by quantum-electronic means. Thirdly, and most importantly is the role that the mind ultimately plays in all things.

In other words, we shall investigate and understand the hidden powers of the mind and how they can be released and enhanced using quantum-electronic technology. Hence, since the mind is capable of imagination – so too may anything within or without the physical universe be accomplished with the application of this power.

Also discussed and explored are a variety of *related* subjects such as perception, illusion, objectivity and subjectivity, - and the meaning of ultimate reality. The two other primary themes of the book dealing with physical technology are; the means to create a viable nuclear fusion generator, - and its application to power a flying craft controlled by an operator within it.

Included in the text also – and importantly, is the subject of a practical philosophy as it relates to the mind and its inter-relations with the physical universe. This is quite necessary since the mind will ultimately be intimately involved in this new technology. Hence the operator(s) will need to understand the nature of this interaction and its effects both upon the material aspect and the operator's mind itself.

THE FUNCTION OF THIS BOOK – 2

The function of this book is several-fold. Primarily it is of course the presentation of a revolutionary and ground-breaking new technology. In addition happily, it has other functions. These other functions and benefits were not included by intention however. They came about as a result of the book's continuing development and as an ever increasing intuitive process. Indeed much of this intuitive process was by way of true revelation and self-discovery.

It will be demonstrated to the Reader how physics at its most fundamental (i.e. quantum) level is little short of philosophy itself. Moreover it will also be shown that physics and philosophy are interchangeable in a very tangible way. For example, the subject of "mind over matter" has long been a field of speculation, experimentation, exploration, and of course philosophical discussion. Indeed there are many cases on record (even accompanied by photographs – as in the case of *Sir Wm. Crookes* - of people levitating; - as observed and recorded by *Sir Arthur Conan Doyle* (of Sherlock Holmes fame)).

If such cases were a common everyday occurrence in society, it would then be accepted that indeed the mind is capable of bizarre and unusual feats. Physics and philosophy would then both be viewed as a single field of endeavor and research. It could then be referred to as "natural philosophy" – as science once was in the nineteenth century.

In addition to this, the Reader will experience a transformation in his mind, his viewpoints, and perceptions. He will find his perspectives and viewpoints shifting so that things which he once considered im-possible will then be viewed to be most certainly very-possible, even probable. Furthermore he will even begin to question his previously held beliefs and viewpoints, - and why we do not commonly have experiences which until now have been regarded as "magic" or science-fiction.

Moreover he will begin to view himself not as a body but as a mind, a will, a non-physical agency, and as a Being; that is - an intelligence - an immeasurable thing, having awareness and having vastly greater potentials than he ever considered possible. It will be shown that in fact such things are not only possible, but that here indeed is a technology which *will* make such things a *reality*.

We may also discuss religion here since there is a close relationship between religion and philosophy. Religion also seeks to find solutions to age old mysteries and how the problems and sufferings of Mankind may be alleviated. Further, it seeks to find the spiritual nature of the Universe and of Man, and how the two may be reconciled. This in effect is a search for God (in the Western tradition), the nature of God, the secrets of life, and how communication on this level may be established.

These problems must also be considered to be philosophical ones. Hence we may now see how religion, philosophy, and science may ultimately be mutually identified, related, reconciled, and merged to become a new way of life and understanding for all.

> "The one man who has more illusions than the dreamer is the man of action", (Oscar Wilde).
> "No man should think himself a zero and think he can do nothing about the

state of the world", (Bernard Baruch).

> "All is a mystery, but he is a slave who will not struggle to penetrate the dark veil", (Benjamin Disraeli).

> "There are more things in heaven and earth, Horatio, than are dreamt of in your philosophy"… Shakespeare.

NOTE: the early articles in this book deal with a real and practical philosophy. Read these articles without omission and with diligence. You will be well rewarded, for your mind is about to undergo a transformation and must therefore be prepared with adequate groundwork. If you should at any time be unsure of the descriptions given herein, always refer to the diagrams, the glossary, or a good dictionary. They will offer guidance and understanding as to the overall intention of the book.

- ADDITIONAL NOTES -
1 ~ THE PHYSICAL UNIVERSE TRAP

I believe that most of my Readers will agree that the world of Man is in a shocking state. Insanity and irrationality are rampant everywhere: [see note]. In the past, the notions of honor, respect, loyalty, diligence, duty, and self-discipline, etc., were given high priority. Today much of this noble essence has been diluted, - even evaporated. So what has happened?

The fault lies within ourselves and our relationship to gross material technology. The acceleration of our downfall began with the introduction of mass-production and the automobile. (Where is the nobility of a Knight or a Prince who drives a car?). After this came the introduction of electronic appliances such as radio, television, computers, and so on.

In today's world, most of our problems are generated by the desire for money and material acquisition. These are allied with a degenerating media and education system, and the abandonment of religious philosophy. Hence the mind and will are becoming ever more stuck in materialism and subject to its failings. Money and acquisition are the hooks by which we are controlled and manipulated.

(It should be noted that it is not acquisition *per se* that deserves the bad reputation, it is the desire for acquisition). These things are not intrinsically evil. The evil is within ourselves and how we permit these things to control us (and how we use them to control others). In part, it is the *difficulty* of acquisition which makes these things desirable. For example, if we as individuals were capable of creating in an instant whatever we wished for, or if anything were free, it is likely that we would place very little value or significance upon them. A primary trap is to make something desirable and then make it difficult to acquire. Obviously, the way to avoid this insidious trap is to not allow the item to become desirable. A new friend is far more valuable than a new car. If we suffer due to our love of material acquisition, we can only blame ourselves! [2].

"Every great advance in science has issued from a new audacity of imagination", (John Dewey).

NOTE 1: the people of one country can look at the people of another country and call them insane, irrational, and misguided. This all due to different cultures and beliefs. Yet when looking at themselves, the majority sees no insanity or irrationality, - their viewpoints, actions, and behaviours are all comfortably justified. It is this *inability to perceive* insanity that is a key reason for this destructive state of mind. This is likewise true for individuals. Man generally is becoming more irrational and so a remedy is needed very quickly before he is sucked into the vortex of oblivion.

When you gaze sleepily out of your bedroom window, all seems calm, orderly, and 'normal'. This situation is a soporific tending to confirm the desired belief that "all is well" with the world. Well quite frankly, all is not well - and it is time for us all to wake up to that fact.

NOTE 2: for example, an expensive European watch may sell for a thousand dollars, yet it can be manufactured for one twentieth of this. The markup is a result of costs for middlemen, storage, transportation, maintenance, and so on, - and not forgetting the element of greed.

2 ~ NOTE TO THE READER

This book is actually several books in one so that a number of texts may be written - each dealing with a separate subject. However as I wrote and continued to write, the book seemed to take on a life of its own, so that these various subjects became inevitably interwoven.

This is due to the fact that all these various subjects are all interrelated and each gives support and sustenance to the others, - as the Reader will discover. (In much the same way that the individuals of an organization give support and sustenance to each other and to the organization itself).

A portion of this book is given to discussions of atomic, nuclear, and particle physics. I will hasten to reassure my Readers that such matters are not to be feared nor avoided as being unfathomable or inscrutable. They are not - as is generally thought by the average population - mysterious, abstruse, or beyond the comprehension of mere mortals. In large part, such attitudes are generated and promoted by an established and educated elite. (See Preface - 5).

To put it bluntly, they are practicing a form of sophisticated academic snobbery, pretending to be superior to the 'proletariat'. Indeed the concepts of atomic and particle physics are simple enough and well within the grasp of the reasonably well educated layman. Ninety percent of particle physics consists of elementary concepts, while the remaining ten percent is concerned with higher mathematics designed to give precision and predictability to the dynamics involved.

For instance, in quantum physics we can see that time is sometimes apparently reversed so that cause becomes effect and vice-versa. In fact this is not so, since it is an *apparency* due to our ignorance of what time really is, and hence is a matter of perception: (explained in the main text). This elimination of a time factor and cause becoming effect is a common theme in quantum mechanics and if one bears this in mind, he will better understand the weird world of subatomic behaviors. [See note 1].

This book contains no such mathematics except a very few elementary and simple "high school" equations which are childishly simple. You may therefore heave a sigh of relief.

The concepts presented herein are described with clear logic and reason. Many of these are presented with simple analogies and sometimes diagrams are provided. A glossary is provided which will explain fully any concept desired for further study and research. References are also made to other articles which contain more data on particular points throughout. *Furthermore, every point made in each article will always be enlarged upon at some other point in the text, so that the Reader will never be left with unanswered questions. Everything will eventually be explained in its proper context.*

Philosophy of course is that discipline which attempts to delineate man's relationship with the physical and non-physical world which surrounds him. This book is therefore an exploration as much of the mind and its abilities as it is of the intrinsic properties of the physical and material universe in which we find ourselves.

All the data presented in this book can be found in any good library or bookshop so that there need be no misunderstanding for the Reader. Again, if the Reader feels that he is unsure of the material, it will be because he has gone past a word or phrase he has not understood. Find out what the word or phrase is and get it properly defined. The material will then once again become clear. [See Note 2 below]. (The space-time continuum will often be referred to in the text as STC, and this concept will be thoroughly explained).

The book is written in a very condensed style so that there are many, often new concepts found within even one single article. It is a treasure trove of valuable discovery, so your reading adventure will be a fulfilling one. Furthermore, if you are an adventurous spirit and have an open mind and a love of technology, you will have in your hands all the necessary information to help build a wonderful new world.

Δ

If you are able to grasp simple concepts like the vibrations of a bouncing ball, the surface of an expanding balloon, a soap bubble, the concepts of electric and magnetic fields, the flow of electricity in a conductor, the formation of smoke rings, a rising bubble in a liquid, the vortex formed by water as it exits the bathtub drain hole, - or the propagation of light in an optic fiber, the planets orbiting the local star, the bow-wave formed by a moving ship, or the behavior of a candle flame, - then you will have no difficulty in grasping the fundamentals of dynamic energy flows, particle behavior, and quantum physics as explained herein.

It is highly recommended that the Reader arm himself with a notebook and pen for he should take notes, - and for his own convenience, will likely find it necessary. Because of the way in which this book was necessarily written and arranged, it will be found that a significant portion of the data contained in articles are interlinked while being dispersed throughout the book. Hence a single article - while being itself a cohesive and coherent expression of thought – will be related to several others, and indeed the entire text. In this respect, the book is much like a net or web in that every node and strand communicates with every other node and strand, - indeed, like the electronic Internet itself.

NOTE 1: It is not difficult to understand that if time is eliminated from a sequence of events, then none of these events can be said to be either a cause or an effect, or indeed sequential. The very concept of 'cause' or 'effect' implies the passage of time between the two events. Since cause and effect will occur in the same instance, either event may be considered to be cause - or effect, - or both - or neither! In this event, we have two things occurring simultaneously, and if one event does not occur, then neither will the other. A simple mundane analogy of this is a pair of shoes. No manufacturer will ever produce just one item, they always come in pairs! This implies of course that two or more events can occur simultaneously and spontaneously. How? This will all be explained.

NOTE 2: (taken from 'The Complete Works of Lewis Carroll', the Nonsuch Press, London, 1939).

"The learner who wishes to try the question fairly, whether this little book does, or does not, supply the materials for a most interesting mental recreation, is earnestly advised to adopt the following Rules:

1. Begin at the beginning, and do not allow yourself to gratify a mere idle curiosity by dipping into the book, here and there. This would very likely lead to your throwing it aside, with the remark 'This is much too hard for me!', and thus losing the chance of adding a very large item to your stock of mental delights…

2. Don't begin any fresh Chapter, or Section, until you are certain that you thoroughly understand the whole book up to that point, and that you have worked, correctly, most if not all of the examples which have been set… Otherwise, you will find your state of puzzlement get worse and worse as you proceed, till you give up the whole thing in utter disgust.

3. When you come to a passage you don't understand, read it again: if you fail, even after three readings, very likely your brain is getting a little tired. In that case, put the book away, and take to other occupations, and next day, when you come to

it fresh, you will very likely find that it is quite easy.

4. If possible, find some genial friend, who will read the book along with you, and will talk over the difficulties with you. Talking is a wonderful smoother-over of difficulties. When I come upon anything – in Logic or any other hard subject – that entirely puzzles me, I find it a capital plan to talk it over, aloud, even when I am all alone. One can explain things so clearly to one's self! And then, you know, one is so patient with one's self: one never gets irritated at one's own stupidity!

If, dear Reader, you will faithfully observe these Rules, and so give my little book a really fair trial, I promise you, most confidently, that you will find Symbolic Logic to be one of the most, if not the most, fascinating of mental recreations! ..."

3 ~ SUPPRESSED KNOWLEDGE

History is replete with instances where knowledge has been suppressed, hidden, and denied to the general population. (The great library of Alexandria was burned to the ground - not once, but twice). In other ages preceding or following such periods, these fields of knowledge have been freely available. There are however two fields of knowledge which have been heavily suppressed, especially in the last two centuries. These two fields interestingly, in a general sense form a *dichotomy*.

The first belongs to the realm of the mind and spirit, and the second to the field of physics and material technology. More specifically, the first target of suppression is the knowledge that the living creature - including Mankind, is a composite consisting of the physical body and a non - physical agency which is what we refer to when we say "I" or "you", etc..

An analogy of this would be to say that the body is like a car which the driver gets into and out of, at will. In fact you are the "driver"/operator of your body/vehicle. This non-physical operator is called a "Being".

The other specific field of suppression is that of knowledge concerning the true nature of the natural Energy known in the Orient as "Chi" or "Ki", and this Energy's incredible potential. It may be said then that this book is a revelation of suppressed knowledge hidden for at least two centuries (with the dawning of modern technology), and probably for the entire history of Mankind, - available to only a few chosen of the privileged.

4 ~ MY GOAL

Many of the important ideas presented in these pages will seem to the Reader to be pure science-fiction. The Reader is of course entitled to say that this work is sci-fi and nothing else: [please see note 1]. The Reader may pronounce this work to be Poppycock. Very well, if this is so then I challenge that Reader to offer more logical, reasonable, and presentable solutions to the major problems facing physicists today. I hereby claim that he cannot, - whether he is, or is not - a physicist. [please see note 1].

However, let us suppose that indeed everything within this work really is sci-fi. This would then seem to imply that all the present day technology that we now possess is the ultimate, and that Man can progress no further. Does this then mean that we have reached an *impasse* - a dead-end in technology - or that revolutionary new methods, rules, and laws cannot be discovered, - or indeed, that existing laws may be invalidated?

Many of you will already understand that this logically and reasonably cannot be so. You will say without hesitation that there may be, or even *should* be in fact, new laws of nature and physics yet to be discovered. This book will then re-affirm this belief held by those progressive thinkers.

> The Energy described herein is the conspicuous agency in Poltergeist manifestations, Psycho-kinesis, and Telekinesis, etc.. It is ultimately a quantum-electronic phenomenon in constitution, and may therefore be generated using purposely-designed electronic equipment. The nature of this quantum Energy makes it the key factor in unlocking other fundamental mysteries of matter, energy, time, and space. Other discoveries are that gravity and inertia are also quantum-electronic in origin and may therefore be increased or decreased using technical means. It further allows that workable *definitions* for gravity and inertia can now exist where they have not heretofore.

Other subjects discussed are Relativity, the inter-relationships of matter, energy, time, space, the "density" of space, and flow rates of time, - all of which may be manipulated using electronic techniques: [note 2]. In other words, along with the presentation of a new form of energy and a new application of electronics, new laws of physics will also be presented. Superconductivity is another possibility revealed herein.

Present day physics is still in its infancy and is undeveloped. It is found that the physical "constants" (K) which we are taught, such as the velocity of light, the flow rate of time, the four 'dimensions' of space-time, specific gravities of substances, atomic structure and oscillation, etc., are in fact not constant at all, but vary according to the quantum-electronic conditions prevailing, - whether they are natural or devised, cosmic or sub-atomic. Also discussed and explained are the static and dynamic influences which bind energy into matter, i.e. particles.

No complete discussion of such interrelationships of matter and energy is possible without examination of the role of the mind and thought. Also covered are "local" or personal space-time continua, - and how they relate to memory, synchronicity, crystal structure, rapport, perception, objectivity ~ subjectivity, and so on. The quantum-electronic nature of "beauty" and aesthetics generally - (a highly subjective study which nevertheless is popularly considered to be objective) - is also discussed. The subject of time is thoroughly examined, with a look at the possibility of faster than light travel.

As elsewhere explained, the subject of quantum physics deals with the very fundamentals of the physical universe, - and even includes excursions into the non-physical. Because q. physics intrudes into the very workings of the mind as well as the material, we can expect that it will influence to some degree the mental constructs of man, especially pertaining to his emotions, thoughts, perceptions, viewpoints, realities, and so on.

Furthermore, because this book inevitably and necessarily investigates all of these aspects, it will appeal to almost anyone in the general population. For example, it should be of more than passing interest to students of physics, engineering, energy production, biology, medicine, the arts of *Chi* and *Ki*, philosophy, culture, tradition, religion, sociology, economics, - even law, and so on.

NOTE 1: they may say that it is science fiction posing as a serious work; conversely they may say it is a serious work posing as science fiction. Such pronouncements are trivialities - merely words - since the Reader is able to read and *judge for himself what he perceives as true.*

(In any event, it's still a damn good book and fun to read*!*).

NOTE 2: we have seen that the revelations of *Einstein* and his ideas of Relativity and so on have surpassed those of *Newton*. This new technology will similarly surpass ideas of Einstein, including those of Relativity.

5 ~ THE FEAR OF NUMBERS & MATHEMATICS

The subject of physics is really not a subject to be afraid of or one to avoid. This "fear" is only an <u>unwillingness to confront,</u> and this unwillingness is due to the existence of *unfamiliarity* and *unknown* factors. For example, many people are afraid or unwilling to approach the subject of physics because of the large amount of incomprehensible mathematics and obscure, abstruse terminology.

Physics itself, without the mathematics - is actually very interesting. If a subject is presented without difficult and mind-curdling number cocktails it becomes quite enjoyable and even fascinating, - especially the field of quantum mechanics.

Ideally, the subject of physics should actually not involve very much mathematics, and there should be a branch of experimental physics which does not require any complex mathematical machinations. What *should* be involved above all is imagination, inspiration, intuition, experimentation, trial and error, and last but not least - a playful inquisitiveness: [see note].

In today's physics arena, physicists rely on mathematics - mainly to predict the results of a theoretical input. This makes it almost unbearably dry and non-productive. It also kills the spirit of imagination.

What they should do is throw away all their brain-scrambling computations and start fresh with ideas and *experimentation*. Instead of trying to figure out on paper what happens when you add a little of this to a little of that, just _do_ it. This is what men like _Nikola Tesla_ did, and he was one of the most productive geniuses who ever lived.

(We should also include _Leonardo Da Vinci_ - and the many alchemists of the Middle Ages who bequeathed a plethora of useful discoveries to us).

Results would be obtained much more quickly and with a lot more fun besides. After all, the purpose of life is all about having fun isn't it? Children and animals certainly seem to know this. If the purpose of life is _not_ having <u>fun</u>, there seems to be no alternative_!_

I believe, as do many others, that science would actually be much further ahead today if it were not for the existing reluctance to dispense with the mathematical crutch. (Indeed, many great discoveries have been made by experimenters and amateurs using only their God-given _intelligence, intuition,_ and _imagination,_ - as we find occurred in the age of the alchemists, and during the _Renaissance_, etc.). Mathematics are certainly useful but should be used to determine the parameters of experiments and the nature of the results obtained. So in a sense, physicists are putting the cart before the horse_!_

Indoctrinated Theoretical Physicists will no doubt strenuously give the reverse argument, - since they have never tried to do science the <u>easy way</u>_!_ They would rather bog themselves down in complex equations – which in fact have often only added to the existing confusion and/or proved to reveal contrary to demonstrated fact (as in the case of the 'Cosmic String' theory)_!_ [Please see note 2].

The entire edifice of the 'String Theory' is based solely upon theoretical mathematics and even now no-one can say for sure whether "cosmic strings" actually exist, how many universes they represent, or even what their structure is_!_ Another mathematical analysis made of Ball-lightning for example completely _fails_ since it denies their structure and function due to faulty reasoning. Because ball-lightning cannot therefore be "explained", it is relegated to the shadowy realm of 'paraphysics', or even the paranormal and psychological phenomena.

Mathematics - like a computer - will only yield answers appropriate to the specific question asked and the manner in which the question is posed. Mathematics doesn't employ fuzzy logic or offer intuitive proposals, it is a purely mechanical system of thought without heart or spirit.

A computer will respond to the precise command given to it. You can program a computerized robot to understand that you always eat only red apples. This will become a passive and dormant datum. If you then tell the robot to "give me an apple" from a barrel of mixed colours, it may then give you a green or yellow apple. It does not make the connection between the two commands or statements. As humans (and animals) we do this virtually all the time without realising it.

The computer responds only to your statement, not your _intention_. 'Fuzzy logic' is where a computer - like a human, _does_ make this connection - but as you may guess, this is a very sophisticated type of programming. To program a computer to respond similarly to all and every situation can be seen to be a virtually impossible task since the computer must be able to determine your unspoken *intentions* in every single case.

This is precisely why a computer *per se* can never become the self-intentioned, self-aware beast of science-fiction. To be such a dangerous beast it must first understand your *intentions* in order to counter them.

Another example is when you ask someone, "would you like some ice cream?" If you were to ask this of a computer, it would not understand the question. You would have to say something like, "do your immediate parameters indicate a requirement for ice cream?"

The state of being self-determined would be tantamount to knowing the intentions of living creatures. Having these abilities, the computer would then become <u>aware</u>, the hallmark of a sentient living creature. (This is not to say however that a computer may not in fact be assumed by a sentient intelligence, just as is your own brain!).

It should be said however that a really good scientist ought to be able to think just like a computer while at the same time being able to employ fuzzy logic. The trick is being able to switch from one to the other and knowing when to do so. A possible rule of thumb would be to use computer logic when dealing with inanimate objects. etc., and fuzzy logic when dealing with sentient Beings, - combining the two when necessary.

> "We have to remember that what we observe is not nature herself, but nature exposed to our methods of questioning", (Werner Heisenberg).

NOTE: experiments are usually conducted on a trial and error basis (which is what experimentation implies). However, this experimentation is based upon a mathematical prognosis which gives an indication as to what might happen during the experiment. Physicists will readily admit that there are many possible *unknown* variables which can affect the experiment. In this respect, mathematics is of less help than a keen intuitive understanding of the forces involved. In fact, if scientists were given free rein to do the experiments as they saw fit, many of them would likely opt for a more intuitive, 'fuzzy logic', hands-on approach!

NOTE 2: mathematics has given us nuclear fission, and this technology may be said to be a purely mathematical creation. It demonstrates the mind-boggling complexity typical of abstract mathematics. Very little of any good has come of this technology and the efforts required to extract usable energy seems almost too much to be worthwhile.

6 ~ CAUTION TO THE READER

The Reader will discover that the articles are not all numbered incrementally, - that there are gaps in the numbering system. This was unfortunately unavoidable due to the way in which the book was compiled. The articles were not originally written sequentially, and some difficulty was encountered placing them in proper order, compiling an index, - and so on. Furthermore some articles were, later on in the compilation – regarded as no longer required and were consequently either discarded or moved to another location.

No doubt the Reader will forgive this small inconvenience since I believe his overall enjoyment and enlightenment will far outweigh this slight imperfection. In any event, the numbering is completely *sequential* so the Reader cannot get lost.

7 ~ OPERATION MIND-STRETCH

Why do physicists - and indeed most people - have such difficulty in entertaining the concept of the possible existence of non-physical agencies? Further, why do people generally hesitate at the prospect of accepting the existence of the unusual powers of the mind, disembodied entities, life on other planets, alien space-craft, re-incarnation, and so on?

It is due primarily to two factors. The secondary factor is because it is difficult for people to accept ideas which are beyond those of his peers, the society, culture, or tradition and system in which he and they have been raised. This can easily be tested by the traveller and the adventurer. More fundamentally though, it is a psychological problem.

People are generally unable or unwilling to entertain ideas and beliefs which upset their notions of stability and predictability, and the "socially acceptable" with which they have been indoctrinated. In other words, they are unwilling to "stretch" their minds beyond what is required. Even when confronted with the reality of such unusual ideas, they are still unwilling to accept them and come to terms with them. The ability to "stretch" one's mind requires considerable Energy generation which most people are not easily able to do. (The mind is composed of Energy flows, and this will also be discussed later).

It is truly amazing how people can easily otherwise justify what they cannot perceive as the truth. (There is a good simile with a balloon which strains to accommodate extra gases. It "does not like" to accept more gas, and resists extra input).

It is profoundly interesting also to note that modern Man with his electronic gadgetry and airplanes, etc., cannot accept the incredible powers that South American or Australian, etc., backward and indigenous tribesmen have over animals. They are able to cause animals to do things which they would not normally do, such as lie down, go to sleep, hunt for food, find a lost friend, etc., - this is real communication between two distinct types of biological creatures. [see note].

Yet these same tribesmen – a few years ago – would have laughed at you if you tried to tell him about your automobiles and computers, etc.. This inability to stretch the mind then is a common human failing and is found among all classes of people, rich or poor, and depends to a degree upon education and experience. (Incidentally, intelligence is positively influenced by education and experience).

Most books are written with the unspoken intention of catering to the targeted Reader's credulity level and the culture and traditions to which the writer and readers belong. Science fiction writers have a license to go beyond this threshold point and people enjoy it since they feel they also have a license to do so, - it being labelled "fiction".

This book – as a serious examination of an extraordinary new physical technology however – is not intended to be cautionary or pandering, indeed it *cannot* be. Likewise this work will stretch your credulity to the breaking point. It may be necessary for the Reader to make an effort of will. Those who are able to read this work in the spirit in which it is written will have strong minds, consequently, they will be richly rewarded.

> "There are many who will view this work as fantasy. This is unimportant in any event, for there will always be those progressive few who will turn fantasy into reality"… (author).

NOTE 1: we have (most of us) had the delightful experience of interacting and communicating with a pet dog or cat. Many people – especially those who live in the countryside, have had the experience – to their wonderment – that it is indeed possible to communicate on a simple level with denizens far removed from our normal experiences of interaction, - such as spiders, wasps, mice, newts, and beetles, etc..

These are not merely bits of mindless protoplasm wandering around. They are living creatures, thoughtful and aware entities capable of reasoning, of controlling their bodies, - and communication. They have families, they belong to groups, and they have systems of etiquette. It can also be observed that they demonstrate a real care for others of their own type. In short, they are Beings who own and operate bodies. (A Being has no wavelength or frequency and therefore has no size or measure, - hence one cannot judge the size of a Being by the size or type of his body).

It should be fairly evident that any creature having a highly organized and structured body must also have a correspondingly well-organized and structured mind with which to operate it. The state of the mind corresponds with the state of the Being creating it. (More on this later).

NOTE 2: Evidently a Being – such as yourself – is actually potentially capable of doing <u>anything</u> he chooses to (through the agency of q. Energy). This is the great secret which has been hidden from Man throughout history by those who wish to retain it for their own uses and the control of others.

If we were to research the history of quantum Energy and its use from ancient times until the present day, we would find many examples of people having unusual abilities. These abilities include moving objects by will-power alone. Indeed we shall also find examples of levitation and objects passing "magically" through walls and other barriers, including matter, time, and space.

An example of this latter is objects which disappear from one location and re-appear at the same time or at some other time in another location, etc.. This is typical of phenomena in the quantum universe and hence these are explained (and per-formed) with quantum technology.

IMPORTANT NOTE:

This book - although not intentionally so, will be highly controversial. There are many who will welcome it and there are some who will not. There are those who will try to suppress it, and this suppression can take many and powerful forms. Ridicule is one effective form. Another form of suppression which has occurred throughout all of Man's history is the burning of books, the acquisition or control of distribution - the loss of friends, money, job, family, - yes, and even one's life - are included in this technique. Let us look at a present day situation; - the automo-bile.

The automobile is the direct cause of more death and maiming than virtually anything else which the common man can get his hands on. Any fool who can read, write, and push a button is allowed to obtain a car and a license and then proceed to cause havoc, death, and maiming. Yet cars on the road continue to multiply in the millions. If the government is so concerned with our safety, why are they not banned? The answer is money and oil. This must demonstrate adequately the power of business and profit over politics.

CLOSING NOTE OF INTRODUCTORY SECTION

Now I should like to say this before closing. I humbly consider this work to be an inspired one. I believe I owe this work to the non-physical forces which are even now accelerating and transforming the world. Most people as yet are not aware of it but as I speak, there is a new movement growing in the West and other parts of the world which is changing profoundly the nature of Man and his ability to im-prove the world's conditions. This is happening because the West especially is in great need of it. The Eastern ("third world") however is now in need of technology.

The technology herein presented is, I believe, admirably suited to the Eastern mentality, history, culture, and temperament. Hence we find again a dichotomy being formed; i.e. West ~ East: Mind ~ Matter.

> I would like to believe that this book will greatly assist in likewise profoundly changing man's conditions, and to usher in a truly Golden Age. This has been man's dream for æons and now I believe with all my heart and spirit that this book repre-

sents a powerful new tool to help create this dreamed of Utopia. [Please see note 4].

It is my intention - to reiterate, that this book should usher in a new and truly Golden Age for Man. If you accept the truths presented in this book, you have accepted a duty to assist in creating this Golden Age. A Golden Age will not happen by itself, - far from it, and quite the reverse. A golden age can and <u>will</u> happen if you accept the responsibility to do something with diligence and dedication. It won't happen with the good wishes of two or three individuals, it will take <u>action</u> - <u>dedicated action</u>, and <u>*group action*</u>, - and groups allying with other groups around the world.

This is not a grim task to accept, for this book offers high adventure and promise of real happiness to those willing to make a reality of the possibilities contained herein. If we accept this duty and willingness, diligence, and dedication, - we can actually pull it off!

> "destiny is not a matter of chance - it is a matter of choice. It is not a thing to be waited for - it is a thing to be achieved". (*Wm. Jennings Bryan*).

NOTE 1: most physicists - and indeed Mankind in general - when investigating problems and seeking solutions, are confronted with a major and unseen barrier. That single barrier is a characteristic of the ordinary mind and it is the fault of assuming or making *assumptions*.

These barriers are often assumptions and preconceptions made by others, but more importantly and often, - the investigating mind itself. Man generally makes many assumptions about the world around him, and this unfortunate trait leads him to arrive at many false conclusions.

Such assumptions are many in number. For example when we visit a foreign country, we usually have the assumptions (preconceptions) that the indigenous people think and feel the same way about a variety of things that we do. Although we see the obvious signs and products of a vastly different culture, we do not use our God-given powers of observation and intelligence as to the people themselves.

Instead we assume that these people have (as examples) the same sense of humor, the same attitudes regarding money, liberalism and conservatism, objectivity and subjectivity, moral values, family values, attitudes regarding the relationships between men and women, attitudes regarding work and duty, or the same ideas about etiquette and politeness, etc.. Such is certainly not the case. The same principle is of course true when investigating physics and science.

It is commonly <u>assumed</u> for example that physicists have a complete understanding of those subjects about which they expound on - whereas in fact they often quite simply do not. They expound and resound with confidence in their glossy magazine articles and diagrams, but when the truth is known - much of it is theory, guesswork, and speculation.

Indeed, one may make well-calculated assumptions, speculations, and theories based on diligence and research (sometimes called "intuition") and it is sometimes a powerful tool to do so, but at the same time one must be aware that he is actually making an assumption, - and further that that assumption may be based on other assumptions (preconceptions) made by himself or others before him.

We do this everyday in social intercourse by assuming that another individual understands precisely what you mean when you say something. (Quite possibly he actually misunderstands you because your (or his) ability to communicate is pathetic. Furthermore, you would not know this about yourself because others are too polite to tell you, - and even if they did, you would assume that it is *they* who are stupid and unable to understand your simple speech and communicating ability).

Another may assume that you mean one thing while you actually mean something else. One final result of this is that when he misunderstands your directions, you think he is an imbecile, while he too will think you are a moron incapable of proper communication. Hence we find that hidden assumptions can lead to misunderstandings, misgivings, poor results, and even hostility. This can happen frequently between different cultures.

(How do we remedy this situation? By learning how to communicate properly! If you're not sure of someone's meaning, ask him to explain it, or ask him to say it again).

This truism is blatantly obvious in the natural animal world where birds of a different feather will almost never flock together unless under a common threat. Why is this? The answer is one reason and one only. The communication between two separate species is very poor (although not impossible). The reasons for this should be fairly self-evident.

(I have however seen seagulls and crows playing games of chase with each other, and even taking turns. No doubt this is because they have affinities. They are approximately the same size, they fly in a similar fashion, they are both scavengers, they steal eggs from each other, etc.).

NOTE 2: another powerful tool in the researcher's arsenal is intuition. Intuition works, and indeed many a physicist, scientist, or researcher has found this tool to be invaluable and likewise responsible for many a great discovery. Physicists however by their very nature and training, do not like it and are reluctant to acknowledge it simply because it cannot be seen, experienced, quantified, or measured in any way. Indeed, intuition is a subjective perception of solutions to the objective problem at hand.

As is well known to many stargazers, some galaxies have the appearance of a rotating mass or collection of stars. They seem to be rotating because it is suggested by their dynamic spiralling appearance. We understand that a fundamental requirement of the universe is that all cosmic bodies should rotate, so we know intuitively therefore that galaxies must be rotating, even though we cannot actually observe them doing so. Furthermore, their spiral forms suggest strongly that they rotate.

(Hence is demonstrated that one may <u>know</u> something even if there is no <u>direct</u> evidence that such a thing is so. Ask an astro-physicist and he will tell you that he *knows* the galaxies are rotating although he cannot actually *see* them rotating. His belief is so strong that he will stake his life on this fact*!* That's called <u>Faith</u>*!*).

There are many things in existence for which there is no *direct* evidence but for which the evidence is implied. Indeed the implication of something is sometimes difficult to perceive so that only those with keen observation or insight are able to perceive that which is imperceptible to others.

> "Because our culture has somehow generated the unsupported and improbable belief that everything real must be fully describable, it is unwilling to acknowledge the existence of intuition", (Sir Geoffrey Vickers, social theorist).

NOTE 3: research as a subject, and physics, etc., has been "corralled" in a restrictive mindset which does not permit any exploration beyond its confining boundaries. Scientists are fond of enunciating new discoveries and researches, but these new fields are tightly controlled and monitored.

These mainstream scientists for example are not permitted to venture into uncharted territories of research such as the fascinating fields of Poltergeist activities, ball-lightning, psycho-kinesis, and so on. These restrictions and 'peer-pressure' do not consist of tangible rules or laws but consist of attitudes and even implied threats. Indeed, "peer pressure" itself may well be a *control mechanism* used by those in control*!*

There is the ever-present awareness that one (a physicist, for example) is at all times walking in an ideological "minefield" where the slightest indiscretion against the *status quo* and established mindset will detonate a barrage of <u>ridicule</u>. It is even possible that this minefield has been laid and encouraged by those who wish to suppress any tendency to free thought and exploration. Ridicule of course is a powerful weapon and is used by those who are either ignorant of the subject they are ridiculing or have ulterior motives and vested interests. Ridicule issues from the very mean spirited and dismal mentality of those who themselves are incapable of helping or of creating anything worthwhile.

It is noteworthy that animals happily have no concept of ridicule. They do not engage in it, neither do they understand it. If you were to attempt to ridicule a dog for example, the worst response you could get is a cocked head and a bewildered look of puzzlement. More likely, he will wag his tail and join in the merriment*!* Unfortunately children do understand ridicule, and they are very sensitive to it. It is a distinctly human failing, probably due to our highly sophisticated language abilities and a mean streak found among certain individuals. Nevertheless, it is evident that more civilized societies do not engage in ridicule as a punishment.

NOTE 4: Much of the material in this book has been inspired by the works of writer/philospher *L. Ron. Hubbard* who created and founded the international mental/spiritual technologies of Scientology and Dianetics. The world owes him a very great debt for his tireless work and incredible genius.

> "Ridicule is the first and last argument of fools": (Charles Simmons).
> "Truth and ridicule are mutual anathema, - they form a dichotomy. Those who love truth, hate ridicule. Likewise it can be said that those who love ridicule, hate the truth". (author).
> "Our wretched species is so made that those who walk on the well-trodden path always throw stones at those who are showing a new road", (Voltaire).
> "The world would rather cling to a wrong idea than accept a new truth", (Seneca, a Roman statesman).

ABOUT QUANTUM PHYSICS: (taken from National Geographic's 'The Science Book'; page 326). ["Quantum physics investigates the behaviour and conformity of matter in the atomic and subatomic realm. The basis of this physics theory - with the physicist *Max Planck* (1858-1947) as its originator - was largely established during the first half of the 20th century. Quantum mechanics is regarded as one of the most important foundations of modern physics"]...

["In 1890 almost all mysteries in physics appeared to be solved. In fact, the future Nobel Prize winner *Max Planck* was advised against studying physics, ostensibly because there was nothing new to be discovered"].

> "Mysteriously, there may ... be a special kind of energy accelerating the expansion of the universe". (From 'The Science Book': by National Geographic, - in reference to so-called 'Dark Energy').

THE ROLE OF THE MIND: physics as it is today, has more or less apparently reached an *impasse*. For example, the automobile over the years has received improved electronics, engine, fuel system, wheels, brakes, aerodynamics, and so on. Yet it is still a car, - a metal box with seats, an engine, four wheels, and obliged to follow a paved road. The engine is still the primitive reciprocating piston petrol burner. The concept of a car has not actually changed an iota in a hundred years since its inception in 1906! The Space Shuttle is apparently a leap forward in technology - but in fact, it is simply an airplane with a rearrangement of already existing technologies and small improvements.

These improvements are merely added colors and shapes on an already existing canvas using already existing brushes. Technology is expanding laterally, its upward thrust has played out. To go further upward with new impetus we need to learn new rules and new laws of physics. We must find new media to construct our works of art.

Until now, the mind as a dynamic factor in physics has been ignored, but inevitably, the new physics - based upon quantum dynamics, will incorporate the mind as a literal integral factor in its structure. The mind creates and projects Energy (known as *Ch'i* in China). The precise nature of this Energy is now understood and explained herein. Hence, it can now be generated in great quantity using quantum-electronic technology.

CONTENTS

430 A ~ STATES OF MATTER & THE RIDDLE OF TIME
430 AA ~ CIRCULAR MOTION AND RELATIVITY (Part 1)
430 AB ~ CIRCULAR MOTION AND RELATIVITY (Part 2)

SECTION 04.2 —- ADVANCED DYNAMIC ENERGY SYSTEMS
42-131 A ~ WAVE TYPES
42-143 A ~ INPUT ENERGY, RESULTS, AND DELETERIOUS EFFECTS
42-143 B ~ RADIO-ACTIVITY & QUANTUM ENERGY
42-144 B ~ MAGNETS AND Q. ENERGY
42-145 ~ MATTER, ENERGY, AND THRESHOLD PHENOMENA
42-146 A ~ MODIFYING ATOMS & ELECTRON RESPONSE
42-147 ~ THE MAGIC OF Q. ENERGY
42-148 A ~ THE MODIFICATION OF MATTER
42-148 D ~ QUANTUM DYNAMIC FIELDS
42-149 ~ LOOSENING ATOMIC BONDS & THE LIFE-FORCE
42-150 ~ SPACE-TIME IS PARTICULATE & INCREMENTAL
42-151 A ~ SPACE-TIME DISTORTION, GRAVITY, & RELATIVITY
42-151 B ~ THE ROTATING ANNULUS EFFECT
42-223 B ~ FLEXIBILITY OF THE PHYSICAL MATRIX
42-223 BC ~ ELECTRONICS, RELATIVITY AND THE TIME FLOW RATE
42-223 C ~ ELECTRONICS AND TIME
42-223 CC ~ MOTION DEFINED
42-223 D ~ RELATIVE MOTION AND THE SPACE-TIME CONTINUUM
42-223 E ~ RELATIVE MOTION AND SHARED SPACE-TIME CON-
TINUA
42-232 B ~ QUANTUM ENERGY AND THE BIZARRE
42-232 C ~ POLTERGEIST PHENOMENA & GRAVITY FLUCTUATION
42-232 D ~ POLTERGEIST AND QUANTUM PHYSICS
44-133 ~ DC CURRENT YIELDS PULSED DC
44-134 ~ HIGH ENERGY FLOWS & SPHERICAL CONDUCTORS
44-135 A ~ ELECTRONIC FLOW AND QUANTUM PHYSICS
44-136 ~ Q. ENERGY ACCUMULATION AND LOSS
44-137 ~ ACCUMULATION IN AN ISOLATED CIRCUIT
44-141 B ~ QUANTUM ENERGY CONDUCTION IN MATTER
44-141 C ~ INTERNAL LATTICE ATOMIC EMISSION OF Q. ENERGY
44-142 ~ GENERATING QUANTUM ENERGY
44-143 ~ ELECTRONIC Q. ENERGY GENERATOR/ACCUMULATOR
49-225 ~ ORGANIZED MOTION AND STRUCTURE
49-227 A ~ TIME TRAVEL
49-227 AA ~ THE VELOCITY OF LIGHT
49-227 B ~ EINSTEIN, ATOMS, VELOCITY, AND TIME

SECTION 05 —- THE NATURE OF GRAVITY AND ITS CONTROL

SECTION 09 —- COSMIC SECRETS

SECTION 9000 —- ADVANCED COSMIC SECRETS

SECTION 01
(A NEW FORM OF ENERGY)

100 ~ OPENING DECLARATION

> "There is hope for the future… and when the world is ready for a new and better life, all this will some day come to pass, in God's good time". (*Jules Verne*: 'Twenty Thousand Leagues Under The Sea').

> "…There must be one overall truth. One truth that blends all into a unified whole. It does not have any contradictions, any falsehoods, or any inconsistencies. Every branch of knowledge must blend into it, harmoniously". (*John Strong* in his book, 'The Doomsday Globe').

< secrets of the universe >

One has often heard this catch phrase used - usually in a jocular vein, as if acknowledging that they should exist but that no-one will ever discover them - ha ha! In fact, the idea that the universe has secrets that still exist has become a stale joke. The secrets I am referring to are the very fundamental secrets of which only a few have been so far discovered. One of these is electricity and its many applications.

Let us list some of these fundamental so-called 'Secrets of the Universe'. There is no reason why these may not be called such since - until their discoveries, they were indeed 'Secrets of the Universe'.

fire and its applications

the power of steam and its applications

electricity and electrons, and their applications

all matter is composed of atoms

matter consists of molecules and crystals and that these are in turn composed of groups of atoms

chemistry and chemical interactions with other chemicals and electromagnetic radiations (photo-chemistry)

electromagnetic radiations, their natures and applications

the electromagnetic spectrum

ultra-sound and X-rays

nuclear fission

nuclear fusion

There are of course many more discoveries which rightly or wrongly deserve the appellation "Secret of the Universe". It will be noted however that almost all of these discoveries have been made in the twentieth and the late nineteenth centuries, i.e. the birth of the technological era.

It will also be noted that the more recent discoveries are also the more fundamental. Moreover, these are ever more concerned with the essential, particulate structure of matter and its behaviour. However - and more to the point, it is commonly acknowledged that the phrase 'Secrets of the Universe' implies a special knowledge that is denied to mortal humans; that it is reserved for the Gods, whoever and wherever they may be.

The simpler the fundamental essence of something it appears, the wider its applications and its power in technology. More recently, physics has become enamoured of the very large and the very small. The latter is exemplified by the branch of science called 'quantum mechanics' wherein matter (and space and time) is found to behave in bizarre and unpredictable ways. Paradoxically, we find that the microscopic and the macroscopic are intimately related in the process of nuclear fusion, which powers the stars.

New discoveries in quantum physics indicate that not only do energies exist which interact with subatomic particles, but further that these very fine energies likewise *interact with the human mind!!* This is astounding. Hence quantum mechanics brings us into an arena which may be thought of as truly belonging to the bizarre, the fantastic, and indeed the very fundamental essence of everything around us: [please see note 5].

We begin to perceive that perhaps all is not what it appears to be, and that the so-called or regarded "solidity" of our everyday experiences has a somewhat unsuspected, shifting, and unstable existence. We perceive further that we are now approaching what may be truly thought of as the "real" and ultimate *secrets of the universe*.

Until the present, we find that science and physics has been bogged down due to its lack of understanding of the full range of the electromagnetic (Em) spectrum. Technology we find now is spreading out laterally instead of upward and vertically; - like the mushroom cloud of an explosion. Furthermore, there are secrets of the universe which have not been discovered for the simple reason that they are *not hidden*.

In fact, they are in plain view for the most part, for all to see - but are cunningly disguised. *The nature of this disguise is their boring mundanity and our own inability to perceive and recognize them*. For example, there are some old movies wherein is shown diamonds being hidden in a sparkling glass chandelier. They are not really 'hidden' but they cannot be seen by those without perception, vision, questioning, or intelligence. In other words, they are *camouflaged*.

In many cases, these "secrets" are so readily seen and observed that they are ignored as being too insignificant. In this regard, they are not analyzed nor recognized, and consequently remain well "hidden".

Electricity is a case in point. Static electricity has been with Man for countless millions of years in the forms of friction and lightning but it took the genius of educated men to recognize a heretofore unrecognized potential. Now we may observe the realization of this potential in today's global technology which is based almost entirely on electricity.

The power of steam is another outstanding example.

The understanding of the power of steam gave birth to the industrial revolution and consequently a changed world. Flowing water, bubbles, balloons, smoke-rings, vortices, etc., yield further clues to these secrets.

The time has now come for the revelation of a new as yet unrecognized power and force. The amazing thing about it is that it has been known and recognized by Man for thousands of years. Its complete and full potentials however have not been understood nor exploited.

Δ

It has already been suggested by two physicists, (the Nobel Prize winning *Dr. Millikan* who discovered the electron, and *Dr. I.I. Rabi*), *that atoms and molecules constantly radiate electric waves, - the frequencies depending on the atom's size or energy level or substance, i.e. the number of electrons and electron shells contained*.

Very simply explained, this book makes the statement that there is a form of Energy which has been overlooked by Western physical sciences, and its potential which has been overlooked by the entire world. Due to its fine wavelength, this Energy interacts with both matter at the sub-atomic level and, perhaps incredibly, the mind or will. Hence, it forms the dynamic bridge between mind and matter, i.e. the physical universe. Further, the nature of this Energy is now fully understood and herein revealed, and thus lends itself to generation in large quantities by present technology.

There is an old truism which states: "if you wish to conquer your enemy, - first understand him": ('The Art of War', - *Lao Tzu*). All of existence may be characterized by a single dichotomy – in fact the very first dichotomy of all. This dichotomy consists of the Physical and the Non-physical: Matter and Spirit. Matter of course is not the only characteristic of the physical universe. There is also Energy, Space, and Time. Matter however may be considered to be the densest and the best representative of the material, i.e. a generic term - followed by the three others in the order given. *Time* then is perhaps the most ethereal and intangible, and the least understood aspect of the universe.

We may regard the non-physical as friendly, and the physical universe as our enemy - since it is the material aspect which ultimately delivers evil, grief, and pain. Proceeding from this understanding we may now examine the universe to fathom its most fundamental elements.

First of all, we can consider that 99% of all matter (and virtually all electromagnetic energy) is involved in and proceeds from the process of *Nuclear Fusion*, (i.e. stars and proto-planets, etc.). Likewise we may say that 99% of all matter consists of hydrogen, helium, ionized atoms, protons, and electrons, etc., - to be found in stars and the interstellar spaces. (Hydrogen is the primary fuel for N. Fusion, and Helium is the by-product of this process). Subsequently, all of this matter is in the state of electronic particles and flows and the produced electro-magnetic energy.

> Note: many of you will disagree with ideas presented here in this work and many of you will possibly agree with them. In the final analysis however, this is not really important. My main intention is to disseminate this valuable knowledge, - to plant the seed of awareness, and to show that the new technology of quantum Energy - albeit bizarre, can be a very useful one, and indeed that it is viable and available with presently existing technology.

Further, it is to show that while quantum physics is based upon physical energies, the rôle of the mind or the observer in this technology is of more than passing significance. Moreover it must always be remembered that when dealing with quantum Energy, we are likewise dealing with a force which does not obey "normal" rules or expected physical behavior. In other words, "one must expect the unexpected". The reasons for these discrepancies will be explained. All will be made clear, never fear!

Δ

If we rest quietly for a moment and examine our natural environment, we will perhaps become aware that the creatures and other phenomena we are surrounded by may in their own right be viewed as strange and bizarre. The denizens of the insect world and sea creatures especially produce many strange and truly bizarre and 'alien' forms.

Tornados and waterspouts can be counted among the particularly bizarre nonbiological dynamic natural phenomena. However we do not normally view these as bizarre or even strange for one important reason. If we were confronted by a beetle the size of a horse, we would call it "bizarre" but at its normal size, we would not give it a second glance.

Most of these phenomena are introduced to us during our childhood. The only time we find them to be startling or intriguing is when we are introduced to them in the first instance as adults. After that they become relatively uninteresting and mundane. Children especially find new phenomena exciting - albeit entirely acceptable. Interestingly, as one grows older, one becomes more cynical and less willing to accept the exciting and the dramatic as being true ingredients of the nature of things.

The same is true of new technology. When electricity was first discovered, it was an exciting subject which everyone talked about, although denigrated by some. Now since we have become "electronically sophisticated", we take all our wonderful and truly astounding electronic technology for-granted, - we casually throw away cell phones, computers, and televisions - yet these are really true marvels of technology: [note 3].

As elsewhere explained, clay is a useful and commonly used substance of which bricks are usually made. Useful as this substance is however, its physical, chemical, particulate, and electronic composition is still largely a mystery. Nonetheless, its practical and observable properties have been in use for thousands of years.

Here is a case of something where particulate and quantum level properties are largely unknown and yet throughout the millennia, perhaps many billions of tons of clay have been produced and used in building and construction. Much of our present technology consists of such elements which are widely used and yet their essential nature is little understood.

Another case in point is of course electricity. In fact, most of our present technology is predicated upon unknown quantities and qualities. Interestingly, the more widely something is utilized, the simpler is its fundamental nature, and the greater is its essential mystery. Two ultimate examples of this are space and time…

Today and for thousands of years, there has existed a recognized form of energy which has been in practical use, and yet its essential nature is virtually still unknown. This energy is known as "Chi" to the Chinese, and "Ki" in Japan.

Now for the first time in Man's recorded history, this energy form is fully understood and explained in detail. Its new practical applications are described, as well as how it may be generated in very large quantities - thereby introducing new and amazing physical phenomena into the arenas of science and technology.

The new technology presented in these pages will at first glance seem bizarre, - impossible, - and the result of an overactive and feverish imagination. Yet a hundred years from now it will all seem commonplace, just as we accept giant aircraft which are true marvels of technology.

This new technology - based upon a form of energy, i.e. 'quantum-electronic Energy', is fundamental to the structure of matter, of energy, of space, - and of time. It is ultimately an electro-atomic energy and hence its primary *physical* source is the atom. Since the electron is the ultimate source of virtually all electromagnetic and atomic quantum phenomena (as will be shown), we may then call the electron a "quantum particle".

The Energy, which is the central theme of this book - may therefore be called "quantum Energy": [please see note 2].

In large quantity, the Energy will have profound and noticeable effects upon matter, - including biological organisms. The body is essentially a chemical structure undergoing constant chemical, electrical, and quantum activities. It is in fact little more than a semi-rigid/liquid mixture of active chemistry. In this context, it may be unkindly compared to an organized compost heap. Moreover it has the 'frightening' ability to move (i.e. to be animated), apparently of its own volition! Chemical reactions and processes are essentially molecular, atomic, and electronic interactions - and atomic and electronic interactions are essentially quantum-electronic!

As will be explained, the body is also capable - on a minimal scale, of actual *transmutation* of elements - such as the conversion of hydrogen to helium, - the same process indeed which constitutes - in a more energetic setting - nuclear fusion in the stars. (Higher elements such as calcium are also transmuted by biological organisms, as will be shown). Q. Energy can be made to create a quantum-electronic field, and objects subjected to this field will be modified at the quantum, i.e. subatomic level.

Now the biological body is surrounded by and permeated by its own subtle quantum Energy field so that there can be an intimate interaction between the two fields. The strength of the E-field is important, naturally - and as is the case with all forms of electromagnetic energy, too much can have undesirable effects. Favorable effects would include regeneration, restructuring, repair, and replacement of vital parts. (To be discussed).

Regained physical and mental youth may be expected. Undesirable conditions such as diseases, the effects of ageing, and so on may be eliminated, - and even reversed! (It has been found by researchers that 80 ~ 90 % of all illnesses are essentially psycho-somatic, i.e. originating in the mind. Indeed there is evidence that in fact, _all_ disorders and ailments, regardless of their nature - are _essentially_ psychological in origin. That is to say that it is the mind which is primarily responsible for the body's condition, and this can even include such conditions as broken bones, severed limbs, and so on! - this too will be amply explained).

The mind apparently and evidently is associated with the physical structure of the brain, the _endocrine system_, and the body, - and since the brain can be modified structurally (including the _Pineal_ and _Pituitary_ glands) with the application of quantum Energy - the mind itself can therefore be beneficially affected.

This Energy is associated with the mind, as already indicated, - and also constitutes a link therefore between quantum mechanics, matter, energy, space, and time, and other phenomena such as Poltergeist activity: [note].

The body's structure and behavior is observably influenced by the mind and the fields and flows of Energy in the body generated by the mind. This has been amply demonstrated by the arts and sciences of Eastern medicine such as Acupuncture, _Shiatsu_, (_Feng Shui_), the martial arts, - and by the more rigid methodology of researchers here in the West.

Furthermore, anyone with a small ability to perceive the personal conditions of others in his society will note that facial form and expressions, and body language will inevitably and always indicate the mental stability, state, and personality of these associates, - and also their social and economic status.

Hence it is possible (albeit admittedly fantastic) that by the use of quantum-electronic fields and energies, the body may be restored to its former health and youthful condition, and the mind to its youthful energy and intelligence. The rate of these healing processes (i.e. involving the _rate of time flow_, indeed) can vary and depend largely upon the intensity and frequencies of the applied field.

These types of phenomena will be recognized by researchers into paranormal and Poltergeist activities, (and writers of science-fiction!). Such interesting data unfortunately are disregarded by the popular media.

SPECIAL NOTE: from time to time reference will be made to energized or 'Energized' (with a capital E̲) atoms or matter. When something is energized, this refers to common electromagnetic energy. When a substance is *Energized* however, this means having its energy level raised by means of the intrusion or infusion of quantum Energy. We will use the term 'En/energy' or 'En/energized'. This is usually in reference to the degree of energization of an atom or mass. (To be further enlarged upon).

NOTE 1: one may see that there is a connection between matter, electro-magnetic energy, and quantum Energy. What we will explore in this book is the relationship between electro-magnetic energy, matter, energy, space, time, quantum Energy, - and the body, mind, will, perception, and thought. Hence we may see that here is a train of linkage from matter and energy - including the body, to the mind through the via of quantum Energy.

Imagine if you can the joy of a child or indeed anyone - and especially the parents of a child who has suffered the absence of a limb or organ and then - through regrowth and regeneration - has regained what was previously absent or lost. This then is the promise of the power of quantum-Energy technology. (This is not something I would mention lightly unless I were totally committed to the truth of it).

This book was written with the primary intention of exploring and explaining the interactions of quantum Energy, electromagnetic energy, and matter, - as stated. As I wrote and researched however, it became ever clearer to me that matter, energy, space, and time represent together an interdependent, interactive, *"holistic dynamism"*. In other words, apparently and evidently no single element can be regarded or examined, - or indeed modified *without the inclusion of these other factors*. [note 4].

NOTE 2: the term "quantum Energy" refers specifically to a range of frequencies between one (1) Ångstroms and that of the electron at rest, i.e. $^{1}/_{40}$ Å. Quantum Energy or 'E-energy' actually exists as an entire spectrum parallel to the electromagnetic spectrum that we are familiar with, having the full range of frequencies from the electron and higher up to long radio-waves. However herein when "quantum Energy" or simply "Energy" (capitalized) is mentioned, the reference is to this specific range of the spectrum. There is no reason why higher frequencies should not extend beyond that of electrons, but we are not concerned with these.

NOTE 3: it is also interesting to observe that as societies and cultures grow older, they too become more cynical and less childlike in their embrace of the world around them. This increasing lack of enthusiasm and of ensuing antagonism then may be said to be some of the indicators of the impending death of the society.

NOTE 4: here is a very simple example. Let us suppose that you have a vertical and cylindrical bottle half-filled with water. The water assumes a more or less cylindrical form. If you tip the bottle on its side, the geometry of the water will change. But note also that the geometry of the enclosed air-*space* also changes. Hence by changing and modifying the form of the water (matter), you have likewise changed the form of the *space*.

NOTE 5: actually it should not be astounding if we understand that the mind is composed of quantum Energy patterns and flows. Likewise, it shouldn't dismay the biology scientists who aver that the mind is only electrical impulses in the brain, - electricity itself being a quantum force.

100 A (1) ~ MAN'S GREATEST FAILINGS

> "the only things we know about the quantum world are the results of experiments", (from Paradoxes And Possibilities).

> "We have to remember that what we observe is not nature herself, but nature exposed to our methods of questioning": (Werner Heisenberg).

Man's greatest failings are not in the fields of physical technology - which he has adequately mastered; they are in the fields of the mind and spirit. One of these failings is the mode of *assumption* wherein Man assumes certain apparencies to be true, but which are not necessarily so.

For example, when one travels to another land and culture he may expect to find similar belief systems and methods of procedure (and indeed assumptions, too). If he has any perceptive ability, he will soon discover that this is not so and consequently becomes confused. This is called "culture shock". He is confused because many of his "stable data" (i.e. belief systems, preconceptions, illusions, and assumptions) have been invalidated – he does not know how to behave in a given situation.

Living in his own familiar environment however, he can continue to retain these false assumptions and his life is consequently unaffected. Nonetheless these assumptions remain with him, and he is thus prevented from knowing the greater truth, and lives constantly with misconceptions. In this regard, he has a lack of understanding and it is this lack which causes a limitation of his spiritual awareness.

Furthermore, when two or more groups or individuals, etc, are in conflict, he becomes a third party (unless he himself is involved). Hence when the viewpoints of one side (A) of the conflict are *first* presented to him he will assume that this is the valid one and he immediately forms a mental bias (unless he is already biased in favor of one or the other (B)). This will occur even though he has not heard or considered the viewpoints of the opposing side (B) in the conflict. This may be called "premature evaluation", or "primary assumption".

Again, if these other viewpoints are presented at a later time or date, he will still cling to his assumption that the *first* party viewpoints are more valid than the second party's, - unless he perceives them to be obviously in error. This is because he has already created a belief system, agreement, and 'reality' based on the first presentation. Hence, he has created a stable datum or "comfort zone" which includes this 'primary assumption'. He has established for himself a new system of stable data corresponding to the first set of viewpoints and is now reluctant to invalidate this new stable datum or belief system. It is this inability to modify his viewpoints and stable data which are one of his great failings. [please see note].

Another of man's failings is his limited scope of vision, i.e. his inability to entertain a greater range of possibilities in a universe not directly perceived. To a large extent however, the general population is at the mercy of social institutions such as the general media, the education system, promoted social values, and peer pressure, etc.. These have generally pooh-poohed any suggestions that valid alternate realities and perceptions may indeed exist.

By doing so, they have created a mental climate or mind-set in the general population. Hence human society has been burdened with such words as "fantastic", "incredible", "bizarre", and so on. The very existence of such words themselves create a cautionary and circumspect state of mind.

NOTE: a very similar sitation presents itself if one side of the dispute is that of a well-known friend. Regardless of whose version he hears first, he will likely side with his friend - even if his friend's version is obviously irrational. Supporting irrationality is of course condoning it and is irrational in itself. It is quite likely however that he knows that he is supporting irrationality but would rather not take a chance on losing a friend. We observe then that friendship - or rather the fear of loss, can support irrationality and through this mechanism, irrationality can *and does* pervade an entire society. Furthermore when bystanders observe this subservience to irrationality, they tend to view it as socially acceptable and hence practice it themselves. Those in society who passively accept irrational thought and behaviour are actually contributing to the eventual demise of the society*!* In fact this is happening now and continually*!*

100 AA ~ THE REASON WHY

Children are very curious creatures and - having a keen insight, are capable of asking the most penetrating and fundamental questions. (I know this because I was one myself). They will ask questions such as, "why is an apple round?", "why do we eat?", or "why do we have trees?", or "why are you old?" You may attempt to provide an answer for those questions and the child will usually respond with another "why?" question. The more such questions he asks, the more fundamental they become.

A possible answer to the second question would be "in order to live". The child will then likely ask, "Why do we live?" this is a very profound question indeed, and an answer is provided herein. The more fundamental the questions become, the more difficult they are to answer. The reason for this difficulty is partly because we, as adults in a technological society, seek answers of a material and quantifiable nature, - partially to satisfy our own developed need for an answer to the overall scheme of things. In fact the more fundamental the questions are, the closer we approach the realm of the quantum level and the nature of mind itself.

Children however - whether consciously or not, are great game players and such a questioning child is seeking from the adult a verification of his own suspicions that things are not quite as they appear to be. The child intuitively suspects that physical existence has an underlying non-physical basis and is seeking a satisfying answer to this end.

The question "why do we live?" may be classed with other such questions, such as "what is space?" – "what is time?" - "why do we exist?" - "what is our purpose?". These are questions which cannot be answered with direct reference to physical or material considerations. In fact, these questions are more easily answered by resorting to non-physical concepts.

Why for example is space full of stars? Then again why should we - ostensibly biological meat bodies, born from meat, and nothing but meat - ask such abstract and non-physical questions? Why would a lump of meat ask such probing and philosophical questions? Indeed, why should a lump of meat compose and enjoy music, - and so on?

We are obliged to answer these with a statement such as, "because that's just the way it is". This is typically the kind of answer a child might readily give if asked the same question. In other words, in matters of fundamental importance, our answers are no more sophisticated than a child's. It is an answer without seeming logic, measure, or practicality, and which therefore does not appeal to the adult of modern industrialized society, - and most especially the dedicated physicist of today.

(Here is an example of the above. All electrons in the universe are precisely identical, and must be so in order for the transmutation of elements to occur in the Sun's atomic furnace. Why are they all identical? No can answer this except to say, "that's just the way it is"!).

This answer then implies that; 1. There is no logical physical reason for the stars' existence and, 2. There is the suspicion that some *non-physical* agency is responsible for it. This further implies that other fundamental phenomena exist likewise due to non-physical agencies. As is usually the case, it is the simplest answers which closest approach the fundamental truths, - and it is the simplest answers which modern techno-Man does not like to accept.

100 AB ~ THE UNIFIED FIELD THEORY

> "only those who attempt the absurd will achieve the impossible"…(M. C. Escher).

> "Most of the things worth doing in the world had been declared impossible before they were done": (Louis D. Brandeis).

> "It is difficult to say what is impossible. For the dream of yesterday is the hope of today and the reality of tomorrow": (Robert H. Goddard – genius of American rocketry).

This theory of *Einstein* is also known as his General Theory of Relativity. Einstein - we are told - spent his last forty years trying to resolve and complete this work. We are told that this theory was never completed by him. This may actually be true, but one must be careful not to read this as an implication that it was never completed at all.

His effort was an attempt to create a mathematical description or formula whereby all the electromagnetic forces and fields found in nature can be unified and explained, and their relationships discovered. These forces include the so-called "weak" and "strong" forces within the atom, electrostatic fields, the magnetic, electromagnetic, and gravity fields.

By completing this work, Einstein would have for example, established the relationships between electromagnetic fields, gravity, and inertia. It would have unlocked all the secrets held by matter, energy, space, and time. Inadvertently, it would also have unlocked the existing links between these aspects of the physical universe, and astoundingly - the mind*!*

Such information is too powerful and beneficial to Man to be suppressed, buried, and forgotten. With this knowledge, all the ills and distress now suffered by Mankind will be vanquished and swept aside. The virtually free energy offered by Nuclear Fusion would become a reality, and the frontiers of space opened wide.

This book then has the power to change history and the world of man.

100 AC ~ THE SECOND BOOK TITLE

The second title of this book is 'A New and Complete Unified Field Theory'. The original Unified Field Theory was compiled by *Einstein*, and we have been told that this Theory was left incomplete. Hence the benefits which may have accrued and been realized are now 'lost' or hidden. A true, plausible, and *complete* Unified Field Theory must be able to explain *every* phenomenon which occurs in nature, as well as within those technical discoveries devised by man: [see notes 1 & 2].

These natural phenomena include fringe areas of research such as Ball-lightning and Poltergeist activities, etc.. Many however do not accept the validity of the thousands of corroborated, personal eyewitness reports of such events (which can be found recorded in most libraries), and generally this type of information is not found in the popular media. Hence most amateurs, scientists, and physicists are discouraged, mainly by peer pressure - from officially investigating such reports and phenomena.

Therefore, if we are to completely understand the universe we live in, we are obliged to examine these bizarre and unusual events, and the evidence which strongly supports the interaction of the mind at the physical (i.e. sub-atomic) level: [please see note 3].

A true and 'Complete Unified Field Theory' then is obliged to include and explain such phenomena, - distasteful as it may seem to the orthodox. (It should be understood by the Reader that the concept of "bizarreness" is a subjective perception, - an adopted viewpoint. Objectively, there is no such thing as something "bizarre").

NOTE 1: Such discoveries would include the simplest; i.e. electricity, the atom, the electron, the proton, the neutron, the photon, radio-waves, and X-rays, etc.. Included also are gravity, inertia, Relativity, Black Holes, space-time distortion, 'Worm Holes', time flow rate variations, the velocity of light, and so on. These can all be precisely described, and their fundamental natures revealed with the concept of quantum Energy and quantum electro-dynamics.

NOTE 2: there is a great deal of data available in easily accessible books which describe research and experimentation of a 'bizarre' nature ostensibly utilizing Einstein's theories of space, time, Relativity and so on. The internet also has many such references. A primary book on the subject is *Charles Berlitz's* 'The Philadelphia Experiment'. Such research is still continuing today. (See also 'the Montauk Project', 'Rainbow Project', etc.).

NOTE 3: if you are interested in such things, you may mention to someone the extant theory that the planet is actually hollow, like a teapot. Many people will simply snicker, turn on their heel, and just walk away.

The strange thing is that these people won't even exercise an intellectual curiosity to ask how it is you accept such a theory. They decide there and then that you're a crackpot, - and that's that*!* I think that this is a distinguishing characteristic defining such people. They have limited curiosity due to limited perception, awareness, and interest generally. We may also say that these factors constitute *general ability* and effectiveness as a Being.

100 B ~ RELATIVITY AND PERCEPTION

Proof for one individual is not necessarily proof for another, - and vice-versa. The world we live in seems on the surface to be quite predictable and comfortably 'solid'. We like it this way and so anything which tends to upset our established notions of predictability and "normality" is shunned and ignored in the hope that it will eventually go away or disappear.

However the laws of the physical universe do not normally permit this to happen. We are stuck with harsh reality. But *what is* reality? - is it the comforting predictability and solidity we have been raised to live with, or is there something else not perceived to this 'cage of unseen bars'.

We are all very familiar with objectivity and the *objective* world, and very little conscious attention is given to the *subjective*. But in fact if one takes the time to consider this, he will perceive eventually that indeed <u>all</u> our experiences in life - including perception - are in the final analysis only subjective. Relative motion is an obvious case in point: [please see note].

Imagine that we have a 'Maypole' device rigged with a weight suspended on a single cord. If the Maypole is caused to rotate, this suspended weight with its cable will revolve around the pole. As the rotation rate increases, the weight and its cable will begin to move away from the pole as it moves with increasing angular velocity. Eventually, a high rate of rotation could be achieved theoretically where the cable and weight should be extending horizontally and perpendicular to the pole.

We can see then that in effect - by the law of relative motion, that the universe itself with the distant stars, and the planet itself, is revolving about the pole. Now suppose we were to have the Maypole stationary with the weight motionless and hanging inertly. If we were to somehow cause the *universe* to rotate about this Maypole we should expect - according to our concept of relative motion - that the same phenomenon occurs. (In this exercise we shall ignore the planet upon which the Maypole stands).

In other words, the suspended weight should move away from the pole and in due course perhaps extend the cable horizontally. So far, the train of events seems reasonable and logical *according to our notions of relative motion*. However, the ways in which universal laws operate *are not necessarily bound to our concept of logic*. The universe is vast and seemingly "designed" to prevent any such experiment.

We can safely say then that this trick is an impossible one. But further to this is the fact that it is also practically impossible to find the precise center or axis of proposed rotation of the universe. The universe - like a Chinese painting, has no defined borders and only exists where you look for its existence. Moreover the universe is in a state of constant flux and flow. Obviously then if we cannot define the borders or limits of something, then we cannot find its geometric center – it would be like looking for the center of a cloud! [see note 2].

(Another way of saying this is that as you approach the "ends" of the universe, the laws of physics as we know them will begin to break down and strange things will start to happen: - More on this later).

If two such *'Maypole'* experiments were conducted in separate locations, the exercise presents a further paradox of logic. The physical universe, like a wheel, could not logically rotate around two separate centers of rotation or axes. Therefore if we have two *Maypoles* or any two such rotating objects, the concept of relative motion is illogical when we attempt to view the universe as a single macroscopic reference point.

To illustrate it another way, - imagine an ice skater rotating (pirouetting) on an ice rink. Relatively speaking, the rink is revolving about the skater. However if there are two or more skaters, each rotating on the rink, we cannot <u>logically</u> see how it is possible for the rink to be revolving (Relatively) about each one of the skaters simultaneously. Yet from a Relativity standpoint, this is precisely what we are obliged to say is happening! In fact for this logic to hold, we must refer to the individual viewpoints of each of the skaters. The logic breaks down when referring to the observers in the stands.

The logic upon which we are so dependent to stabilize our concepts and perceptions of the universe around us becomes shaky indeed! Concepts like Relativity and relative motion, although seeming to be physically <u>objective</u> superficially and "solidly" founded, become little more than abstract <u>subjective</u> ideas.

If there are a dozen pirouetting skaters on the rink, each one observes the rink revolving around her. This then indicates that while Relativity is not always valid *objectively*, i.e. for the bystander, it is completely valid for each skater involved *subjectively* in the process. This suggests then that Relativity is not an objective reality but *is only valid for each observer's point of view*. (>*Thomson Experiment*).

In other words, Relativity is a purely <u>subjective</u> phenomenon. In the same way, the rate of time flow for each individual is not objective as we like to believe. It is in fact again wholly subjective, - including our observation of clocks, the turning of the Earth, and so on. Likewise, a given *space* can be constrictive for one individual and too big for another.

The point being made here is that the physical universe can be convincingly physical (objective) but when closely scrutinized and analyzed, actually becomes quite elusive, intangible, subjective, abstract, and in fact, - not quite what we expected! (please refer to the *Rainbow* analogy; 300 A). This becomes profoundly true and observable at the quantum physics level.

Hence it appears that Relativity, the space-time continuum, and quantum Energy share a whimsical relationship, - and an attempt to understand this relationship must be made by the committed space traveller and all others who would understand time, space, and energy. This book then is just such a dedicated attempt.

NOTE: this concept of relative motion actually has much to do with Relativity as proposed by Einstein. We are discussing relative motion in reference to the universe which has an immeasurably large mass and consequent gravity component. Relative motion concerning high gravitational fields is the essence of Relativity. Any motion in space must be *de facto* <u>relative</u> motion or it is not motion at all.

This motion is in the final analysis relative to the universe as a whole. The physical universe then may be regarded as the <u>base</u> within which all motion occurs; - in fact very similar to the moving wheels within a clock. It is the entire clock which becomes the base for the relative motion of the wheels! Moreover, the clock need not be fixed; it may be falling from an airplane. ("the whole is simpler than the sum of its parts"… Willard Gibbs).

NOTE 2: a cloud is an interesting example of how the organized universe works. We may say that the cloud is an organized entity. It appears to be a discreet object and indeed, one cannot say that it isn't.

A cloud consists of water droplets arranged in a given volume. However, these droplets are constantly forming and evaporating within seconds of time, yet the cloud remains seemingly intact and discreet.

Some quantum physicists think that even a piece of matter behaves similarly in that the constituent atoms continually form and disperse as energy, - the energy being used to create replacement atoms. Yet the mass remains apparently un-changed.

100 C ~ A NEW REALITY

> "It is a fool's prerogative to utter truths that no-one else will speak"… (Wm. Shakespeare).

> "So-called reality is but an illusion; - illusion however is the reality" (*author*).

There can be little doubt that through the ages, Man as a social group has been blinkered and blindfolded. Until the late nineteenth century, he was greatly inter-ested in philosophy and the spiritual universe. In the twentieth century with the ad-vent of industry, mass production, commercialism, and material acquisition, we have become even more blindfolded. The materialism we find ourselves immersed in keeps our attention focussed on objects (and bodies), and the means to acquiring these and having a better one than the neighbour.

The ancient Egyptians were a very civilized people and their entire culture and society was structured around purely spiritual, abstract, and noble ideals. This was both literally and figuratively a "Golden Age". (And it is very interesting to note the evident and apparent connection of the two forms of 'gold'). During the ages when the great pyramids were built, their culture was extraordinarily high. They were ruled by a priesthood and ruling class originally attuned to lofty and spiritual ideals. This is likely why many people are drawn to and fascinated by this ancient civilization.

They were concerned with their origins from space, the Sun, Moon, and stars as sentient beings, - and longed to be re-united with them. Their pyramids were built with this consideration in mind, to be tools or devices by which they might both communicate with the gods, and at the end of their Earthly lives, - join them for eternity in their heavenly abode.

Since then, the world has descended ever more deeply into a swamp of war, anger, deceit, terrorism, drug-dealing, destruction, corruption, anti-social behavior, inflation, pettiness, and so on.

The question is often raised, "what is reality?" This question is frequently asked by Man because he has been so completely isolated from the truths concerning the mind and the physical universe. These truths indeed constitute our unknown existence. If Man should ever discover these truths, he will at last discover the intimate nature of ultimate reality. These truths lie in the mind and indeed, for our immediate purpose, - quantum physics.

In the old black and white '*Flash Gordon*' movie series of the 1940's, one can find all manner of bizarre creatures created by the fertile mind of the script writer. Such entities were the 'Golem'-like "Clay Men". These humanoid creatures were able to *merge* with the clay walls of certain caves. When I saw this, I was fascinated and struck by it, thinking it was a marvelous concept. (Indeed, in Medieval Jewish lore such a creature, i.e. the 'Golem', was reportedly created by a *Rabbi Loew* of Prague using long forgotten and hidden magical techniques found in the 'Kabbala').

Bizarre as it may seem to our own experience, this is an example of the possibilities in a new reality and the coming age. "Time-travel", "parallel spaces", the merging of objects, space-travel, and other strange ideas excite and stimulate our imagination. These bizarre concepts all received their genesis immediately after the cessation of World War II when science fiction writers began expressing such ideas. (Credit must also be given to those important writers of the late nineteenth century Victorian era such as *Jules Verne* and *H.G. Wells*, etc.).

From what source did these futuristic ideas suddenly spring? The likeliest source is an experiment conducted 'in secret' by the US Navy during the war in 1943, as described by *Charles Berlitz* in his book 'The Philadelphia Experiment'. Soon after these experiments, the bizarre details leaked out to the public and were adopted by writers as fertile material for new ideas.

Evidently, during the course of the experiments, a ship (the 'USS Eldridge', later renamed the 'USS Andrew Furuseth') became invisible to Radar, the naked eye, and the camera. Not only this but further, it was claimed that the ship became the source and terminal of a spatial-time anomaly, sometimes referred to as a "time-tunnel" or "worm hole".

Reportedly, some crew members were transported along this 'tunnel' into another time-flow or continuum, - or as sci-fi writers like to call it, another 'dimension', or alternatively, a 'parallel universe'. Other members of the crew became *merged* on an atomic level with metallic parts of the ship, and after the experiment consequently had to be cut free.

Δ

Be forewarned therefore, this new technological reality is not a wishy-washy whimsical idea. Neither is it for the faint-hearted. With this new physics of matter and mind, this new technology becomes a fierce new reality whereby the weak can be made strong, the strong made stronger, and the distressed made happy.

With this new technology, Mankind may once again return to his former state of grace in which his primary concerns are social values, ethical conduct, joy of living, and love and caring for one another.

The desire for the acquisition of material objects which fuels our present mundane existence will come into disrepute since - in the new reality postulated here, such things will firstly not be required, - and secondly, will be so easily obtainable as to reduce them to insignificance.

100 CC ~ THE ORGANIZING INFLUENCE
> ("the *organized* whole is greater than the sum of its parts"… author).

The entire physical universe, including matter, energy, space, and time may be said to be a seething cauldron of quantum mechanics, with q. Energy as the organizing influence. Called by many great natural philosophers including *Plato, Aristotle, Mesmer, Kepler, Galvani, Reich, Hieronymus, et al*, the "Organizing Energy" or something parallel, - it is responsible for the dynamic organization and structure of all that exists.

This includes sub-atomic particles, atoms, the crystalline and non-crystalline structures of matter, - vortices and tornados, sand dunes, beach corrugations and so on. Moreover, all the organized structures of nature including trees, and animals, etc., are the consequence of this organizing influence.

In the case of Mankind, we find this influence strongly manifested with structured activities such as tradition, culture, education, law, engineering, medicine, chemistry, and electronics, etc.. Indeed many of these fields are not the exclusive domain of Man, for we find in many levels of the animal kingdom the equivalents of tradition and culture, engineering, law, education, and even medicine. Hence, we find as life organisms that we have much in common with our little furry friends.

In fact, in every field of endeavor which characterizes Man and life generally we find *organization*. Indeed all life, and likewise the entire physical universe are characterized by organization! The unifying common thread to all of these systems of organization is quantum Energy.

Motion at all levels is characterized by incremental steps of organized energy. When we walk in the park, we are expressing this fundamental law literally "step-by-step". All clocks work by generating successive pulses of energy and motion. The wind blowing across the surfaces of water and fields of wheat create organized waves through time. (It is interesting to note that fields of wheat are composed of discreet and individual stalks which can be considered to be *particles* responding to the dynamic energy of wind. Each of these particles in turn is caused to oscillate like pendulums but they do not actually wave. It is the particulate group action which gives rise to the wave phenomenon). [See Note a & b].

As the Earth turns by rotation, this motion itself is characterized by the incremental exposure of land surfaces to night and day, cooling and warming, sleeping and waking, contraction and expansion, off-shore and on-shore breezes, - and so on.

Man is born and dies according to cycles. All of these cyclic activities are expressions of organizing influences which are inherent to all levels of life and the universe. The most fundamental influence of all is quantum Energy and the *primary* source of all q. Energy is, as we shall see, - the non-physical agency, mind/will, or Being.

NOTE a: being of an obviously particulate nature, these ears of wheat behave according to quantum rules of behaviour. Hence, we can use an equation to describe this behaviour and the energy contained in the waves. This equation is $(E = pv)$ where (E) is energy (p) represents particles in motion and (v) is velocity. (Note that (v) is not (v^2)). Please also note that in this equation and other similar to and derived from it, that *the symbol (=) does not necessarily mean equal to; rather it means equivalent to, which is not the same thing*.

NOTE b: We observe that a field of wheat provides a medium for waves created by the wind. It should be noted that firstly, the medium of wheat is particulate. It is this particulate nature of wheat or any medium which permits wave energy to manifest. Without a particulate nature no medium can transmit or support wave energy. In a sense, we may say that the wheat becomes a dynamic fluid at the moment an energy wave passes through it. The same may be said for any substance no matter how rigid. (Hence, a solid mass plus a *suitable* energy form yields a fluid).

The wave creates a fluid-dynamic situation within the mass. Secondly, we should note that the particles (i.e. the ears of wheat) are behaving as a group. The important distinction here is that the ears of wheat are not bonded to each other, yet they are still able to function as a group, - like the molecules of air, oil, or liquid helium. In solid matter, the situation is different since the particles (e.g. atoms) are bonded together with electronic forces.

It is these forces which coerce the atoms to act together as one discreet unit. The ears of wheat by contrast have the freedom to "disobey" the conformity necessary to contribute to the wave energy. Another thing to note is that the ears are suitably juxtapositioned near each other to create the wave phenomenon. For example, if each ear were placed several meters from each other, an energy wave could not be said to exist.

Similarly if the ears were each positioned so that it was touching its neighbour, the friction created would tend to nullify any wind-wave action. Hence, we see that any medium which supports waves must be of a density which corresponds with the type of wave energy of interest. The wave action in a medium of unbounded particles is more of a quantum phenomenon, whereas waves created within a solid mass are more of the mechanical variety. This then constitutes the difference between the mechanical universe and the quantum universe; i.e. bonded particles, and unbonded particles - operating together as groups.

It is also interesting to note that while the general behaviour of the wheat-ears constitutes the wave action, there may be individual ears which do not obey in a given frame of time. In fact, there may be an entire population of ears that do not obey in this time frame. In another time frame however, they do obey. (They may be considered to be 'taking a break'). This small percentage or population is unimportant in a short time frame. It is the conforming majority that constitutes the phenomenon.

100 CD ~ NATURAL AND ARTIFICIAL

When we state that something is <u>natural</u> or has a natural source, - what do we mean by this word? Natural means "of nature" and when we refer to nature we immediately think of biological organisms, stones, water, planets, stars, galaxies, and so on. We must also include other natural energy forms such as lightning, rotation, and gravity, etc..

Further, we must also include so-called "un-natural" occurrences such as Poltergeist phenomena, ball-lightning, visions, ghosts, *Doppelgangers*, etc., - and other weird or bizarre phenomena which inevitably get reported to various agencies, etc., (and never heard of again!).

Now many believe that such manifestations do not occur and that they are the results of "imagination". But then what is imagination? - and from whence originate visions, ghosts, Doppelgangers, shadows, and what-not? Obviously, the individuals who report such phenomena are having a real experience. When others do not have the same experiences, they claim these experiences are not real.

The question then is, are they real or are they not real? Certainly they seem to be natural occurrences, and they are certainly <u>real</u> to the individual having the experience, - this we must admit to.

If an individual were capable of, let us say, moving an object with his mind (e.g. psychokinesis), are we not obliged to say that this is a *natural* occurrence and phenomenon? Likewise if he were capable of causing objects to appear and disappear, would we not also ascribe these phenomena to natural sources? Should we call them artificial? What then means 'artificial'?

We commonly believe that something which is "artificial" is man-made, such as a machine or a skyscraper, etc.. But Man himself is also a natural occurrence, or so we believe, - so surely the products of a natural creature are themselves natural, just as are the nests of birds and squirrels.

What about the natural constructions of other animals - such as the lodges and cleverly built dams of beavers, or the warrens constructed by badgers and rabbits, etc.? Are these to be considered natural or artificial?

Like subjectivity and objectivity (and indeed any dichotomy pair), the distinction becomes blurred when examined under the strong light of reason and logic. The superficial stability and the apparent dichotomy, i.e. subjectivity/objectivity -

begin to demonstrate in-stability, elusiveness, and dissolution. In other words, they begin to "fade out", like the borders of a Chinese painting: [please see note].

"Natural" and "Artificial", like all dichotomies - thus are only constructed concepts by which we are able to stabilize the apparent "solidity" of our physical experience. The purpose of this and other articles then is to demonstrate that the mind is a powerful tool, and that <u>any reality</u> may be constructed. However, I get ahead of myself…

(Note: the word 'solid' unfortunately has several disparate meanings. It can mean *rigid* as opposed to fluid; it can mean the opposite of 'hollow' or porous like a sponge. It can also describe rigidity as opposed to flexibility. Sometimes the word is used to describe relative density or heaviness.

(Moreover the word is also used to describe the 'physical presence' (pp) of an object. This is a condition opposed to that of an *ethereal* state wherein an object shifts into another space-time condition, much like a cloud. In other words, something may be "solid" or "ethereal". This word 'solid' is of such fundamental import in the language that it is recommended that its various meanings should be indicated by other new words. 'Rigid' is one such, 'ethereal' is another. Hence, care should be taken when reading this word. The term (pp) is found in the equation $(E = {}^m/_{pp})$ which will be explored later).

NOTE: the Chinese style of painting presents us with a wonderful metaphor for the universe itself. If we imagine the painted picture to represent reality, then the fading out at the picture borders represents an entirely new unconsidered reality - in the context of the picture. This is the canvas itself - which is the true reality of the painting, although we ignore it because it is more aesthetically comfortable to do so. We would rather imagine ourselves to be living in the beautiful picture than outside it. We do not wish to be confronted with the harsh reality of a manufactured and paid for canvas, itself demonstrating that the beautiful picture is actually nothing more than an illusion.

102 A ~ LIFE ENERGY AND ENTROPY

< please refer also to 100 CC > According to the science of astronomy, the universe - with its countless trillions of tons of plasma, rock and metal, etc., exploded out of a pinpoint billions of years ago in an event called the "Big Bang". Ever since that moment, the universe has been in a state of expansion and decay - we are told, undergoing a process called entropy, - which means that all organized systems in the universe move toward disorganization and chaos, and that all energy forms eventually degrade into heat and microwave radiation. [see note].

(This certainly begs the obvious question; if the universe is on a one-way trip to chaos and disorganization, then from whence came its original organization, and how can it contain any organization at all?).

If this is true (and some astronomers now suspect that it is not), then it is interesting that we should find any order at all – and especially interesting to find clusters of millions of stars (pulsars) all pulsating together *in unison*. (Such a group is called a "Quasar"; - Quasar 3C273 is just such a group. Its overall diameter is seven light years).

It is also singularly amazing - if we are to believe this teaching, that a purely mindless mass of matter and random energy moving inexorably toward entropic oblivion should by itself give rise to forces (i.e. sentience, life, and the 'life-force') which are capable of bringing directed order out of lifeless matter. And not just simple directed order, but order of such sophistication that it is capable of putting Airbus 380's into the skies of our planet, and instant global communication into the laps of our children.

This cannot be by any stretch of the imagination, a step in the direction of entropic oblivion. Are we to believe then that "mindless matter" is capable of such a feat?

Wilhelm Reich himself declared that his discovery, - a new kind of energy which he called 'Orgone' was an *organizing influence* - even on a cosmic scale, - "hence", he claimed, "the universe can never 'run down'".

NOTE: how could one judge whether this "pinpoint" is big or small? The estimation of the size of a 'point' within a Void is impossible since there would be no reference dimension. This point could be vast, but no-one could know since the concept of "vast" would have no meaning. When physicists use the term "pinpoint", it too is meaningless.

102 B ~ THE FUNDAMENTAL NATURE OF THINGS

Technology today is well developed and serves to fulfill many of our needs and desires. It is however largely a result of trial and error since there are still many unknown qualities possessed by the physical universe. We use technology but to a large degree, we do not know the nature of the components we use. For example in televisions and computers, etc., we employ the energy of electricity, but in fact we know very little of its fundamental nature. Even though we understand how to use it, all this has been learned by *trial and error* - not by understanding and prediction.

Likewise we know how to use and apply matter, energy, space, and time, but of their true and fundamental natures we know virtually nothing. This is especially exemplified in instances of Poltergeist phenomena and ball-lightning, etc.. We can reduce all technology, whether it is mechanical, chemical, or electronic, to the interactions of atoms and sub-atomic particles. (In mechanics for example, we are confronted with problems of friction, heat, lubrication, inertia, and momentum, etc. - all essentially quantum-electronic phenomena).

Scientists experiment with and investigate quantum physics to determine what occurs under various circumstances. However the more fundamental their researches become, the closer they approach a point in their investigations where they can no longer say, "This or that happens because of these or those conditions". They must all eventually accept the fact that something *is* simply because it *is* ; it exists because it exists*!* Accordingly, we may also ask why a red car is in a parking lot. Obviously it was put there*!*

Again, we may ask why all electrons are evidently identical in structure and function, why do they all possess a negative electric field, and why are they all supremely interchangeable. This is conformity of mind-blowing dedication. There can be no explanation or "why" for this, - they *just are*. This of course is precisely the simple but profound wisdom a child would give*!* There is however one explanation remaining for those who are willing to accept it. That is that this situation is a <u>created</u> one*!*

Physicists will balk at this concept, - the *simplest of all explanations*. Yet truth itself is founded upon the utmost simplicity. Likewise simplicity is founded upon Truth. Physicists do seek after truth in their work, *but only as far as it agrees with their preconceptions, experience, peer pressure, and education*. Beyond this they will not accept it - unless one can offer proof of a very convincing nature. They will not accept it because they are physicists, - both by education and by personal nature.

103 A ~ THE POWER OF SIMPLICITY

> Truth is founded upon simplicity…simplicity upon truth. Likewise, real beauty is founded upon simplicity and truth. (author).

> The simplest hypothesis which promises to resolve the greatest number of quandaries is probably closest to the truth… (author).

> In the "good old days", life really was better. Life was simpler and people were willing to give more and receive less. Paradoxically perhaps, this is the way to true happiness. (author).

What is the single and commonest element of existence which causes us to enjoy a walk in the park, to play with a dog, to be attracted to children, to prefer life in the countryside, or to travel to less advanced cultures? That single element is *simplicity*. It is also evident that true simplicity is also beautiful: [see notes 1 & 2). (It is also valid that true beauty is simplicity. We may therefore equate beauty with simplicity).

If we observe children and animals, we can see that they delight in very simple things. Playing with a ball or a stick, splashing in a stream or pond, digging a hole in the soil, and so on. Indeed if nature could talk, she would say she never intended for us to have automobiles, watches, computers, airplanes, televisions, and so on. If you stand back for a moment and look carefully, you will see how incredibly bogged down we are with all this technology and the unseen infrastructure which supports it.

A child finds great joy in being permitted to carry a shopping basket, for example. Mankind will someday rediscover his hidden inner child-like self, and likewise find simple enjoyment. Like children/animals also, he will find himself less interested in complexity and things bought and sold.

The fundamental nature of Man and life everywhere is to gravitate to and to seek out simplicity. We are aware - if not consciously or analytically, that herein lies *the real secret of happiness*.

There is a device which has been in existence for many years and in fact, various forms of this device have been in use for thousands of years in places such as ancient Egypt, Rome, Greece, China, Japan, etc.. It is perhaps one of the first ever man-made "automatic" devices. Automaticity is the underlying and essential principle of the device.

This automatic device does not produce anything and does not "do" anything except perform a very simple automatic function. It has no practical value as would be appreciated by a farmer of the stone-age, bronze-age, iron-age, medieval, or feudal lifestyles. It is not manipulated in any way as a 'tool' would be. In fact its sole function is simply to be *observed*, - and yet it is probably the most used, the most manufactured, and the most sought after automatic device in the entire world.

Furthermore, this device does nothing more than measure or record an abstract and somewhat subjective concept which Man continually creates with his mind. This concept cannot be seen or indeed sensed with any of the senses and yet it is of utmost significance to every living creature.

I refer of course to <u>clocks</u> and <u>time</u>. A clock is not necessarily simple in construction but its concept and *raison d'être* is the simplest thing imaginable. This utterly simple thing - this abstract concept - has the power to create a multi-trillion dollar industry which in turn has produced and sold no doubt trillions of these devices globally. This is a classic example of the power of an idea. [see note 2].

>"The whole is greater than the sum of its parts".

(This should be better said, "The *organized* whole is greater than the sum of its parts").

(It has also been said that, "the whole is simpler than the sum of its parts"... Willard Gibbs).

There is probably no individual in the world who does not own a timepiece or does not have access to one*!* It would seem then that Mankind is, because of his obsession with the material world, also inevitably obsessed with <u>time</u>*!* This is observably a peculiarly adult human trait. The more technical and mechanized a society is, the more concerned it is with time. Primitive societies which live almost as an integral part of nature (such as the Australian or New Guinea aboriginals) have no need for, and are not interested in the measuring of time, nor the concept of timepieces, - except maybe as novelties and decoration, or status symbols.

Observably also, animals are completely divorced from any need or desire to measure time, - as are children. Children - it is observed, have only the very simplest of notions regarding time, - there is "now" and there is "sometime", or "just a minute" - and as we can observe, they are very much concerned with "Now". Interestingly also, children seem to have no interest or concept of past events. They can be taught to understand 'a few minutes', or 'tomorrow', but beyond this, everything is a subjective dream world.

There is yet one more splendid example of the power of simplicity, - and that is electricity. Electricity itself is as simple as rubbing two surfaces together. Its earliest beginnings with Wimshurst machines and Van De Graaf generators were in the Victorian era. (Both of these machines operate on the principle of friction). [See Note 3].

In just a hundred or so years, we have seen how the sciences of electronics have revolutionized our world and enabled Man to voyage to the Moon. Indeed today we cannot conceive of life without this amazing yet simple power. It has insinuated itself into every tiny crevice of life.

Astro-physicists tell us that over ninety-nine percent of all matter in the universe is in the state of <u>nuclear fusion</u>. A simpler dynamic and automatic energy process truly does not exist. The process of nuclear fusion requires that only two parameters be in juxtaposition. They are hydrogen, the simplest and most abundant element, - and electronic flow, the simplest and most ubiquitous energy manifestation. The hydrogen represents matter, the electronic *flow* represents energy, time, and space.

>"It turns out that beauty in science is something much closer to the notion of simplicity", (K. C. Cole writing in 'Sympathetic Vibrations').

NOTE 1: anyone who has taken an interest in the so-called "Mandelbrot Sets" (created by *Benoit Mandelbrot*) will be familiar with "Fractal Geometry". These computer generated geometric forms are truly infinite in form and extend into the cosmic or the quantum level. It is proved for instance that all natural forms in nature such as trees, animals, mountains, rivers, and so on can be described and actually pictorialized using this generated geometry! In other words, a computer using this program can create virtually any natural scene one can imagine!

Furthermore, this powerful form of mathematics is based upon a very simple mathematical equation which is: ($Z = z^2 + c$). This is further evidence that the simplest elements will effect the most all-encompassing and effective creations. And we are all familiar with *Einstein's* ($E = mc^2$).

(Actually I don't have a clue what Z stands for, but the answer is out there, - known to 'Fractal' mathematicians.).

NOTE 2: it is possible to construct a triangle, each side representing in turn, - Simplicity, Beauty, and Power. These three are interdependent and increase or diminish together: (more later).

NOTE 3: If we were to make a list of the simplest things imaginable, we would find that all these things are closely associated with quantum physics and the electronic states of matter. Such a list would include for example the states of solid, liquid, and gas, coil springs, pendulums, sharp edges, stretched piano wires, bells, light and heat radiations, flame, sound waves, etc., - and of course electricity itself.

The simpler something is, the closer to the fundamental (quantum) nature we approach. Complex structures such as airplanes and computers, etc., are merely agglomerations and combinations of crude matter and simpler elements. We see then that the entire physical universe is constructed essentially of elements based on quantum physics and electronic principles, i.e. 'quantum-electronics'.

> "Everything should be made as simple as possible": (*Albert Einstein*).

103 B-2 ~ THE POWER OF AN IDEA
> "There is one thing stronger than all the armies in the world, and that is an idea whose time has come", (Victor Hugo).

Try to think of the simplest thing imaginable. Is it the bow and arrow, a sharp edge, a pointed stick, fire, the wheel, an explosion, paper, ink, the pen? These are all very simple concepts and their power has been demonstrated throughout history. Wars have been won and countries have been conquered due to these important contributions to technology and science. Is there anything simpler than the things listed?

There are simpler things, but they are things which we do not usually consider when looking for such. We have not yet considered the <u>mind</u> nor the products of mind, such as thought or emotion, speech, vision, hearing, taste, and smell, etc.. Evidently and apparently the simpler something is, the greater power it accrues. We may here also consider the numbering system, the alphabet, fire, electricity, and the concept of organization. These are probably five of the most powerful forces in existence, and anyone is able to admire both their simplicity and far-reaching power.

The numbering system, writing, speech, and organization possess perhaps more power than all those things yet listed, and yet they exist not as solid objects but only as concepts, products of the mind and sustained by the mind. Without the mind, they do not exist. What could be simpler than these last four?

Simpler than any of these things is an <u>idea</u>, which consists in its basic format of a single image picture formed in the mind. It is *ideas* which have captured the imagination of men and which have produced all manner of technologies, both physical and cerebral. The religion of Islam for example is nothing more than an idea and yet it had generated the largest land empire the world had ever seen before the British Empire.

(The British Empire quite possibly was never conceived of or planned intentionally. It simply grew from a small beginning and was fed by the desire to obtain wealth for England).

103 BB ~ AMBIGUITY

When we examine the material world, we find that 'solidity' is relatively un-ambiguous. In other words, it is easy to find agreement upon the existence, form, and function of solid objects. Solid objects can be seen, felt, weighed, measured, tasted, smelled, and subjected to various kinds of tests and measurements - both direct and indirect, with relative ease. Solidity is the easiest thing for any group of individuals to agree upon.

Liquids however are slightly less subject to accurate analysis due to factors such as absorption, evaporation, chemical combination, shape-changing, and so on. Like-wise, we find that gases and electronic plasmas become increasingly ambiguous and less subject to human control. This is even truer with electromagnetic energy and phenomena. As we move away from the material aspect then we find that am-biguity increases. In the case of quantum Energy (i.e. 'Chi', etc.) ambiguity in-creases further as is demonstrated. For example, - how many people do you know who would even agree upon the very existence of this energy form?

There are those who believe in its existence and there are those who do not. It is interesting to note that everyone agrees upon the existence of the mind, and yet it is not subject to any kind of acceptable *direct* scientific analysis. Any analysis made of the mind has to be a subjective one. Everyone knows that he has a mind (in most cases) and can therefore accept as proof the evidence or apparency that others have minds also.

If those matters which are initially deemed ambiguous become personal expe-riences, they are no longer viewed as ambiguous. Hence we may say that things which cannot be agreed upon are ambiguous, and those things which *can* be agreed upon are accepted as unambiguous, i.e. 'real'. This ambiguity however does not mean they are non-existent, or that they cannot be 'proven', - for example radio-waves.

There are those who claim that spiritualism is nonsense and that those who claim to be spiritualists are in fact charlatans. Yet there are those who claim that spiritu-alism is very real - that there are real unseen forces at work. The same is true of fortune-telling methods such as the Tarot Cards, the I-Ch'ing, the Pendulum, Dows-ing, Homeopathy, and so on.

Hence we find here another ambiguity which resolves itself when it becomes a personal experience. The subject of infinity is rather ambiguous since on the one hand it is a useful mathematical concept, and yet nothing in our experience seems to *verify* the existence of an infinity. However if one were to personally experience an infinity, this ambiguity would cease to exist. Because mathematicians find this tool useful, they have fallen into a trap of believing that it actually exists some-where. In fact it *does* exist, but not in any physical form.

The point being made here is that anything in the physical universe, i.e. matter, space, energy, or time, may be measured and calibrated in some way and is therefore not ambiguous, - especially to scientists. Anything beyond the bounds of the electronic and electro-magnetic phenomenal universe may be therefore classified as ambiguous. Hence if one encounters a situation which has unresolved ambiguity attached to it, he may then regard it as beyond the realm of the physical. We may say then that the physical universe is not infinite in extent, but it is ambiguous: [see note].

This is especially true with the application of quantum Energy technology. Matter, energy, space, and time may be made to behave in ambiguous and strange ways, i.e. not according to the acknowledged laws of physics. We find this ambiguity already displayed for physicists in quantum mechanics and in various situations and experiments such as the *Thomson Experiment* : (otherwise known as *Thomas Young Experiment* or the '*Double Slit Experiment*').

NOTE: There is a simple example which may be given here. The value of (Pi) - i.e. the ratio of a circle's perimeter to its diameter, has not yet been precisely calculated. It has been calculated to well over a million decimal places, and yet shows no sign of being resolved. Hence, it is an ambiguous value having a tenuous and precarious perch in that which we call the physical universe. It is almost as if the circle is not quite real, and it is easy to see why the ancients regarded it as a symbol of the spirit; - the spirit being regarded as immeasurable and ambiguous).

The nearest approximation to the value of *Pi* is achieved with the fraction ($^{355}/_{113}$). Any greater accuracy is unnecessary and impractical.

103 BC ~ THE APPLICATION OF THEORY

What is a theory? The major themes of this book must be theoretical since they are concerned with principles of the physical universe never before tried. How then do we define a theory? A theory may simply be said to be a statement deduced and extrapolated from known and 'proven' facts: [please see Notes 1 & 2].

The truth of a theory should be self-evident, even if unproven. When a theory is 'proven' to be true, this 'proof' lies in its general acceptance as a fact. Indeed this can be the only 'proof' which is sustainable. It is still called a theory even though it is accepted as 'fact'. There are many theories extant which indeed all society is thoroughly grounded upon.

A theory seems to go through a process of transmutation whereby it becomes an established 'fact'. It is an ambiguity which becomes a 'fact' when it withstands the rigors of repeated practical applications through time. Such theories include the theory of electricity and electrical flow. No-one is really sure of what electricity is or what its flow entails, and yet electricity is now the major sustaining force on the planet (together with oil). It has been 'proven' by practical application. Indeed, repeated successful practical application is only one of a few criteria which truly defines the 'Proof'.

There are many other theories upon which modern life is predicated. Some of these include Einstein's Special Theory of Relativity and his General Theory of Relativity (also known as the 'Unified Field Theory'). In the modern parlance of physics, Theory with a capital 'T' is said to be a proven body of data, however the 'Theories' discussed herein will be treated as 'theories' since there are still many unanswered questions surrounding these data. There are also the theories of atomic structure, of crystal structure, of gravitation and inertia, of atomic fission and of fusion. In fact almost any scientific statement of the cosmic, the mundane, the molecular, chemical, crystalline, atomic, and sub-atomic levels is theoretical. This is even truer when explaining quantum mechanics.

It would appear then that virtually all modern technology is essentially based upon <u>theory</u>. We must also include the Theories of Creation and Evolution. These two theories however do not lend themselves to practical application and hence remain more or less conditions of faith and conviction - that is, states of mind. In other words, these two theories are more akin to religion than to science! In the minds of many however, Creation has a religious connotation while Evolution is decidedly materialistic and humanistic. These two Theories therefore may be said to form a dichotomy.

The Reader may say of this book that it is largely theoretical; *but of course it is*, - it could not be otherwise! The theories contained herein have the advantage of being grounded upon already 'proven' information. Some statements may be said to be leaps of faith, but they are in fact intelligent deductions and extrapolations, many of which are based upon known and accepted facts, corroboration, and observations. Accordingly, the "Theory of Quantum Technology" has a far sounder technological base than do the already accepted Theories of Evolution or Creation.

NOTE 1: the 'fact' that the Sun is powered by nuclear fusion is actually only a theory. Physicists however have developed this theory of fusion by careful observation of the Sun's behavior, spectrographic analysis, and so on. This theory is so much accepted as a fact that science has spent billions of man-hours and trillions of dollars in research and attempts to duplicate the fusion process. Indeed, the concept of nuclear fusion is also so philosophically satisfying that it will never be abandoned.

NOTE 2: a theory is somewhat like an illusion, or the result of "*Legere de Main*". An illusion (> 104 BB) is an apparency created in the mind. It resembles something we imagine to be 'real' yet to an 'objective' observer, does not consist of those criteria which we associate with reality. A theory likewise is an abstract mental creation attempting to become a 'reality'.

A wonderful "illusion" is the *Mobius Strip:* (see 300 AD). Does it have one side or two? This can be hotly debatable and is a good example of ambiguity wherein perception can be either subjective or objective.

In passing, it is interesting to note that stage magicians create illusions which pass for 'reality'; for example, the reflection in a mirror. The joke is that "reality" itself is an illusion, so that the magician's illusion or the reflection - are alike illusions within illusion. Hence an ugly woman who sees a beautiful reflection in a mirror is creating an illusion of an illusion of an illusion! This is a pitiful state to be in as the Being has become so entangled in illusions that he no longer has any grasp on reality at all, and has become completely lost, i.e. mad.

104 A ~ REALITY, SUBJECTIVITY, & WHAT IS PROOF?

> "The notion of absolute truth is shown to be in poor correspondence with the actual development of science...", (David Bohm, British physicist).

> Sanskrit 'Maya' meaning "the illusion of reality received by the physical senses".

There are three kinds of experience which we may accept as proof; - they are; practical application - corroboration - and subjective analysis. Many of you will ask, "but what about objective analysis?" In fact as will be shown throughout the book - ultimately, objectivity itself must be *subjective*, and ultimately all 'proof' is subjective - as will be explained.

There is an old English saying that "The Proof of the Pudding is in the Eating". The meaning of this is a description of the true nature of proof. Eating something for example supplies the 'proof' of whether the pie, etc., is enjoyable or not. One sees that this is describing an entirely *subjective* experience. In fact this saying is suggesting that proof itself, in the final analysis can only be subjective. If I say that the pie is good, would this be an objective analysis? – no, of course not – it can only be subjective since I am stating the results of my own experience, no-one else's.

It can also refer to the use of tools, systems, and so on. If someone creates or invents a new system, etc., to do a certain job - the proof of its efficacy lies not in theoretical analysis, wise debate, detailed drawings, mathematics, discussions, or arguments. The proof lies in its practical application, and this has always been true. It is called, "the acid test".

You may tell someone that you like ice-cream or the color yellow, etc.. He may then ask you why you like ice-cream or yellow. How does one answer such a question?; - you like it because you like it; it is because it is; - there is no reason or logic involved. The liking of something is entirely subjective and is not prone to analysis. Liking or disliking, etc., is therefore not a condition of the mind, since the purpose of the mind is to store information, to analyze it, to make comparisons, correlations, etc..

It is therefore a condition of the *will* (or spirit, if you prefer) which is not subject to any form of analysis, reason, or logic. Reason and logic are mechanisms of the mind, not of the spirit: (explained further elsewhere).

Is anyone able to prove that you like ice-cream? The only evidence that you like ice-cream is the fact that you say so: [see note 0]. Is it necessary for someone else to *prove to you* that you like ice-cream? Of course not; your own subjective reality is 'proof' enough to you that you do.

Let us take a simple example of perception and the elusiveness of this concept we call 'proof'. A tree exists in your own garden or your local park, - i.e. in your own objective/subjective universe - because you perceive it. In other words, you can see it, you can touch it, smell it, and so on. If you could not perceive it in any way, then of course it would not and could not exist in your personal universe. So although we regard a tree as an objective experience, it is also a subjective one if you perceive it. [note 3].

In other words, you only perceive it if you create it, (in much the same way that a clock 'creates' time) - and axiomatically you create something if you perceive it. [see note 10].

In fact perception may be considered to <u>be</u> creation*!* Furthermore if *no*-one perceived it, then it could not and would not by any definition, exist. It is perhaps a difficult concept to get one's mind around, but only because we have our shared reality constantly re-affirmed (corroborated) by those around us with whom we are in <u>agreement</u>. [see notes 1, 6, & 7].

We are so used to saying something is real or objective, etc., and having others agree with us. It therefore does not occur to us that this thing we call reality is in fact created by each and every one of us as we agree. It is a lot like everyone in a painting class - all painting with the exact same style. If this is done repetitiously in class after class, it will be found after enough repetition that no one will conceive of painting with any other style, or that other colors can be used, or that another medium can be used, etc., except perhaps one or two independent individuals with insight, perception, and an adventurous spirit. An entire society could be constructed this way. The Chinese style of painting is an example.

It is an observable fact that virtually everyone likes to conform, regardless of their beliefs and ideologies. This means being in <u>agreement</u> with the *status quo*. Anyone who disagrees is shunned and ostracized, and this is an intolerable situation. Society and groups therefore *coerce* us to conform and agree, - to stay in agreement with whatever the present agenda happens to be. For better or worse, the universe demands and operates on conformity. He who does not conform will ultimately perish, - in a social sense: [see note 5].

It doesn't really matter what the subject of agreement is - so long as you *agree*. The subject of agreement may actually be entirely false and even mad, but this doesn't matter as long as you <u>agree</u>*!* It is this agreement and its enforcement that creates that which we may laughingly call "reality". This is one mechanism which enforces the existence of the physical universe.

(Apparently this is the mechanism of tradition and the consequent creation of culture. According to this mechanism, we see that the Chinese have a traditional style of Chinese painting. If the Chinese knew nothing of the outside world they would probably know nothing of other styles of painting! Hence they have established a reality of their own, as indeed do all people and all cultures in one way or another…

(It is only the true artists and creators of the world who are actually able to "step outside of the box", - to realize that their chains are in reality only shadows, - and to look further than most - and beyond. They have that enviable ability, - which is a true spiritual quality)…

The concept of proof is ultimately *subjective*, and although we strive to achieve objective proof, - there is nothing completely objective about it. In fact, it is a common failing in Western Man that he constantly tries to reduce everything to an entirely objective analysis and quantity, - (the subject of time is a case in point), - whereas the universe (as "solid" as it seems) is ultimately a whimsical and subjective construction. Of this we shall see ample evidence: [see notes 8, 9, & 11].

It will be seen in fact that the universe, as "measurable", analyzable, or gaugable as it may seem in all its aspects, is in constant flux and flow. It annoyingly and irritatingly seems to elude us at every effort to "pin it down" to some ultimate formula such as so-called 'constants' (K) including the 'constant' velocity of light, - (which by the way does not actually exist!). (This will become clearer to the Reader as he progresses). This will be seen in discussions of quantum Energy phenomena later on.

The very word 'proof' itself suggests the notion of perception, insubstantiality, apparency, and conceptuality. When we ask for the "proof" of something, we are essentially asking for a demonstration of that thing which will satisfy *our own concept* of reality. (Like a taste of pie!). When we accept that demonstration as having done so we then call it "proof". If we do not accept it as being in agreement with our concept of reality, we say it is not proof.

It becomes apparent upon analysis therefore that 'proof' is entirely a subjective and hence a personal phenomenon of perception. One man's 'proof' then is not necessarily another's. This can be readily experienced by those who visit cultures which are vastly different from their own.

For example, if a rice farmer in Bali prays to a local god for rain, - and then it rains; that is proof to him that his prayer has been heard and answered. So it seems evident that proof is surely a very personal thing. To ask for proof of something therefore seems inadequate.

(Furthermore anyone within that culture will state that this occurrence is an objective reality, while an outsider is just as likely to claim that it is 'only' a *subjective* perception. So is it a subjective reality or is it an objective one? Who is correct? The answer of course lies with the individual and how he creates his own personal universe and perceptions).

Inevitably then the only way to find the true 'proof' of something is to look for it and find it oneself. To accept someone else's 'proof' is to cheat yourself and be untrue, - especially if this proof does not correspond with your own perceptions. Above all else, be true to yourself and establish your own realities, for indeed - this is the only reality there ever is*!*

Many who read the articles and statements in this book will declare that "such and such is unproven", or that, "known physics does not support that contention". My rebuttal will be that first of all, the central theme of this work concerns a portion of a spectrum parallel to the electro-magnetic spectrum which has been completely unknown and unexplored (or ignored) by modern physics - due to the elusive nature of this Energy.

Therefore there are very few - especially in the world of science, who can claim to be a complete authority on the qualities or effects of this Energy. (See 'corroboration').

Secondly, there exists a vast plethora of things which are 'unproven' yet which physicists today accept as everyday "facts". Examples of these are the structure of the atom, and that electricity is a flow of particles, - that the mind cannot interact with matter or energy, - or that an anti-gravity force cannot exist, or that all stars rotate, or that solid objects cannot pass through walls unscathed, or that cats cannot see anything which we cannot also see.

It is interesting to observe animals and to note what constitutes proof for them. An animal such as a domestic dog or cat will recognize its master returning home by his voice or by his footsteps, and yet will always <u>smell</u> him as a final 'proof' that he is who he appears to be.

For an animal, this subjective, personal, and abstract whiff is the ultimate proof, and if you are his closest friend, - he will attack and attempt to kill a threat on the authority of this whiff. Observably, animals live in a world which is fluid and flowing, - a world which is abstract, composed of ever changing energies, where nothing is measured but is accepted for what it is at the moment. They care nothing for material possessions, - only for the enjoyment of freedom and movement. [note 2].

Humans in turn will accept the identity of someone from merely looking at his face, or hearing a voice. To most people, this abstract exercise of perception is 'proof'. And yet what makes one face different from another? It is an intangible, abstract, and indefinable thing since most faces are physically, virtually identical. (It is evident and apparent that animals are generally quite unable, - or at least have a poor ability to recognize faces, thus demonstrating that humans apparently and evidently have much finer visual recognition abilities).

It is something of interest to note that as humans, we readily accept others into our group, our families, organizations, etc., merely on the strength of our recognition of their facial features. We do not ask them for their I.D. cards in most cases, so this acceptance into our intimate fold is based entirely on our subjective recognition of their faces. The same is true of course for animals and their sense of smell.

In the final analysis then, the Reader must weigh the evidence given and decide for himself what constitutes <u>proof</u>. [See note 4].

There is another form of 'proof' which is highly convincing and that is *corroboration*. For example, if you wake up one morning at an unusual time and you suspect that your clock is in error, you may be able to compare the time given with that of another clock. If now both clocks are telling the same time, this is highly convincing evidence that the correct time is shown by both clocks, and that they both corroborate each other.

The second clock may be called a "backup system". When the time is corroborated by yet a third clock, you will be completely and unquestionably convinced that you do have the correct time. This indeed may be accepted as 'proof' in this instance. What has transpired here to make our assessment of the time so convincing? The answer is <u>agreement</u>*!* The more clocks we have telling the same time, the greater is the agreement, - and it is extremely difficult to disagree with a hundred clocks all telling the same time*!* (It can be done, but it's difficult: - see note 12).

This is the unfortunate quandary which Mankind finds himself bogged down in. He is in such awfully deep agreement with his environment and his associates that he cannot extricate himself from this trap; - a trap it is*!*

Some things in this book are "proved" by the simple expedient of multiple corroborations. For example, there have been many sightings of *ball-lightning* by witnesses who claim variously to have heard or seen corroborated behavior. We now know for example that such entities pulsate and have an internal structure - as does our own star, the sun.

> "Strictly speaking, *no hypothesis or theory can ever be proven*, it can only be dis-proven. "When we say we believe a theory, what we really mean is that we are unable to show that the theory is wrong – not that we are unable to show beyond doubt, that the theory is right", (physicist Gerhard Robbins). (my italics).

NOTE 0: interestingly, most people will actually accept this as "proof" that you do, since they too have experienced this situation. In other words, their experience and yours are mutually corroborating.

NOTE 1: the fact that quantum Energy exists by no means invalidates this assertion. This Energy form has a long evidence in history and even today there are researchers and so-called "metaphysicists" who readily affirm its existence.

NOTE 2: Man will attempt to measure his fellow-man by the material products and advantages which he possesses, controls, administers, or has authority over. These are attempts to know and understand each other. Animals and children on the other hand, know nothing and care nothing for these considerations. They are ready and willing to accept one for what he shows himself to be, - and surely this is the greatest treasure - to be yourself and to be accepted, - and willing to accept the world at its face value, without judgment!

NOTE 3: the following is another example of how perceptions can be altered by simple suggestion. (From 'How to Think Clearly': *R. W. Jepson*; London, Longmans, Green & Co..). "I was present some years ago at a lecture by a professor of psychology. He began by talking to us about *Napoleon's* campaigns and referred to the battle of Marengo, Hohenlinden, Austerlitz, Jena, etc. Suddenly, without warning he produced and showed for a second a piece of white cardboard with a word on it printed with large capitals. He asked us to write down the word we had seen. The majority of us wrote BATTLE. As a matter of fact, the word was BOTTLE!"

NOTE 4: when we say that something depends upon the viewpoint of an observer, we are implying something quite profound. For an observer to adopt a viewpoint requires an intention or decision to subjectively perceive something in a certain way. Hence the dynamics of a spiritual, mental, emotional, or physical situation depends upon the subjective perception of the observer! As explained, this is amply demonstrated to physicists in quantum mechanics: (*Thomson Experiment*).

NOTE 5: even self-styled 'non-conformists' actually conform to a way of life which is approved by other so-called 'non-conformists', since Man is a social animal. To be truly non-conformist one would have to deny his humanity and reject the conformity to dress, conduct, speech, food, housing, walking, use of his hands, the use of amenities such as the toilet, the telephone, and so on. In short, the very fact that one lives and decides to survive in a human society with a human body is proof that he is conforming! One who espouses the benefits of non-conformity is actually preaching a doctrine of "counter-culture", - a euphemism for 'anti-social'. The only way one could be a non-conformist would be to become a bird which flies upside-down and backwards!

NOTE 6: actually of course, the clock creates nothing. It is the one using the clock who creates time by postulating it into existence. This is why he has a clock, - to measure the time he has postulated.

NOTE 7: I may know that there is a blue triple star system on the other side of the galaxy. I may know this if I have been there. You however likely would not know this. For me then this system would be *real* as I have perceived it, whereas for you this system does not exist since you have not perceived it.

This system exists for me but not for you, - it is a subjective reality. If on the other hand you have perceived the same system, then we have both had the subjective experience of perceiving it. The experience is now a shared one and we may now both say that this is an *objective* experience. The more individuals there are who agree with this perception, the more objective it becomes.

NOTE 8: it is a tribute to Man that he has done so much with clocks. With this simple device and its sequential and incremental division of a completely abstract phenomenon, he has 'harnessed' time and constructed an entire global industrial society and culture. It is very likely that the incremental measurement of time is the first prerequisite to establishing such a culture.

NOTE 9: Man has been able to devise a complete technological civilization based on the intelligent and knowledgeable use of matter, energy, space, and time. This civilization may be said to be a culture of objectivity. The products of all this technology and objectivity however, such as automobiles, computers, airplanes, television, and so on are - interestingly enough, designed for *subjective* use. The pleasure one derives from watching TV for example, or from going for a ride in the country is subjective, - and this is what these items are ultimately designed for. Other types of systems, etc., are designed to expedite this ultimate goal of subjective experience.

NOTE 10: let us be abundantly clear on this very fundamental matter. Obviously if you create something, you perceive it. What is not so clear however is that if you perceive something you create it. The Being (you, or I) is continually creating, - it is a non-stop creating "machine". In fact, the Being continually creates, every moment of every second of every day of every year etc, etc. indeed for eternity; he never stops creating. Likewise, the Being cannot help but perceive; it is his nature; it is how he is defined. The only thing in existence capable of *creating* is the Being, and the only thing in existence capable of *perceiving* is the Being. The Being is 'continually' and everlastingly creating and perceiving.

The one aspect is the dynamic mode and the other is the static mode. These are the two aspects of the Being. We may say that the Being can also decide to be both the Cause and the Effect. To *create* is to be the cause while to *perceive* is to be the effect. Now cause and effect only exist in the physical universe and this is simply because in order to have cause and effect one must have *time*.

Cause is the *primary* manifestation of something while Effect is the *secondary* manifestation. Between cause and effect therefore, there is the passage of time, - and this is how we must define Cause and Effect.

The Being *does not* exist in time, he exists and he is, - but he does not *persist* since this must introduce the concept of time. He exists in the eternal 'Now'. It is the physical universe which persists through time. Therefore if a Being creates something, he perceives it simultaneously in the same moment of 'Now'. Likewise, if he perceives something he creates it in the same moment of 'Now'. Creating and Perceiving become one and the same phenomenon so that to <u>perceive</u> something is likewise to <u>create</u> it; it is the same thing*!*

Hence, if you see a tree in the park, you are creating it in your own universe. "But", you say - "how can this be, the tree is already there?" Yes, it is there because it is *agreed upon* by many other Beings. You perceive it only because you have *gone into agreement* with those other Beings. This means that we perceive and create something if we are in agreement that it is there. But always remember that we create it and perceive it in our own universe, no-one else's. You can also create/perceive a tree that no-one else can*!* (This will all become clearer as we progress).

The power of agreement can be demonstrated in many ways. We have all heard about dangerous intersections where car accidents are frequent or common. How does this situation develop? At one time the intersection had no such reputation until it became known that it was a bad one.

One day through carelessness, a driver causes an accident. His argument (perception) is that the intersection is poorly designed in some way. To any observer however it appears (is perceived to be) just like any other intersection. He tells his story to the police, to the insurance agents, to the newspaper reporters, to the traffic court judge, and to his friends at the local pub. The story gets around and people start to believe it, even though they themselves have never had any difficulty at this location.

Then one of these believers has an accident which *he himself has created through his agreements*. This accident is reported in the same manner and the intersection then begins to gain a reputation.

Thus is created a situation of cause and effect where effect becomes cause and vice-versa. The accidents create a reputation and the reputation creates accidents. Which is the cause and which is the effect? It is neither, - but the source of it all is sentient Beings and their agreements. This is where the situation is *created* and comes into being.

Thus is a situation created which is very akin to quantum mechanics, where cause can be effect and vice-versa, or both, or neither! It is akin to quantum mechanics because Beings and their agreements are involved.

NOTE 11a: as discussed elsewhere (108 DE), Western man especially has been trained to be a creature who operates predominantly on the Beta wavelength of analytical thought. Likewise - and in association with this, he is a very left-brain animal - especially the male of the species. Females (human) generally are more right-brain types and likewise are more Alpha-frequency types. Indeed one may draw the conclusion that the right brain lobe is associated with Alpha wave activity while the left-brain is of the Beta persuasion!

One may note and observe that building bridges and designing and constructing aircraft, etc., is a decidedly left-brain activity while raising children, homemaking, caring for others, and group activities are definitely right-brain oriented activities.

This idea will be fiercely disputed by the politically correct who would have us all living in a bland, colourless, concrete world, where nothing happens, where no flags are waved, where no pride of origin is permitted, where real music has vanished forever, where everyone wears grey, and where women have finally achieved 'their' goal to become men!

More so since they understand the reality of right/left-brain, - an issue which they cannot bear to confront, and which they refuse to discuss - along with all the other obvious differences in gender and race, etc..

It should be of great interest for all to note that a truly civilized society will be composed of people who possess right and left-brain lobes which are equally well developed. This should be self-evident in any event, and in fact, the Japanese - albeit of the Homo Sapiens persuasion - had come very close to achieving this through social engineering, education, and the introduction of Western methods of production and business.

Now, with the application of quantum Energy to developing minds of the "quantum Age", mankind will achieve a true step in 'evolution'.

> "You cannot teach a Man anything, you can only help him find it within himself": (Galileo).

NOTE 12: it is interesting to observe that within a city, a society, or indeed an entire country, it is possible to have every clock and watch telling the same time, - second by second, minute by minute, hour by hour, day by day, and so on. This is called _agreement_!

I have lived in Japan for several years. This is a culture dedicated to precision, good order, reliability, and good manners. Every intercity train is punctual, - leaving not just on the minute - but _on the second_! This is reliability! The same may be said for the Germans and the Swiss.

104 B ~ STRUCTURED PERCEPTION

"They do not want my music any more than they want your science. What they want is lackeys to their system". (_Mozart_ speaking to _Mesmer_).

A Structured Perception is quite simply a perception which has been structured artificially by the environment. For example, people raised in different cultures perceive the world in different ways. This is the way we actually perceive ideas and concepts, and the resulting physical perception (which of course is only one form of perception).

Structured Perception naturally results from pre-conception and assumptions, although this is only one source of structured perception. S.P. can originate with verbal or visual suggestion, etc., (as in advertising), television, demonstration by peers and superiors in a social or working environment, belief systems, etc.. (refer to Note 3, & 140 A).

The perceptions and preconceptions of superiors and authorities passed on to others is a powerful and persuasive force in maintaining ideas and beliefs, i.e. the '_status quo_' - and indeed traditions in a society. The full spectrum of the media and the education system are especially strong in today's technological society and are a powerful influence on re-structuring the perceptions and belief systems of the masses of people.

A very simple example of structured perception are the ways in which we view a glass of water. If the glass is half-full, one may say that it is half full or alternatively, he may say that it is half empty. We have all heard this little conundrum, but apparently few have given it much thought. It is the life experiences and resulting perceptions of the viewer which determine for him whether the glass is half empty or half-full.

One may say immediately that he who views it as the latter is an optimist - while perceiving the former, he is a pessimist. This of course is not necessarily true, but may be taken as a general rule. So you see there really is a difference, - it's all a matter of perception and adopted viewpoint. Moreover, there is also a difference in the quantum universe.

Furthermore, you can see that this example is a relative situation where perception plays an important role and determines the perceiver's reality. (Recall the pirouetting ice skater).

Here is another endearing example of perception. In the winter, squirrels will come around the house looking for food. You may put five or six nuts on your window sill. The nuts will remain there for a few days being politely ignored by the squirrels. Eventually however you will find one missing. The squirrel believes that you may not miss one out of several. This is the squirrel's perspective (and system of etiquette) regarding food - which after all is of primary concern to him. He doesn't want to take food which he believes to be yours, but at the same time he wants to eat it, and so he draws a compromise - *assuming* (without analysis) that your ability to count is fortuitously no greater than his. Likewise, it is very difficult for us to conceive of Beings having greater abilities than our own, and further we cannot imagine what such abilities might be. [see note 3].

One of the reasons why a society slowly decays is that when perceptions and preconceptions are passed on to inferior intellects and children, the lesser adults and children - lacking intelligent and proper understanding, then take these ideas as they themselves understand and define them. In this way then these noble ideas - which may also be clever, intelligent, and wise, - tend to lose their razor edge and become duller through multiple translations and misunderstandings, - and the adoption of inferior ideas. They then become less noble, less clever, less intelligent, and less wise, - and ever more so through the passage of time, ages, mis-translations, and social upheavals.

An example of this is religious ritual. At the beginnings of a new religion, - a ritual, a belief system, a ceremony, rite, or practice etc., may actually have a very real and meaningful, practical, even technological significance. As the religion continues through the ages, the true meanings of these practices are lost, modified or altered and re-interpreted. They are also lost simply because they lose their own intrinsic meaning, relevance, and value in a changing society: [see note 1].

This may be due to the changes inherent in a society becoming more reliant on technology or other changes such as climate and demographic changes, the invasions of different cultures, the increasing reliance on money, trade, manufacturing, transportation, modes of transport, and so on. Indeed one cannot discount the influence of those in power with evil intent; and these are usually those 'behind the throne', so-to-speak.

The perceptions of those who constitute this religion have gradually and constantly been re-structured so that the religion as perceived by them today is not the same religion that existed those ages ago. (Christianity is a case in point; - In the early days of Christianity, re-incarnation or rebirth was an accepted fact, just as it is today in Buddhism. However early on, probably during and after the 'Dark Ages' this was all to change, - whether by design, by resignation, disavowement, or by unfortunate happenstance: - see note 2].

NOTE 1: To illustrate a crude example, we can see that a sailing ship with brass cannon is completely useless in today's venue of warfare, so that the traditions of the men who built and maintained those ships, cannon, and equipment are obsolete. Their skills - which may have taken years and great diligence to learn, have no value in today's world. Other examples are of course the wagon wheel and the marvelous ships called "Triremes" built by the Mediterranean Romans and Phoenicians, etc..

NOTE 2: there is strong evidence to suggest that the so-called "Dark Ages" was a severe climate change - a mini-Ice Age which began in 535 A.D. and lasting for about two years, killing off a major percentage of the population around the entire globe; - a literal 'dark age' in fact! Evidence for this occurrence can be found in many disciplines, including climatology, botany, core sampling, vulcanology, and mythology, etc..

The evidence is strong that this climate change was precipitated by the huge eruption of a vast volcano in the area of the Mediterranean Sea, i.e. now Santorini. The volcano's crater above sea level is estimated to be about fifty kilometers wide, - a really, really, big bang! In fact according to reports of the time, the explosion was heard around the globe.

Hence the Western notion of a supreme God being the embodiment of goodness, love, and grace is severely shaken in times of natural disasters and plagues, etc.. During such periods, life must certainly seem like hell on Earth, - the abode of Satan himself (her-self, for the politically 'correct'!).

NOTE 3: As humans, we are easily able to distinguish at a glance the difference between four nuts and five. How do we do this? We know at a glance that a nut is missing and we immediately count them to be sure. The fact is that we already know one is short but we count them anyway because we have been indoctrinated into a society which cannot conceive of perceiving something without some form of rational, logical analysis.

If now we had a barrel full of nuts and a squirrel took one of them while being unobserved, would we then be able to tell if one nut was missing? The answer is probably not. But suppose that Beings of greater ability existed that *could* tell the difference. How on Earth could they do this? For such Beings, it would not be necessary to count the nuts, yet somehow they would be able to know the difference.

So here we see on the one hand analytical analysis of a quantity, and on the other something which parallels analytical analysis but which surpasses it in effect and results. By *knowing* the difference without actually counting the nuts or doing a rational analysis would be tantamount to knowing something without supporting data.

We actually have a term for this and it is called 'intuition'. What is intuition then? It is naught else but simply _knowing_. Is this possible? Well, if it were not possible, we would not have words like 'intuition'. There are many who will argue against the concept of the existence of intuition. The biggest joke however, is that we use intuition every moment of our lives.

For example when we walk from our house to the supermarket, we use intuition to tell us that walking will get us there. We do not calculate this with mathematics, nor do we engage the arts and sciences of statistics. We do not require others' opinions, neither do we consult logarithmic tables, or the passage of the Moon. Quite simply, we *know* that walking will take us from A to B. Since there is no logical analysis or reasoning involved, we are entitled to call this knowingness "intuition".

104 BA-1 ~ CYCLES, CIRCLES, AND DECAY

The entire universe operates on the principle of cycles within cycles within cycles within cycles - *ad nauseam* (and illusions within illusions within illusions, etc.). If we wish to describe a cycle, we may refer to it as a recurring process of action through time, i.e. "a cycle of action". We find that there is probably no machine or tool in existence which does not rely in part on the cyclic principle, i.e. cycles of action, either in its use or in its manufacture.

We can observe for example the cycle of fashions and dress codes through time in a society. Other cycles are evident, such as the rise and fall (i.e. decay) of civilizations, the birth and death of organisms, the cycle of buses as they travel around town, the passages of the Sun, and phases of the Moon, and so on. [See note 1].

It will be observed also that cycles generally get either longer or shorter, - they virtually never remain constant or static. Generally speaking however, it is apparent that the greater social cycles called 'civilisations' are becoming shorter and of less duration as humanity ages.

We can observe a parallel of this for example in the decaying orbit of a satellite and then its eventual plunge to Earth. Hence, although these cycles exist, they become shorter. A cycle whose turn requires the same length or duration completes a 'circle'. However, if it is shorter either in length or in duration than its previous turn, it cannot complete the cycle/circle precisely or fully. It thereby becomes the segment of a spiral or helix - like the segment of a coil spring.

(In this case, the turn of the coil does not meet itself 'head to tail' and is therefore an 'open' turn - and thus incomplete; see note 3]. This is one reason why the helix/spiral form has such significance. Furthermore, if each succeeding cycle is smaller than before, the circle becomes a spiral and the helix becomes a "dwindling spiral" (actually, this latter is a helix, not a spiral).

If we represent this as a three-dimensional diagram of dynamic vectors through time, it takes the form and geometry of a vortex! This is easily and simply demonstrated by a helical coil spring which is larger at one end and smaller at the other. This then represents the "dwindling "spiral".

We have often heard this term used to refer to and describe a diminishing social situation, or to the gradual decay of a business locked into unfortunate circumstances. For example, if social conditions cause the sales of a business to diminish, the profits are reduced and the company cannot be competitive. It is then forced to cut corners and costs to produce lower quality products and services. Furthermore, it finds it necessary to eventually increase prices: [see note 2].

The overall result is poor quality and overpricing. This may occur over several years and during this time period, we will see the company experiencing a gradual decay, i.e. a "dwindling spiral". An entire country can have the same experience. In a strong economy therefore, we can expect high quality goods and services at low prices. This of course stimulates spending, sales, production, - more spending, more sales, and more production - with the result that everyone is happy!

NOTE 1: the Moon is a good example of these ever changing cycles within cycles. As we know, the Moon revolves around the Earth every 28~29.5 days - more or less, depending upon which lunar "month" you are referring to. (There is the 'Solar Month', the 'Lunar Month', and the 'Sidereal Month' - i.e. the Moon's month as measured by its motion against the backdrop of stars).

Its points of setting and rising on the horizon change every day, and these points gradually move further South and then North again.

This incremental motion forms a range of rising and setting points over the horizon. However, even this entire range expands and contracts and is itself moving first North and then South again. As it does so, the backdrop of stars likewise slowly changes in a great cycle which lasts for 2160 x 12 (i.e. 6 x 6 x 6 x 6 x 20) years, - the so-called 'Great Year'. It is probable that even this Great Cycle also goes though its own cyclic periods, including expansion and contraction, and so on, again and again.

NOTE 2: we find this especially true of government-operated businesses which have no interest in competition, even in a competitive market. They are eternally financed by the taxpayer, whether he wants it or not! The result is inferior and overpriced service. It is for this and other reasons that the government should bypass business altogether and get on with the job of governing! This state of affairs is especially true in a socialist or communist country. In fact communism is defined by all businesses being controlled by the state.

NOTE 3: the circle/cycle is not perfectly duplicated in the physical universe and so continues to exist as a spiral or helical segment through time. If it were perfectly duplicated, the turn would become a complete circle, self-enclosing and discreet. This happens to be of great philosophical importance. It is of further import to note that when a cycle of action is perfectly completed, it no longer exists in space and time. When it is *not* completed, it continues to persist. This therefore is yet another mechanism which tends to perpetuate the physical universe, i.e. the misduplication or incompletion of cycles of action.

We may use the rotating propeller to demonstrate this principle. The propeller is designed to create relative motion between it and the air (or water). Hence, in order to produce a desired phenomenon, there must be the dynamic force (the propeller) and the static medium (the air or water). When the propeller rotates, it describes a helix in the static medium. If there is no medium, the propeller simply describes a circle. In this event, no third phenomenon is produced.

Hence in this latter condition, the cycle/circle is a perfect one producing no effect upon the physical universe. If however a cycle is to produce an effect upon the physical universe, the cycle must be *imperfect* and become a helix. A helix then is a dynamic circle through time. The helix form therefore may be taken as a symbol for time.

Hence we see that the physical universe is maintained by compromising perfection.

104 BA-2 ~ DUPLICATION AND PERSISTENCE

If something (e.g. an object, an energy form, or a space) is perfectly duplicated, it ceases to exist in time, and therefore ceases to exist!

If however, it is improperly duplicated it continues to persist - thereby creating time, since time is created by persistence. Any cycle of action is ultimately an attribute of the Physical Universe and the P.U. must contain (misduplicating) cycles of action in order to persist. Furthermore, this persistence is brought about by slight imperfections in the duplication of each cycle and this creates time and persistence!

This imperfect duplication is precisely why things decay and age. An object or body, etc., will eventually decay because this is the mechanism of the universe, i.e. "dynamic imperfection". The body, etc., exists from moment to moment in increments of time, and in each moment it is being created anew. However if it is misduplicated each time, it becomes slightly less new and actually goes through a process we call ageing. Ageing is neither good nor bad; it is simply a necessary mechanism of the universe and its persistence. In other words, misduplication (or dynamic imperfection) is a mechanism 'designed' to promote persistence of the universe, even though it also promotes decay.

Nothing is ever perfectly duplicated because if it were, it would cease to exist, and the P.U. cannot tolerate this heresy, - it <u>must</u> persist! We find this mis-duplication everywhere. We find it in the mis-duplication of instructions given by the boss to his subordinate; we find it in the copying and translation of documents; we even find it in machine controlled mass-production where each article is ever so slightly less perfect than the one before it. Eventually the die or tool must be replaced or re-engineered.

This is why boilers must leak, why shoelaces must wear out, why machines must be maintained, why cultures decay, and so on. Mis-duplication is why things go wrong, and since mis-duplication is an ingredient of the physical condition, we can always expect things to go wrong, - it is inevitable, especially in the human condition where mis-duplication is always lurking just around the corner!

104 BB ~ COMPLEXITY, ILLUSION, & PERSISTENCE OF THE UNIVERSE

The entire physical universe is structured upon the created concepts of the dichotomies. One may select many dichotomy pairs and by careful reasoning and analysis deconstruct these pairs one by one. Let us take for example the concept of left and right.

It is quite easy to see how left may become right and vice-versa simply by adopting a different *point of view* or perception. In space of course left and right, up and down, backward and forward, etc., are meaningless unless reference is made to a stable datum such as a star or planet, etc..

Apparently and evidently also the more complex something is, the more physical it may be judged to be. Complexity then is a "trick" employed to establish the apparent solidity or permanence of the physical universe. [see note 1]. What then is complexity? (And likewise, what is simplicity?).

It is simply (!) the agglomeration of simpler elements. For example, an organization (e.g. a government or industry) can be extremely complex but upon analysis, it will prove to consist merely of a vast quantity of simpler elements. The "mortar" which binds these elements together is the cooperative intention of sentient entities, e.g. people, – or more fundamentally - their collective intelligence and will. (Hence we may find that the entire physical universe is supported and maintained by a collective will! - See note 1].

Let us take a human body as another example. Its complexity is vast, ranging from the electric waves comprising the electron up though the atom and through the range of atomic elements, molecules (including of course the DNA molecule), chemicals, genes, chromosomes, cells, organs, bones, flesh, - and indeed, gender. The simpler and more fundamental dichotomies are the easiest concepts to reduce by analysis. [see note 2].

Note: It is perhaps a good idea to clarify here what an <u>illusion</u> is. An illusion is something which is <u>not</u> real - posing as, and pretending to be something which <u>is</u> real. For example, the reflection of your face in a mirror is an illusion. Furthermore, an illusion is created by the perceiver.

It is the perceiver who chooses to arrange the various frequencies of light reflected from the mirror into a form which he then chooses to recognize as his face. Illusion then is something which is created by the perceiver himself. In other words, he is creating his own reality*!*

(It is observable that this recognition ability is also true for animals of higher intelligence. In fact, this may be a tool for determining the relative intelligence levels of various animals. Indeed one of the definitions of intelligence is the ability to recognize similarities and differences. It should be noted here that intelligence and sanity are not related).

Likewise, time is also an illusion created by motion. As will be explored further, it can easily be reasoned that matter, energy, and space are also illusions. This will become apparent as one reads further.

Another excellent example of an illusion is created with the use of earphones. If we put one earpiece to the ear, we recognize instantly the source of the sound. However, when we put on a *pair* of headphones, an amazing phenomenon takes place. The sound fills our heads and seemingly has no direction. This is an illusion - but it is a very convincing one since we no longer recognize its source or true nature. This is what we call reality (i.e. the physical universe); it is something which exists, and is tangible and convincing - yet we know not its source or origin.

Note: the exploration of reality and illusion may seem irrelevant to some Readers. However such matters cannot be ignored in this new technology of quantum physics. It will be seen that objectivity and subjectivity may be separated at times only with some difficulty.

Quantum physics by its very nature is intimately involved with subjects like perception, illusion, reality, subjectivity, objectivity, viewpoint, and so on, as will be seen. In any event, such a technology as presented herein will likely severely strain the new Reader's credulity so that he should prepare himself in this context for material to come*!*

NOTE 1: another such "trick" is the existence of <u>cycles</u>. It is generally recognized by virtually everyone that the universe and indeed, even social and organized life is composed of and governed by cycles. In a social context, this is largely governed by the cycle of night and day, i.e. the revolutions of the planet. Other important cycles are of course eating, sleeping, working, and so on.

This particular cycle, which mainly consists of "work-eat-sleep", is widespread and most people accept it as a comfortable way to live and survive. It has the advantage of being highly predictable and therefore "safe". However truly creative individuals feel the need to step beyond and out of these cyclic "merry-go-rounds", and to accomplish something which has never been done before.

These creative people feel the "aimless futility" of such cyclic existence, since on a broad level it is hardly progressive and does not seek to discover answers to existence. Hence the universe is constructed like a complex machine with cycles within cycles within cycles - like an intricate system of gears, - indeed like a clock. It is this cyclic nature of the universe which sustains it and maintains as a physical reality. Most people are content to simply ride along on the turning cycles - indeed, just as we are content to ride along on the spinning planets.

The phrase "cycles within cycles" also refers to the fact that cycles themselves have a multi-cyclic nature. It is this seemingly endless complexity which creates the illusion that the physical universe is permanent and here to stay!

Another subtle "subterfuge" - combined with the sheer immense vastness of the universe - is the way time behaves. When an object (like a space-ship with an occupant) accelerates to Relative velocities the rate of time flow for that object slows down. Hence time becomes a 'trap' which every Being with a body wishes to avoid.

This change in time flow rate is only experienced locally, i.e. in the craft/object's own local space-time continuum. Hence friends, relatives, and business associates, etc., will belong to a separated space-time continuum not shared by the object. In order to share the same space-time condition with these associates, one must conform to the same velocities and time flow rate, - hence the enforced trap-like condition.

This is one reason why we live and behave as groups, i.e. in cities, in societies, and even on one planet. This grouping is a mechanism which protects us from the vagaries of time. Many of us live in a city and this way of life is comfortable because it affords predictability and an ultimate form of safety (and recreation). It also provides the protective envelope of a shared space-time continuum which we can all agree upon. It gives us a feeling of "belonging". This means that not only do we belong together in the same space, but also in the same space-*time!*

The existence of danger is yet another form of entrapment. The universe generates fierce energies and dangerous places. Bodies and constructed machines are frail and easily damaged – often with resulting pain (yet another device designed to enforce agreement with the material existence). This is quite a cunning trick since danger is both an attractive element and yet one designed to guarantee permanence. Danger is attractive for indeed, the presence of danger generates excitement and drama in many endeavors and for these reasons, brings pleasure.

Pain is also necessary as a continual coercion for a Being to take care of his body, thereby ensuring that he stays in it! In other words, the constant danger keeps a Being continually focussed on his physical body, thereby ensuring his continued association with it. Danger and pain then are necessary components of the physical universe to ensure its own survival.

By keeping our attention focussed on danger and resisting it, we are in fact focusing our attention on and resisting the physical universe itself. Anything which is focussed upon (i.e. given attention to) and resisted - then tends to persist. In other words, our constant attention to the physical guarantees its persistence. Hence if we resist danger, it persists; if we resist pain it persists; all this resistance guarantees the persistence of the physical universe!

Hence we have a number of elements associated with the physical universe which tend to guarantee its continued existence and persistence. These are: Danger, Pain, the Dichotomies, Complexity, Vastness, Cycles, Rotation (i.e. relative motion), Bodies (the ownership of), Social interaction, Social structure and agreement, Relativity and the flexibility of the space-time fabric. The *denial of truth* is yet another mechanism to be discussed.

Thus, it may be considered and perceived that indeed the physical universe is a cunningly conceived <u>trap</u> which coerces conformity and agreement. As long as one obeys its coercion, one will be reasonably safe - but will also be subject to its entrapment. It is a little like being in your house while it is surrounded by wild tigers. You can't go out to visit your neighbours or enjoy a picnic, but you can be warm, well fed, and safe!

NOTE 2: as elsewhere described, many dichotomies may be considered to be three-part triads or 'tri-chotomies'. For example, not only do we have the dichotomy of left and right, we also have the 'middle'. Hence, if male and female form a dichotomy there must also be a middle element. This element would of course be "no-gender", or "both-genders".

It is commonly known by biologists that there are many creatures which exhibit both genders in one organism and these are called 'hermaphrodites'. There are also bio-organisms (usually sea creatures) which indeed have <u>no</u> gender. These are for example bacteria and amoeba, etc., which reproduce by splitting in two. It is even conceivable that there are higher forms of life somewhere in the universe which also have no gender and which "reproduce" by *electronic* means. Such possibilities will be described and discussed.

104 BC ~ WHAT IS TRUTH?

> "Man occasionally stumbles over the truth, but most of them pick themselves up and hurry off as if nothing happened": (Winston Churchill).

> "The greatest homage we can pay to truth, is to use it": (James Russell Lowell).

> "Philosophy is the science which considers truth": (Aristotle).

> "Ridicule is the first and last argument of fools": (Charles Simmons).

> "Resort is had to ridicule only when reason is against us": (Thomas Jefferson).

> "to speak the truth is to drop the Anvil of Reality upon the delicate china of illusion": (author).

Truth and ridicule are mutual anathema, they form a dichotomy. It may be said that those who love truth, hate ridicule. It can also be said that those who hate the truth, love ridicule (i.e. to use it against others). (It may also be said that the dichotomic twin of truth is *lies*. However there is a difference. The utterer of lies acknowledges the truth, while the ridiculer has no conception of it. In other words, the ridiculer denies that which he knows not of. This of course demonstrates that one who ridicules is a lesser Being than one who merely lies).

(I am reminded here of a line heard in the movie: 'Lawrence of Arabia', starring *Peter O'Toole*. "A man who tells lies merely hides the truth, but a man who tells half-lies has forgotten where he put it")*!*

It is an observable fact that people generally do not like truth, although they will strongly profess that they do - especially in the arena of finance. They will do all manner of things to hide it, obscure it, avoid it, and deny it. Many people will say they prefer the truth to its obfuscation, but these people for the most part only wish for the truth when it suits them or their purposes. (They will also say it because they dare not profess the opposite*!*). The cosmetics industry for example (a multi-billion dollar enterprise) is based upon this simple fact.

Many ladies feel that they need additional color or padding, etc., to enhance their appearance. This psychology (which has been largely inculcated by the industry) is in effect a created desire to hide and conceal what they imagine to be their inadequacies. In fact one may say that in today's culture, aesthetics are used largely to hide truth, - i.e. to tell lies.

In business and politics, one finds a great deal of denial, - the obfuscation and denying of truth. (Politicians especially are adept in the art of denying truth, twisting truth, confusion of truth, telling lies, ridiculing, and so on - while at the same time managing to sound plausible. Hence denial or diminishment of truth is another mechanism which ensures the continuation of the physical universe. (see 104 BB).

The denial of truth and its substitution with fabrication is actually the creation of illusion. This continuing denial is a contributing factor to why civilizations collapse: [see note 1]. Before they collapse completely however - we find increasing apathy, dishonesty, the promotion of sex without love, vandalism, theft, drug abuse, crime, corruption, the perversion of gender and traditional roles, etc., with the consequent increase of laws, barriers, and restrictions to perpetuate the insanity, etc..

Indeed, one of the techniques used to control people, society, and countries is to destroy traditional values in that society. The social values become confused, the people lost, and then easy to control by substituting new and perverted values. These are the techniques used by evil men.

We also find as a further consequence that life, its enjoyment, and a strong economy become ever more difficult to attain. Truth becomes more difficult to find. Life in such a society becomes duller and less interesting, -and even frightening. The denial of truth in a society thus creates a state of hostility, boredom, disinterest, fear, and resentment. It also generates a reliance on material things to replace those important missing elements.

This mechanism of degeneration is precisely why real truth is difficult to <u>confront</u>, - since the only truth available is a diluted mixture of half-truths and lies. Hence the real truths of the universe may not be confronted by Man since he lives in societies which specialize in the denial of truth: [please see note 2).

When <u>real</u> truth is found, Man is more or less programmed to deny it, ignore it, hide it, ridicule it, look askance, and so on, but he will not confront it. Hence, although many truths are in this book revealed, there are many who will seek to ridicule and deny them; - this is not guesswork, it is a predictable fact. It is therefore to the strong and able to which this book is dedicated.

> "…There must be one overall truth. One truth that blends all into a unified whole. It does not have any contradictions, any falsehoods, or any inconsistencies. Every branch of knowledge must blend into it, harmoniously". (*John Strong* in his book 'The Doomsday Globe').

NOTE 1: hence, we can see an interesting correlation between the continuation of the physical universe and the inevitable collapse of societies. Apparently the cause of social decay (and physical death) is a mechanism designed to perpetuate the physical universe*!* If this sounds irrational, it is, - for the universe thrives on irrationality.

The physical aspect is anathema to sanity and rationality. Its business is lies, insincerity, pain, death, and anguish*!* Hence we may further deduce that the material condition is anathema to the spirit, and that the combining of the two is an imperfect experiment fraught with spiritual danger and false ideology. (Which is why life is so difficult*!*). It is also - paradoxically, why life in a physical body is actually an *unnatural state!*

NOTE 2: thus we may observe that there is a series of stages through which this degeneration progresses. First there occurs a denial of truth in the ruling classes in order to gain advantage in commerce and control of the lower classes. This dishonesty then percolates down into societies' upper classes and then on down into the greater populations.

People become generally disillusioned and trust begins to evaporate. Along with this begins the abandonment of honor, self-discipline, loyalties, affection, and all the nobler aspects of life. People then - seeking for more meaning in life - cling to material possessions, since they have nothing else to cling to - and indeed, are *offered* nothing else. With this comes widespread avarice, greed, and the desire for more of the material. Hence we can see a dwindling spiral of degradation.

"a paradox is a truth standing on its head to attract attention"… (Nicholas Falletta)

We have no doubt at some time looked at an astronomy magazine and seen pictures of stars, galaxies, and clusters of galaxies, etc.. Astronomers write articles about their discoveries and pictures, and we believe them since the evidence is very convincing. After all, these magazines are expensive, glossy, well known, and they are published nationally and internationally. [See note 5].

It is brought to our attention on rare occasions by astronomers and writers that there are stars which are greater in diameter than our entire solar system. This may be difficult to imagine and believe, and yet the evidence is abundant enough that they do indeed exist.

We may hear from some obscure source that there exists an individual or a family in the world whose net worth is in the *trillions* of dollars, and yet we cringe in disbelief at this news; - but why should we? After all, someone has to be the richest: [see note 10]. We may hear hundreds of stories about things which we cannot accept as true simply because they are so "incredible". But what is incredibility, - why can't we believe?

Why can we not believe that there are vast undiscovered cave and cavern systems many miles deep below in the solid rock below the Earth's surface? Why can we not believe that there are people living on other planets, or indeed - inside our own planet? Why can we not believe that there are visitors from other planets flying around our skies in quantum-electronic spaceships? And further, why cannot we believe that the liquid planet we call Earth is actually hollow like a tennis-ball?

Why can we not believe that there may be new laws of nature and physics as yet undiscovered? Why can we not believe that there are men who are able to fly, or walk through walls with the power of their wills? I am not here claiming these things exist, simply asking why we find it so difficult to believe them. [see notes 3a & 3b].

There are perhaps four fundamental reasons at present why we cannot stretch our minds to encompass the "incredible", i.e. the "unbelievable". One is the fact that we are reluctant to appear foolish in accepting something which is outside of our experience and therefore "unproven".

This is of course a purely psychological problem. Secondly, science has not yet discovered the potentials of quantum electro-dynamics and thirdly, we have been denied the truth about our own reality as Beings occupying material bodies. Fourthly, we have established all about us 'stable data', that is points of reference (sometimes called 'anchor points') with which we are able to negotiate the vagaries and perils of life in general.

For example, our place of work, the bus route, the medical system, the education system, the media, arts and entertainment, our street, our dwelling, our bicycle, and so on. These are stable landmarks by which we are able to orient ourselves. These all form intimate parts of our lives and they are almost as important to us as are our bodies. These four factors then keep us imprisoned in a false, restricted, suppressive, grey, and mundane "reality" which we call "normal" life here on this planet.

We usually fail to analyze such tales in the light of reason and logic taken step by step. Man, - the "technological animal" - has been on this planet (we are told) for something like 50,000 years. (The evidence for high technology involving metals has been found in <u>coal seams</u> dated at *hundreds of thousands of years*, however.*!*). It has been estimated that there are millions of planets in existence capable of supporting not only life, but also *intelligent* life having the ability to create technology comparable to our own. Many of these planets have - reasonably and logically - been inhabited by intelligences for perhaps much longer than our own Earth.

Most Readers would agree that this analysis thus far is logical, reasonable, and acceptable, - at least possible. Our own present technology has existed in its evolution for about 5,000 years. In another 5,000 years from now we shall have developed new and more advanced technology, without doubt voyaging to other stars, solar systems, and even galaxies. This too is reasonable and not unbelievable.

Another such race, on another planet elsewhere possibly has already developed technology for 10,000, 20,000, 200,000, perhaps even a million years. We should then expect such an advanced race to visit other inhabited worlds such as ours; this is likewise logical and reasonable. We may also expect them to modify their bodies to withstand the rigors of space and time. We may expect them to be capable of modifying space-time itself. We may also expect them to land and take samples of flora and fauna. [see note 6]. Furthermore it is likely that they have developed a far higher form of civility.

We may take this scenario a step further and suggest *reasonably* that several such technological planets would form an alliance or federation in the galaxy. This would be a natural progression of technology and social development. Indeed such an alliance may already have existed for hundreds of thousands of years. The Reader at this point is surely obliged to admit the logic and reason of each of these independent suggestions.

It is also reasonable (and logical) to further suggest that this federation of planets has criminals, rebels, and "politically incorrect" individuals in their societies, so that consequently there exist penal colonies [note 9].

All of these isolated factors must each one independently certainly appeal to logic and reason, and yet taken together and woven into a natural, sequential, historical progression of events, certainly gives us pause for reflection and the temptation to dismiss it all as a bizarre tale of "poppycock". Yet even so, it must be admitted and conceded that one is not able to dismiss it all as *<u>impossible</u>* however much it may strain *our* level of credulity. And if we are able to say that something is *possible* in a universe as vast and unlimited as this one is, may we not also say that it is also feasible, - if not downright *probable*? [see note 1].

Our inability to assemble and compose a series of logical and reasonable facts or propositions into a rational and intellectually acceptable composition is surely nothing more than a psychological problem; not a problem related to sequential and possible physical events. In the world of *Homo Sapiens*, such suggestions invite ridicule rather than intelligent investigation and analysis. [see note 8].

Let us look at yet another example with the personal computer. It is a highly complex device yet it is composed of perhaps hundreds of thousands of very simple components, each one evolved from simple research and discoveries over many years: (e.g. the resistor, the LCD, plastic, the transistor, the microchip, and so on, etc., etc.). The wonderment of it is how we have learned to place these components in juxtaposition in order to have a computer. This is technological creativity at its finest. (And now we throw them into the garbage can*!*).

NOTE 1: a brief look at our own Western history will illustrate this admirably. In just nine hundred years we have come from the sword-swinging Crusades to building Boeing 747's and putting men on the Moon. One has to admit upon reflection that this seems fantastic and unbelievable, and yet today it is just a ho-hum fact of every schoolchild's life. This in itself must indicate Man's tremendous potential*!*

NOTE 2: this lack of randomity is of a similar order to that experienced if we were to drop for example, a cupful of dice onto the floor. We should expect the dice to be clustered closer together on the floor at the drop point and increasingly further apart as the distance from this drop point also increases. Hence, there is a relationship between the distance of dice from the drop point and the distances between the dice themselves. (In a manner reminiscent of the well-known d^2 formula for Em radiation.

This distribution is also embodied in the familiar 'distribution bell curve graph'). (A dynamite explosion in a pile of pebbles will give the same distribution). It will also be seen that the number of dots on each upturned die will also correspond to a statistical, orderly arrangement. It may be noted that spiral galaxies also tend to conform to this pattern.

This then creates a pattern of predictability (i.e. a distribution pattern) which is subject to mathematical analysis. This pattern is therefore obviously not a random pattern, and being not a random pattern must therefore display the presence of intention or intelligence in its formation, - including quantum Energy, which is the physical expression of this intelligence. This will no doubt be difficult for some people to grasp, but in fact the pattern of 'randomity' is itself an effect of creation and planning. (This concept will become clearer as one progresses).

Statistical behaviour of particles and objects is one of the founding principles of quantum mechanics.

We may look also at life on this planet. Insects, animals, and plants all fall within a range of size and outside of this range, we do not find recognizable life. It is interesting to note that almost all life forms except bacteria and viruses fall within a range of measure comparable to the range of microwave and infra-red radiation, - as does also the human body itself, from the thickness of a hair to its full height.

NOTE 3a: indeed there are unbelievable occurrences here on our planet which we accept as commonplace; for example, the Cepahalopods (meaning "head-foot"). This is the class to which belong the octopi, the squids, and cuttlefish, etc.. If no-one had seen such a creature, who would believe that an animal could ever exist whose head consists of eight snake like arms having suckers? These are truly bizarre creatures. Furthermore, the relative size of the cephalopod eye belies a superior intelligence in the animal world.

NOTE 3b: there is also a notion abroad that there are "holes" at the Polar regions of Earth extending from the outer surface to the inner surface and hollow cavity. Now the idea that the planet is hollow like a tennis ball is perhaps outrageous enough, but to suggest along with it that there are also Polar openings is surely going overboard beyond all reason?

(A little comforting analysis and reflection however will suggest that such a "bizarre" construction as a hollow planet *is bound to be accompanied by other "bizarre" peculiarities and characteristics*. In this context, and from a fluid dynamics viewpoint, what would be more logical than Polar openings? [Explained elsewhere]).

However, supposing hypothetically for a moment that indeed the planet is hollow, we should then admit that the planet must have been formed this way by natural and dynamically predictable forces. Let us not forget also that the planet at present is actually not rigid - it is composed of liquid magma. Those same forces which may create a spherical hollow space or cavity within its sphere must also be responsible for the circular holes which join the inner and outer surfaces – (if indeed they exist).

Hence, we find that two outrageous and incredible ideas may in fact be dynamically, mutually, and *necessarily* linked to form an even more bizarre interdependent reality, - one feature being contingent upon the other*!* Those who would dismiss out of hand such an outlandish conception, do so out of ignorance and lack of understanding of the simple dynamic forces acting on plasmas, gases, and liquids in a special gravitational/inertial situation. (see diagram 101).

Let us take a slightly parallel situation such as a water-spout (a water-born tornado). This entity is bizarre enough in its own right, but to suggest that it has the form of a *tube* is to take this bizarreness further. Yet a water-spout (or tornado), because of its form and function, cannot exist except by having this structure. The same reasoning may be applied to any "bizarre" entity such as a squid with its many peculiarities, and so on. One may also consider that a "normal" type of animal with just one bizarre feature is bizarre in itself (such as a six-legged dog). Hence we may consider that an animal (or other entity) with one "bizarre" feature is likely to have other "bizarre" features also.

NOTE 5: this is a good example of cause and effect being reversed, so that cause becomes effect while effect becomes cause. In other words, the presence of "Authority" on an interesting subject stimulates the production of quality magazines and books. In this case, the 'Authority' is the cause and the magazine production is the effect. Likewise in reverse, the production of a quality magazine or book, etc, lends 'Authority' to the writer and the written article. The result is the same; - more sales, more profit. We may then ask which is the Cause, and which is the Effect?

This 'reversed-cause-effect' situation is a perfect example of how we interpret illusion as reality (or reality as illusion), and how we perceive the world; is it real or is it illusion? In fact, it is both and it is neither - it is one or it is the other, it all depends on how you *choose* to *perceive* it! As we shall see in quantum mechanics, cause and effect are likewise interchangeable and indistinguishable from each other. Furthermore, one effect can have *several* causes just as one cause can have several effects.

NOTE 6: we should *expect* such a race to have advanced spiritually to a degree unimagined by Homo Sapiens. They would have developed beyond the need for artificial technology to accomplish 'miracles'. In our sight, they would truly be *gods* having incredible powers.

NOTE 7: the range of sizes is large to be sure but it does have limits, and these limits are set seemingly with reason. In the same way, we may compare all multi-footed animal life-forms on this planet which range from several nanometers in length (e.g. nematodes) to perhaps a hundred feet or thereabouts, (e.g. the whale).

NOTE 8: after all, we do this daily using material components to construct complex jet airliners. Each of millions of components used to build the plane are brought together to produce it. These components are for example, computers, clocks, screws, dials, gauges, valves, pipes, wires, metals, plastics, other exotic materials, radios, radar, levers, switches, lights, buttons, knobs, and so on, and on; - truly a mind-boggling venture.

Consider also that each of these items has undergone years of design, fabrication, integration, testing, manufacture, and consist of even smaller components which also have also been through the same processes.

A simple gauge for example is the result of many years of development, discovery, and improvement. Not only this, but consider also the vast, hidden infrastructure ranging throughout the entire society needed to support the design, manufacture, and integration of these components.

When everything is considered, one realizes that a modern marvel such as a jet airliner is hence the product of perhaps a million man-years of technology development, continual refinement, and increasing complexity. On top of all that is the incredible infrastructure and logistics required to maintain the craft, repair it, replace parts, flight control, emergency control and contingencies, training and instruction, - and so on and on.

Think of that when you next travel by air, - and then wonder why the 'plane doesn't crash!

NOTE 9: Australia was at one time a penal colony for English criminals and today most White Australians are descended from these criminals. It is even possible that our very own planet is just such a penal colony for criminals and the insane. (Does this sound reasonable and acceptable?). (It has been said by some that this is the maddest planet in the universe!).

NOTE 10: since this family controls or influences, directly or indirectly most of the world's news and other media, very few know about them or their existence.

104 BN ~ ABOUT KNOWLEDGE & DICHOTOMIES

> "Every great advance in science has issued from a new audacity of imagination", (John Dewey).

> "Science does not know its debt to imagination", (Ralph W. Emerson).

> "It is easier to perceive error than to find truth, for the former lies on the surface and is easily seen, while the latter lies in the depth, where few are willing to search for it". (Goethe).

By definition, a list of dichotomies is a list of opposites and their pairings. Hence on one side of the list we may for example find the concept of "Big". Its paired opposite would of course be "Small". Listed under 'Big' we can find other concepts such as Strong, Powerful, Fast, Causative, Superior, and so on. Their paired opposites would then be Weak, Slow, Effect, Inferior, etc.. The test of a true dichotomy is the fact that one of the pair is dependent upon the other for its existence. [see note 4].

We can see then that there is a definite relationship between items on *one* side of the list and that these items may function together as a harmonious group. For example, an army which is Bigger is also Stronger, more Powerful, and is at Cause on the battlefield. It is therefore a Superior army. The same is true with any other machine such as an airplane.

If an airplane of a certain class (e.g. a fighter) is more powerful, it must follow that it is bigger and stronger in order to carry bigger and more powerful engines. It is also faster and able to carry a greater load. In its class therefore it is a superior craft if it can perform a given task more effectively. We can apply this dichotomy list to virtually any situation or energy system in the physical universe. [see note 4].

We may also apply it to people and societies in general. Hence we can admit that the application of dichotomies must be a fundamental mechanism of the physical realm. In most classes of endeavor, we can therefore safely say that "Bigger is better", - and indeed nature seems to affirm this principle. In nature we see that the survivors are those with either superior wits, cunning and intelligence, or those with superior physical force. Force and intelligence then are then a winning combination and the intelligent use of superior force is an unbeatable combination which may conquer matter, energy, space, and even time!

The entire body of knowledge as a vast accumulation of facts is like a net where every 'node' or knot is connected directly or indirectly to every other node or fact.

Hence one fact - assimilated - can lead to another by means of research, and all are thus in communication. Just as one piece of a jigsaw puzzle owes its existence to that of all the others and becomes a part of the entire matrix, - just as every tiny root owes its existence to the tree, and hence by extension to every other root, and just as every knot in a net owes its existence to the net - which is itself simply a collection of other knots, - so all existing facts or data are similarly inter-dependant and interconnected - however remotely. In the same way also every atom in a mass is in communication with every other atom either directly or indirectly. (Evidence for this is seen in quantum mechanics; see *Bose-Einstein Condensate* 430 A). One may term this interplay and inter-dependency of seemingly unrelated factors, - "Holistic Dynamism".

It is then possible to take one or several isolated and independent known facts and then extrapolate or deduce an additional potential fact. For example, if we find a square of string having four knots, we may deduce that another such square may be constructed and attached to it. By further extrapolation we may deduce that an entire series of squares may be so constructed and attached which then introduces (and creates) the existence of an entirely new concept, i.e. that of a "net". ("The organized whole is greater than the sum of its parts").

There are many ways to arrive at new truths and data: - by observation, by research, by inference, by extrapolation, by empirical reasoning, by insight, by deduction, by 'intuition', by revelation or 'cognition', by perception, by the recognition of existing assumptions and preconceptions, by recognizing similarities and differences, and so on. One can even postulate a fact which may hint at its own existence, and by careful testing and analysis determine whether it is valid. [see note].

Or one may simply set it aside until such an opportunity arises which will allow this datum to fit into a matrix of other proven data, in a way very similar to that in which a piece of a jig-saw puzzle may be set aside and later inserted into a partially constructed picture. (In fact this simple "jig-saw-puzzle method" is a valuable tool, as are both intuition and logical reasoning, - and one which has given success to many a scientific researcher). The jig-saw puzzle in fact is a very clever tool, and is the mechanical expression of reason, logic, and intuition. [see note].

Some great men who made important discoveries by intuitive creativity were; - *Andre Ampere; Nicola Tesla, Friedrich Gauss*, and chemist *Auguste Kekule*, who discovered the spherical carbon molecule ('Buckyballs');.

Note: the jig-saw puzzle also introduces us to a fundamental aspect of the construction of the physical universe, i.e. that everything - including matter, energy, space, and time, is in fact *incremental, particulate, and digital*. This will be enlarged upon in due course.

< dichotomies >

Quality and quantity are a dichotomic pair and it is interesting to note that wherever one looks - whether to nature or to sociological situations, that each one is usually found objectively (or subjectively) interdependent on the other. In a *Valley*, one may find a *Wide* river having a vast *Quantity* of *Slow, Passively* moving *Muddy* water. Alternatively, we may find a *Narrow Mountain* stream of *Energetic, Aggressive*, and wildly *Rushing, Fast* moving *Clear* water. The Quality (i.e. the Velocity) of this water rarely has the Volume (Quantity) of the slow moving river, - and vice-versa. Only in rare situations is one able to find the one associated with the other.

An example of this latter would be Niagara Falls and other such places where quantity (*volume*) is associated with quality (*velocity*) of flow. In electronics this could be equated with high amperage or *density (quantity) of particle flow* and high voltage or *velocity of particle flow*.

A transformer will transform a current of one Amp and ten Volts into a current of ten Amps at one Volt, or two Amps at five Volts, etc.. In any transformer, both input and output energy can be expressed as Volts times Amps, i.e. Watts, - so that (theoretically) the output energy is equal to the input energy, (input Watts equals output Watts). Volts and Amps then are directly related and interdependent. (in such situations however, output energy is actually never exactly equal to the input due to internal resistance and consequent energy loss, usually by heat exchange. It is interesting nontheless that although energy input is converted, the total energy output is *equal to* the total energy input).

It may therefore be said that a Volt is the dichotomic twin of an Amp in spite of the curious fact that both units are derived as purely arbitrary conceptions or quantities. The only relationship they have is their direct conversion one to the other in the transformer. They have no other relationship than as simply measures of different aspects of electrical flow. They become dichotomies because they are essentially the only pair of data which describe such a dynamic flow, and also because they can be associated with other dichotomies such as quality and quantity, velocity and density, etc.. In many instances then, quality of flow is synonymous and identified with velocity of flow, e.g. voltage, and vice-versa, - so that quantity equates with amperage and "slowness" of flow. In electronics, this is precisely what a transformer does by converting one to the other.

The same dichotomy may be found in, for example - manufacturing and business. One may exercise an option to purchase an expensive, quality car such as a *Rolls Royce, Mercedes,* or *Ferrari,* etc., or one may option to buy several (volume) cheaper, lesser quality cars with the same budget. Of course there are sound technical reasons for this separation of Quality and Quantity. However the main point here is that the dichotomic mechanism (e.g. quality and quantity) here and everywhere seems to be a law of the universe. In manufacturing and trade, these two are almost always governed by their exchange value measured in units of money.

Another good example is the way in which batteries may be connected in Series or in Parallel. If they are connected in *series,* the result is a current of increased Voltage (i.e. quality of flow) but low Amperage. If connected in *parallel* the amperage is increased, the voltage remains low.

It is therefore seen that Series and Parallel are a dichotomy pair as are Voltage and Amperage, and also Quality and Quantity. Furthermore, it is also seen that *parallel connection, quantity,* and *amperage* are on the same side of the dichotomy list while *series connection, quality,* and *voltage* are on the other. Hence there is evidently then a relationship not only between items on both *sides* of the list but also between items on the *same* side.

In many cases, quantity may be converted into quality and vice-versa, - just as a transformer converts a quantity flow into a quality flow. In nature, one can find the same dichotomy in electronic emitted wave forms. At one end of the electromagnetic spectrum will be found radio waves (which have large wavelengths but dispersed and diluted energy values). At the other end will be found the very fine wavelengths of quantum E-energy and electrons, Gamma rays, and so on - with concentrated energy values. The former is a volume or <u>quantity</u> form of energy while E-energy is a fine, or <u>quality</u> energy. In this latter case a lot of energy is "squeezed" into a very short wavelength.

By allying both quality and quantity, useful and interesting technological feats are possible. In the case of Niagara Falls, this combination is used to drive high energy producing electrical generators. Similarly, by harnessing the fine electronic nature of E-energy and generating it in very large quantities, - <u>many truly new and wonderful feats may be accomplished.</u>

It may be perceived also that concepts on one side of the list may be directly or indirectly related. For example, people have always known intuitively that words like female, moisture, darkness, Moon, static, passivity, and receptivity, etc., are associated - while their opposites; male, fire, dryness, light, sun, dynamic, and aggressiveness are likewise associated. (Apologies ? to the Politically Correct).

Dichotomies are found everywhere in the universe and in the laws of physics. Indeed, dichotomies are the fundamental basis of our physical existence. A very simple example of a dichotomy pair is left and right. The very concept and existence of a 'left' <u>cannot exist</u> without a 'right'. The concept of 'left' is completely meaningless without its dichotomic twin. Even in space where left and right are ambiguous, one is *obliged* to arbitrarily assign a 'left' and a 'right' to an object in order to perceive the object in relation to oneself. If it is rotating this becomes easier to do since then, you can relate it to the position of your body, - if you have one*!*

(If you don't have a body, the entire concept of visual perception becomes very subjective, abstract, ambiguous, and arbitrary).

The concept of quality implies a superiority, which further implies other examples of comparison. Lesser quality implies inferiority. Other examples of course imply quantity. It seems therefore that the concept of quality too, cannot exist without that of quantity.

(The inter-relationship of dichotomy pairs is a good example of this. Examples of dichotomic pairs are black and white, up and down, left and right, Sinister and Dexter (meaning simply left and right). (We obtain the word 'dexterous' from Dexter). (It is interesting to note that the left hand in ancient lore has always been associated with darkness, the unhallowed, indeed the "sinister", etc., hence our modern use of the word).

It is easy to see then from these few examples, that the left side of the dichotomy list contains terms like: sinister, darkness, downward, unhallowed, under-handed, evil, demonic, heaviness, covert, hidden, and so on. On the right side of the list we would have: rightness, goodness, upward, sanctified, angelic, lightness, overt, extroverted, and so on.

It may become apparent to the Reader that the concept of dichotomies and indeed dichotomies themselves seem to possess and hint at a guiding force beyond that of the purely physical.

In the world of electrical science, one may consider the metals Iron and Copper to form a dichotomy since iron is invariably associated with magnetism - while copper is always the electrical conductor: [see note 3].

(In ancient cultures such as the Greek, - One (1) (and all odd numbers) was considered to be a masculine number while Two and all even numbers, including 0, were considered to be feminine; (for those interested in anatomy, this is easily understood). Zero also can easily be considered to be an even number and feminine, so that we can have (1) masculine, and (0) feminine. One and Zero of course also represent on and off in the digital world - and by extension, active and passive, dynamic and static. The possible lists are seemingly endless).

(In the same context, a triangle (*three* sides*!*) was considered to be masculine - being associated with fire and activity (and interestingly also, pyramids, arrowheads, bullets, etc.), - while the square was feminine, being associated with earth, stability, matter, structure, bastion, house, or home, - and by extension the physical world and the universe).

Of further interest also, the circle was associated with the spirit. The two fundamental dimensions of the circle, i.e. the diameter and the circumference, (the *straight* and the *curved* if you will, - and by extension the direct and the indirect, the one and the zero, the definite and the vague, the true and the perversion, etc.), are related in an irrational (i.e. un-measurable) proportion, i.e. the value of Pi.

This conveyed to the ancient geometricians the magical idea that the circle was "not of the material world" - the physical existence - and not measurable in the rational sense that a square might be measured and calculated. The diameter was of course rational and measurable while the circumference itself was ir-rational (as is the area). [see note].

It is also interesting to see the extraordinary lengths that Man will go to in order to "rationalize" the non-rational, - in this case, to calculate the area or circumference of a circle. It is as if Man has been programmed to detest anything that cannot be analyzed with computer-like precision.

(The real reason - if one cares to consider this - is that Man has become so enamoured of his material surroundings that he no longer is able to perceive the non-physical and the true reality; no longer able to perceive the world as did the ancients and the magicians. He has therefore become focused on the tangible material in the desperate hope that this will provide him with the answers he so longs for. Moreover, this is the reason why he has endeavored to calculate Pi to several million decimal places. This is an attempt to subdue, control, and to rationalize the irrational, to conquer the universe, - which Man considers to be a form of enemy).

The circle is magical because it is intimately associated with natural and biological structures, i.e. the "created" world of life and creative forces. Naturally occurring structures such as bubbles, fruits, stars, and planets are of course spherical, - the material expression of the circle! The Sun, Moon, and Earth are also spherical, and we naturally assume that atoms and electronic particles are too. (There is mounting evidence in fact that atoms are not all spheroids, and this will be explored).

It is interesting to note that the closest thing to a true sphere is probably a soap bubble. Here then we have the inevitable association of the spirit, creative forces, and magic. 'Magic' by implication is of course a quality associated with the mind, will, and spirit. In fact Magic, almost by definition - is that phenomenon which demonstrates the interaction of the mind and will with the material universe.

(Incidentally, the reason why children delight so in soap bubbles is because they closely represent the lightness, insubstantiality, uncontrollability, abstractness, transparency, perfectness, fickleness and playfulness of the pure spirit. Children can easily identify with them).

The ancients' penchant for trying to "Square the Circle" geometrically was their attempt to express in a magical way, the allying of spirit with matter, the irrational and the rational, - thereby representing life - and by extension, the interaction of life with the physical universe. (One of the primary functions of the *Great Pyramid's* geometry was to square the circle. (The height represents the radius of a circle whose circumference is equal to the perimeter of the base). This, by magical intent would make the pyramid truly a creature having life, - or at least embodying a principle relating to life!. (> section 1.6).

By using this method of reasoning, it becomes apparent that "ordinary" life is indeed an expression of true Magic since we are obliged to admit that life is a manifestation of the principle of mind, spirit, or non-physical agency interacting with matter. Spirit and matter - Nothing and Something - Zero and One - ('Off' and 'On') are also dichotomies, as exemplified in the binary system of computers. Electric fields may also be dichotomized, i.e. negative and positive, in the same way that magnetic fields are polarized 'North' and 'South'.

A negative terminal emits (outflows) electrons while a positive terminal receives (inflows) them, - (reminding us again of male and female functions). The wheel rim and axle are also dichotomies, and it is interesting to note that the wheel rim is geometrically equivalent to a zero, a circle, and by extension, a 'hole', - while the axle represents a 'one' or a 'shaft'. Furthermore a wheel rim is curved, (and by extension indirectness, a female quality) while the axle is straight and therefore represents directness, a male quality. A shaft is open ended and therefore *finite* while the circular rim is endless and *infinite*. [note 2).

The first level of ionization of an atom is where it has lost or gained a single electron, thereby resulting in an electrically positive or negative ion. It resembles in this respect the binary "on – off" switch of computers (1 & 0). The hydrogen atom may be "*bound*" and therefore constricted into the form of a single electrically neutral atom, - or it may be "*unbound*" to form two free and electrically charged particles (the electron and the proton).

We can now extend the dichotomy list to include Voltage & Amperage; Electricity & Magnetism; Wheel & Shaft; Copper (or aluminium) & Iron; Light & Heavy; Odd & Even; Male & Female; Negative & Positive; Outflow & Inflow; etc.. Interestingly also, a magnetic field always tends to intensify at an *internal, core* area (e.g. in an electro-magnetic coil), whereas an electric or electro-static field always tends to intensify at the *external, surface,* - (of a conductor). This is because opposite magnetic fields are inherently and internally mutually attractive and *contracting* while electric fields are externally and mutually repelling and *expansive*.

This is an important feature in the "new physics" as will become clear.

Here now is a list of dichotomies which may be seen to provide a connecting link throughout: -

Uranium	Hydrogen	Condense	Rarify
Old	New	contract	expand
Heavy	light	low	high
slow	quick	subjective	objective
Weak	strong	inflow	outflow
Dim	bright	proton	electron

Passive	active	hidden	exposed
Confused	clear	deep	superficial
Opaque	transparent	quantity	quality
Inactive	active	material	spiritual
Dead	alive	cloudy	sunny
Degenerate	regenerate	night	day
Destructive	constructive	oblate	prolate
Chaos	order	hub	rim
Course	fine	static	dynamic
Rigid	fluid	particle	field
Solid	plasma	resistant	permissive
Confused	perceptive	closed	open
Lethargic	dynamic	push	pull
Nucleus	electron	force	entice
Magnetism	electricity	sour	sweet
Infra-red	ultraviolet	ugly	beautiful
Resist	agree	follow	lead
Contract	expand	female	male
Inertia	energy	tail	head
Gravity	levity	insulate	conduct
Matter	electronic field	base	noble
Apathy	enthusiasm	dangerous	safe
Curved	straight	devious	direct

NOTE 1: anyone who has learned to type or drive a car, etc., will testify to the fact that these tasks at first require analytical , i.e. rational thought and concentration. After becoming proficient however, rational thought is abandoned and control of these devices becomes a procedure which no longer requires analysis. Control then becomes a "thoughtless" process, i.e. it becomes "ir-rational", so that it is then a process directly linked to the will, Beingness, or 'spirit'. In other words, we may say that driving, etc., becomes a *subjective* experience as opposed to an *objective* one. This why some race car drivers are so good. They almost literally become one with the machine; i.e. mind and machine become a single living creature.

This of course is the technique sought after by the schools of martial arts. One becomes "at one" with his weapon or instrument (or body). In this event control of the device (or body) becomes so skillful that it causes wonderment to all who are unfamiliar with the technique involved. The attainment of such skill is no wonder at all, it is simply one learned so well that analysis and thought is bypassed, - just as with a skilled typist.

NOTE 2: interestingly, there are two fundamental types of magnets and magnetic fields, and these will be examined at great length because of their fundamental importance in the technology discussed herein. The first type is the simple bar-magnet (or electro-magnet) with open ended polarities. The second is the <u>toroid annular</u> magnet which resembles an automobile tire (or wheel rim) and has no open-ended polarity. The polarity exists but is hidden within the annular structure.

NOTE 3: one may note also that the atomic number of copper (29) is an odd number while that of Iron (26) is even. An even better pair in this regard would be Iron and Aluminium (13) since Al. is used extensively as a conductor where weight and cost are important factors (as in overland transmission lines). (Aluminium - as well as being light, is a good electrical conductor second only to copper in conductivity, i.e. a ratio of 1 to 1.6, - or in other words - volume for volume, Aluminium's conductivity is 62 % that of copper. Furthermore aluminium is actually a superior choice since it is more abundant, cheaper, and lighter.

The only undesirable characteristic of Aluminium is the fact that it is weaker than copper and after repeated vibration and flexing will become brittle. It is therefore unsuitable for many applications. (Aluminium was used extensively in WW II for the electric coils in aircraft to both save weight and for its cheapness and availability. Further, it was not expected to be in service for an extended period).

Iron is used predominantly as bulky elements in electromagnets, (e.g. transformers). These devices are therefore heavy, whereas Al. electrical coils are light. (Interestingly, the atomic weight of Iron is precisely double that of Aluminium with a ratio of 26:13.

The physical universe is constructed of matter, energy, space, and time, - but equally important on all fronts for its existence are the *Dichotomies*. They represent the quintessential binding force holding everything together. The list of dichotomy pairs may be likened to the two sides of a zipper. The two sides together constitute the complete system and if one side is missing, the zipper as a functioning entity ceases to exist. Likewise due to the existence of the dichotomies, the two sides of the 'universe-zipper' can be merged together to form a complete and discreet functioning unit.

Like a zipper also is the fact that every item on one side of the list is intimately related to its neighbor above and below it. Every such item depends for its existence on <u>all</u> the items in the double-list. If only one single item should be removed, its paired opposite will similarly cease to exist, - e.g. white and black, up and down, left and right, etc., - and the entire physical universe will become structurally unstable. Its very existence would be compromised, i.e. it would become "unzipped".

One interesting example is the pair "good and evil". Apparently the entire physical universe is likewise contingent upon the existence of even this fundamental dichotomy. If we remove 'evil' from the list of dichotomies as a fundamental structural element of the universe, the entire list of dichotomies is likewise severely compromised.

If we remove this pair then we also invalidate the many other co-dependent pairs which form the dichotomic list. (We may say in effect therefore that the dichotomy list is constructed as a "holistically integrated structure"), - indeed, very much like a zipper! (It may be said that in the way that a zipper holds a coat together so also does the dichotomy list similarly hold the universe together!).

For example, 'good and evil' are co-dependent upon such pairs as 'Dexter and Sinister'; 'right and left; light and dark; Sun and Moon; male and female; day and night; sky and earth; rich and poor; active and passive; fluid and solid; - and so on. One can easily see then that the entire dichotomy list is - if thusly compromised - thereby rendered structurally unsound and likely to collapse like a stack of dominoes, - (taking with it the physical universe into oblivion, - poof! Or should I say "foop!"). In a strange and perverse way then, the entire physical universe and its perpetuation is equally dependent upon the existence of good - and of course, evil!

NOTE 4: now we may grant to the 'politically correct' that "superiority" is only an opinion. But if we look closely at the dichotomy list, we find that "superior" and "superiority" are indeed valid concepts, and that things really can be superior or inferior in spite of how we may feel about this truth and reality. The same can be inferred for male and female and their various attributes and roles, whether they are physical, emotional, or psychological.

The words "superior" and "inferior" exist in virtually all languages and have functional meanings. Words which have no function or meaning simply do not exist. To deny that something is superior or inferior is to deny a concept which the world's population has agreed to and which has a valid and practical application, - and has had for Lo, these thousands of years. ("Political Correctness" is an attempt by the confused to deny reality; to create further illusion, doubt, ineffectuality, and confusion; to weaken the male and destroy femininity. In other words, PC is destructive to the good order and dynamism of society).

NOTE 5: let us take as a simple example how it is possible to deduce many new data or possibilities from a single fact. Suppose a salesman in a shop sells you a copper kettle. You trip delightedly home believing that you have a bargain. When you are happily boiling water for your hot cup of tea, you notice by chance that a small magnet has attached itself to the kettle. This seems odd since you know that copper is a non-magnetic metal. You instantly deduce therefrom that this kettle must employ a magnetic metal in its construction. This is likely to be iron or steel - but not nickel or cobalt, since these latter two - while being magnetic, are not common nor cheap.

Since we already know that steel is far more common and cheaper than iron - and further, is used in vast quantities in industrial manufacturing. We may now deduce that steel kettles are very common, and are likewise obliged to assume that the kettle is indeed made of steel.

What else does this tell us? It tells us that it is possible to create steel kettles which look like copper. This tells us that manufacturers will create illusions to sell their wares and that retail industry is aware of this and have bought into this purveying of illusion. Further, it tells you that people will to some extent either buy into this purveyance or are victimized by it.

It also tells us that this salesman is unscrupulous and a liar. We may further deduce from this that if one salesman is a liar then there are likely to be other lying salesmen - and further, that you don't have to be a salesman to be a liar, (and conversely that you don't have to be a liar to be a salesman). This means that other items which you have bought may also possibly be less than you assumed them to be. This further tells you that the profit motive is often more important to the seller than is honesty.

It also tells you that he cares little for your feelings or for the feelings of others, and that there consequently must be many others who don't care about other's feelings. This will eventually take us into politics, international business, law, medicine, media, education, finance, - you name it, by golly. So you see how this chain of reasoning, deduction, and extrapolation can escalate almost to infinity in all directions, - all from one small incident, a little knowledge, and some thoughtful analysis.

104 BP ~ DICHOTOMIES, TRIADS, SUBJECTIVITY, & THE ABSTRACT

A dichotomy is a matched pair of opposites. A 'trichotomy' is a group of three elements - that is, a triad, triune, or triplet. Such a triplet can be found in the cycle of action, i.e. Start, Change, and Stop. Other triplets are Create, Survive, and Destroy; Inflow, Processing, and Outflow; Initiate, Sustain, and Terminate; Birth, Life, and Death; Oblate, Sphere, and Prolate, and so on.

(Not all dichotomies can be thusly analyzed; some are *pure* dichotomies which stand alone as simple pairs, - such as dead/alive; curved/straight; subjective/objective; hidden/exposed; quantity/quality; closed/open; follow/lead; push/pull; and so on): [see note 2].

It can be seen that in all of these triplets, two form the dichotomy pair while the third element occupies an area in between them. Many dichotomies therefore form a triad when considering a third element. The third element however does not form a well-defined concept and is more abstract and subjective, - having no definite limits. It does however contain what we may consider to be a transition or threshold point.

For example, a rocket is fired up into the air. It follows a curved, parabolic trajectory and then falls back to earth. When the flight of a rocket reaches its zenith, we may consider that there is a very definite transition/threshold point where it is not rising or falling. This point is a moment of "now" and actually has no duration in time even though the rocket is still in motion : [see note 1]. Briefly then we may say that many a dichotomy is actually a triad, a triune, or a triplet consisting of a pair of opposites and a third element being a transition/threshold point.

To summarize all this preamble then - we may state that firstly, the physical universe is constructed on a foundation of dichotomies, i.e. it is bi-polar. Many dichotomies form, and are a part of a triad or 'tri-chotomy'. The third feature (i.e. the transition/threshold point) joining the dichotomic pair is an abstract and subjective construction (i.e. a philosophy) which exists mainly in the mind of the observer.

We see then that the dichotomy/trichotomy is a reality born out of a subjective and abstract concept. Hence, according to this line of reasoning - it is apparent (and evident) that dichotomies are themselves abstract and subjective constructions, or rather attempts to "solidify" and render physical the abstract.

Since the physical universe is constructed upon, and consists of these subjective and abstract creations, we are forced to concede therefore that <u>the universe is itself an abstract and subjective creation</u>. In other words, while being apparently and evidently a material existence, it is rather a thing born out of subjective and abstract concepts, - it is a created thing!

NOTE 1: It may be assigned an arbitrary duration however, since a camera shutter at $^1/_{125}$ of a second using a high-speed film can capture the motion in a still photo. We may then consider for practical applications that a moment of 'now' is $^1/_{125}$ sec.. This however is only an attempt by Man to assign a numerical value to an abstract thing which is immeasurable. Furthermore, on cosmic and subatomic levels, scientists must create moments of 'now' having different postulated durations.

NOTE 2: even here though, we may find that some concepts are more abstract and indefinable. For example the pair - 'objective/subjective' is not easily definable since it can be shown that they often merge, becoming indistinguishable one from the other!

For those who are interested, the third type of dichotomy may be called the 'relative dichotomy' and includes concepts such as; old/new; heavy/light; weak/strong; quick/slow; chaos/order; dim/bright; opaque/transparent;conductor/insulator; dangerous/safe; ugly/beautiful; and so on. Hence we perceive that there are at least three types of dichotomies. There is the "tri-chotomy", the "relative dichotomy", and the "pure dichotomy" - (e.g. 'quality and quantity').

SPECIAL NOTE: Dichotomies are a way of defining the Physical Universe. If we can eliminate a dichotomy pair in a closed system, that system may be said to no longer belong completely in the PU. A pair of values such as 'heaviness' and 'lightness' is a dichotomy symbolic of the physical universe.

If we can eliminate 'heaviness' from this pair, the pair itself can no longer exist since lightness as a concept or perception depends for its value upon the existing concept of heaviness (and vice-versa). An object may be "heavy" or "light" depending upon one's subjective appraisal. This appraisal delegates the concept as a measure of the physical universe.

Hence if we eliminate its weight (i.e. "heaviness"), the object - technically speaking, will no longer be truly a part of the P.U. as we perceive it. It will therefore no longer obey the physical laws we are familiar with. In other words, the object becomes a part of another space-time (i.e. parallel universe, etc.) with which we are as yet unfamiliar.

105 B ~ INTRODUCTION TO (QUANTUM) LIFE ENERGY

> It has already been suggested by two physicists, (the Nobel Prize winning *Dr. Millikan* who discovered the electron, and *Dr. I.I. Rabi*), *that atoms and molecules constantly radiate electric waves, - the frequencies depending on the atom's size or energy level or substance, i.e. the number of electrons and electron shells contained.*

> "The discovery of the energies associated with psychic events will be as important if not more important the discovery of atomic energy", (*Dr. Leonard L. Vasiliev*, psychologist and winner of the Lenin Prize).

> "At this point we are then, it seems, faced with the need of another order of energy, not radiant", (*Prof. J. B. Rhine*, pH. and researcher at Duke University, Durham, N.C., (1935~1966).

> Biologist *Kammerer* stated, "the existence of a specific life force seems highly probable to me, —- "it is present certainly in the formative process of crystals".

There has existed in all cultures and peoples around the world since prehistory a knowledge of an unusual type of energy existing in nature. This energy to date has not been isolated nor objectively measured nor demonstrated except in a few cases, - and only to a limited degree by private individual researchers, such as *Anton Mesmer*. Notably and in more recent times, they have been *Baron Von Reichenbach;* and *Wilhelm Reich*, 1897~1957 - (who has the unenviable distinction in 1935 of having had his books burned by government agents of the U.S.A.). Even more recent are: *Pavlita, Drbal, Uri Geller*, and many more.

H. G. Wells, Nicola Tesla, George Bernard Shaw, Bulwer Lytton, Wm. Crookes, Sir Arthur Conan Doyle, and many scientists and researchers were all convinced that a mysterious form of energy existed all around us in nature called by some, "the Life Force", or "Life Energy".

This Energy has been discussed, researched, and examined by and large on a mostly subjective level, - except for some brave attempts at objectivity by a handful of adventurous fellows, and a few others as mentioned above. This is partly because this energy form is very subtle and also very pervasive, – like heat, capable of permeating all matter, but unlike heat, not lending itself to being technically isolated so that it may be objectively studied. In sufficient quantities however its effects may be objectively observed if not measured. (Indeed, it is possible that in the future, methods may be devised with which to measure it). [See note].

The Energy has been known by many names and has been researched scientifically and exhaustively on a subjective and an attempted objective level in the West for hundreds of years.

Psychics, sensitives, dowsers, Chi ~ Ki practitioners, and Yogis, etc., when describing this energy, use words like 'vibration', 'subtle energy', and so on. Other names by which it is or has been known are, *'Orgone', 'Odyll', Vril-Power, Animal Magnetism, Life Energy, Bio-electricity*, - and so on, (see following list: - see also *Formative Energy*).

These are all attempts to describe an actual form of energy. Other practices claiming to utilize this energy are the Chinese "Feng Shui", so-called *"Radionics, "Radiesthesia, Dowsing, Reflexology, Homeopathy, Acupuncture, Accupressure,* and *Shiatsu"*, etc.: (as we shall also explore, various types of divination systems such as the *I Ch'ing*, the *Tarot Cards*, the *Pendulum, Dice*, the *Ouija Board,* etc., also employ this energy).

Science fiction writers and readers perennially amuse themselves by indulging in speculations about the existence of a new form of energy and the miraculous wonders it is capable of performing. The existence of such an energy form has therefore always been regarded as a figment of writer's imaginations. But what if such an energy form really exists, - and what if I told you that indeed it actually *does* exist!? And furthermore - that this energy form has always existed, has been known and understood by the initiates of ancient civilizations, and is today widely taught and used in Asia and the Orient? In the East, this energy is known as "Chi" and in Japan as "Ki". It is generally acknowledged as being a very subtle form of energy, and its effects are therefore usually not directly observable.

However it has now been determined to be an electronic energy form and can even be located in (but not <u>of</u>) the electro-magnetic spectrum. Where this energy comes into its own as a genuinely applicable new form of energy is when it is generated in phenomenally large quantities. This can be done using presently existing technology.

< names of life energy >

Other names given to Life energy throughout the world and in different times and locales are:

Prana (Hindu)
Nous (Plato)
Yesod (Cabalists)
Formative Cause (Aristotle)
Dark Energy (Modern Physics)
Mana (Polynesia)
Munia (Paracelsus ; 1490 ~ 1541)
Facultas Formatrix (Kepler)

Animal Magnetism (Anton Mesmer ; 1734 ~ 1798)
Life Force (Galvani and Reich)
Etheric Formative Force
N – Rays (Blondlot)
Elan Vital (Dr. Nandor Fodor)
Etheric Force (19th century physics)
Eloptic Energy (Heironymus)
Bioplasma (Grischenko and Zdenek Rejdak)
Psychotronic Energy (Pavlita)
Psionics
Vital Magnetism
Vital Energy
Bio-magnetism
Para-electricity
Magneto-electricity
Bio-energy
Vril Power (Bulwer Lytton)
Magnale Magnum (Von Helmont ; 1577 ~ 1644)
X – Force (L. E. Eeman ; 1889 ~ 1958)
Mitogenetic radiation, (M – Rays)
Pyramid Power (Drbal)
Serpent power (Hindus and Celts)
Aether and 'Dark Energy' (modern astrophysics)
Formative Energy and 'Life energy' (W. Reich)
'The Force' (The Star Wars Trilogy)

We can see then that this unusual form of Energy, which I choose to call "Quantum Energy", or 'E-energy', has been known about for a long time and in all corners of the world. The question then which one is invited to ask is: "is it possible that so many people from all cultures and at all times can be mistaken about this ubiquitous Energy?". The answer to this question must be answered by the intelligent Reader, - and of course it is the fundamental theme of this book.

NOTE: *Reich* is perhaps one of the most adventurous and demonstrative of all the researchers (second to Nicola Tesla), having built several apparatus for the generation and manipulation of his 'Orgone' energy. It was he who contrived a variety of experiments to measure and quantify this 'Orgone'. He found for example in a carefully controlled experiment, that a liquid alcohol thermometer could detect the presence of Orgone by showing a slight rise in the thermometer's indication. This demonstrated an expansion of the liquid not due to heat energy.

> "... But the history of science is strewn with the ruins of mighty edifices toppled by an accident, or a triviality"; ('Eaters of The Dead', *Michael Crichton*).

Because of its nature and the manner in which it is generated, I have given it the long title "Electro-Atomic Field-Wave Emission Energy". It may also be referred to as "E – wave energy", "quantum Energy", Quantum-E, or simply "E – energy", and sometimes even simply "Energy" (capitalized). The 'E' could handily be applied to other terms such as "Etheric energy", "Elusive energy", "Electronic energy", "Emotion energy", "Electro-atomic energy", "Elemental energy", "Ephemeral energy", "Essential energy", or even "Excellent energy"! Further - if you enjoy word games, you may note that E is the most prolific letter in the English language and hence the most abundant letter in the entire world of languages. It is therefore a suitable symbol for the ubiquitous nature of this Energy. [see note 4].

'Quantum' means "package" or "precise measure", so that a quantum of energy is a precise measure of electro-magnetic energy. Quantum physics refers specifically to that branch of physics which deals exclusively with the behavior of energies and particles at the atomic or sub-atomic level. (In fact however it may be applied to any types of particles which behave and interact as groups. These particles may be as big as stars, as in galaxies, or indeed even as galaxies themselves. Statistics is one such mode of behaviour).

Since E-energy is of a wavelength corresponding to atoms and sub-atomic particles, this Energy thus qualifies as a quantum level phenomenon, and that is why I have termed it thusly. (An electron at rest for example has a wavelength of $1/40$ Ångstrom. An Angstrom is usually taken as the diameter of a hydrogen atom. E-energy therefore belongs properly to the field of quantum physics. [see note 1 below].

Quantum physics as a subject of study contains the answers to many commonplace and sometimes unexplained phenomena such as biological growth, healing, and the regeneration of body parts, - especially as witnessed in lowly life forms such as star-fish, salamanders, and worms, etc.. This is usually mistaken as being a function of bio-chemistry, which is undoubtedly included. However, chemistry and molecular bahaviour are but tertiary and secondary levels to the fundamental processes which are atomic, electronic, and quantum-electronic.

It can also further offer solutions to other elusive fields of research, such as nuclear fusion and superconductivity, etc.. Beyond these exciting possibilities are many others, - some perhaps beyond belief. We shall see!

One of the peculiarities of this ultimately physical Energy is that it is capable not only of being generated with electronic technology, but also of being influenced by the mind - or more precisely, patterns of thought and will, - especially emotional thought. This Energy is in fact the intermediary - the dynamic connection between the mind and the material body it controls. I am referring to you and your body! It is the 'Ki' or 'Chi' of Oriental healing philosophy. This is not difficult to understand or believe, it is as simple and as ordinary as the steam used to drive a steam loco-motive, or the electricity used to power a flashlight.

The mind uses emotional thought to control and operate the body and its im-mediate environment. Animals and children operate almost exclusively on non-an-alytical and emotional thought, (i.e. Theta and Alpha-wave thought, - as opposed to analytical Beta – wave thought). The mode of operation is therefore as follows: the will > higher, analytical thought > lower or emotional thought > quantum life Energy > electro-chemical energy (i.e. the glandular, blood, and muscle systems, and body movement) > - and environmental influence: [see note 2].

This Energy is the underlying force in nature which allows animals for example to function in harmony with their natural surroundings. These same surroundings, which are intimately connected with creatures' lifestyles, are in turn physically in-fluenced by animals' needs and unconscious desires. This is done by means of this subtle Energy, - the link between the mind and the physical, both bodily and envi-ronmental.

(In China the philosophy of Taoism (*Tao* = 'The Way') is based on this ani-malistic concept of the avoidance of intervention or artificial modification. Another more modern way of expressing this philosophy is to "resign oneself completely to one's fate", to "take life as it comes", "to live one day at a time", etc.. It could also be called 'The Way of the Child'). An example of this is given here.

A small animal such as a bird or squirrel etc., finds a big pile of delicious bread crumbs or seeds, etc.. The animal eats until his appetite is satisfied and then he is on his way. He does not carry a bag with him in order to pack away the extra bread-crumbs for a future meal. He gives no thought to the remaining food, he lives day to day, moment to moment, confident in his *ability* to find more food when he is hungry again. This is the way of *Tao*; it negates greed, selfishness, and the fear of death or loss; it is the ability to create. *Tao* then may also be termed, "the exercise of ability".

(This resignation and passivity is also embraced by Hindus and those who be-lieve in 'Karma'. This in fact is the life-style of animals and children, and for them it is observably successful. Children and animals however are extremely curious and explorative, and this is likely an admirable quality.

(Millions of Chinese for thousands of years can't be wrong, it would seem - so that this is apparently a very workable philosophy. (Not creative, not stressful, - but apparently workable to some degree, if not on a modern material level). Indeed it is this type of philosophy which has been shunned by industrialized Man, since it does not contribute to his ideals or way of life)…

The Energy forms a natural dynamic connection between mind, energy, and matter. The natural need or desire to explore recesses, tunnels, groves, hollow spaces, caves, pools of water, waterfalls, streams, trees, woodlands or marshes, and biological matter, etc., is in fact a natural tendency of this Energy form which by itself seeks out its own natural accumulations, (*W. Reich*). *This Energy therefore is itself in essence the 'instinct' or "programming" we assign to animals*. Indeed you yourself are probably also wont to seek out these delightful and quiet sanctuaries. [see note 5].

This is not an analytical deduction, it is something you simply enjoy - without rationalization, - and it is the natural function of quantum Energy interacting with the mind/body combination*!*

In other words, 'instinct' is an Energy form allied with the creature's awareness, and protoplasmic body - that is, a life form. Furthermore, animals - like children, because of their exclusively Theta and Alpha brain wave activity, are keenly attuned to this Energy and its survival dictates and coercion. The Energy tends to seek out or be drawn towards other accumulations and sources of its type. ('instinct' is another word for 'knowing'). (All matter gravitates because all matter emits a gravitational wave force. In a similar way, quantum E-wave energy "gravitates" towards its accumulations). *Life energy* therefore exerts a force on the awareness of the biological organism. [see note 3].

Life-energy is generated by several natural source types. They are, moving water (or other liquids) and the created friction, - bio-organisms, chemical activity, heat activity, fire, sound, and objects impinged upon by other forms of electro-magnetic energy and electrical activity. (*Von Reichenbach*).

> "...the most profound scientific insights are based on imagination and inspiration rooted in a firm foundation of fact...", (K. C. Cole, - 'Sympathetic Vibrations').

NOTE 1: there is actually no reason at all why electric waves of this wavelength should not exist. It is not beyond the logic of physical laws. Neither it is beyond the capability of the physical universe to generate, emit, and radiate such frequencies, - any more than it is capable of generating, emitting, and radiating particulate electromagnetic energy.

The only thing lacking is our acknowledgement and understanding that such Energy *can*, and therefore *must* exist - and consequently, a lack of technology designed to produce such Energies.

NOTE 2: if the Reader has any doubt that the mind has a power over matter, he may observe his own body and how it moves under his own volition, - not by any other force. Likewise observe the faces of your friends. Every face is different, but more than that - each face is molded and formed by the personality and mind lurking behind it. The face then becomes a reflection or refraction of that personality. In other words, you can see the personality engraved upon the dynamic 'mask'. This is actually so common and mundane that it hardly needs mention, except to bring it to the Reader's immediate attention. This then is "mind over matter". (Refer to 305 CC: *The Amoeba*).

NOTE 3: Awareness may be defined as the ability to perceive specific events and the details associated with them. These details would include the emotional levels of others, their body language, and implied and unspoken communications. You may test your friends and associates if they are willing. Ask one to walk around the city and look at various items, say ten clocks. After he has done this, ask him how many had circular faces, how many had second hands, how many had Roman numerals, how many had no numerals, and where each of these clocks was situated.

If he is very alert and aware, he will be able to give a good account of himself. Alternatively you can ask him to go into a coffee shop for a few minutes and then ask him how many people were standing, and how many sitting, the location of the clock if there was one, how many staff members there were, how many were male, how many tables and chairs there were and were they square or circular, were the chairs padded, and so on.

NOTE 4: as we shall see later, this Energy can be described by a very simple equation: $(E = pv)$, where (p) and (v) are parameters of a flow.

NOTE 5: you can prove this to yourself very simply. If you are not completely "brain-dead", you probably have some modicum of curiosity. Nowhere is this curiosity more often experienced than when we stop to peer into open boxes, open pits, tunnels, recesses, holes, secluded spaces, and so on. This natural curiosity is quantum Energy at work in the living bio-organsim. It is also a survival mechanism since recesses, etc., can harbour danger, or alternatively can provide sanctuary *from* danger. There is also the aspect of cause and effect. When we find a mystery we wish to discover it. The mystery puts us in the position of effect while discovering it puts us at cause. We all wish to be at cause, - not effect.

108 A ~ THE ORGONE ACCUMULATOR

"Our wretched species is so made that those who walk on the well-trodden path always throw stones at those who are showing a new road", (Voltaire).

Wilhelm Reich is well known for his researches into quantum Energy (which he called "Orgone" - or "Life Energy"). He conducted many experiments based on the concept of Orgone as an independent type of energy form subject to unique laws, - and according to these laws could then be regulated and manipulated: (see *Reich's Cloud-Buster*; 123 B). One device for which he garnered fame was his so-called 'Orgone Accumulator'.

The purpose of this accumulator was to serve as a bio-medical device which would concentrate and increase the intensity of Orgone in the patient's body. This in turn would bring about an improvement and reportedly, even remission of conditions such as cancer, etc..

Evidently Reich had great success with his accumulator, managing thereby to stir the wrath of the powerful American medical community. This wrath finally succeeded in causing the burning of Reich's books, records, and manuscripts by a government agency. He himself died in jail where he was sent as a result of false accusations and charges.

Orgone (q. Energy) is absorbed by electrically non-conductive organic materials such as wood, wool, cotton, leather, and so on. Books, magazines, and newspapers especially are excellent absorbers since they are composed of many layers of absorbent paper. In this regard they are superior to wood and cloth, etc.. In fact it is likely that a stack of books would be a superior insulator against grounding of the Energy.

Metals however – being electrically and "quantum-active" substances - are attractive to q. Energy. If a sheet of metal therefore is placed against a sheet of wood or paper, etc., the ambient Energy in the local environment absorbed by the non-metals is attracted to the metal. Several such composite layers may be arranged in a sandwiched or laminated affair so that the Energy is absorbed by the outside of a box thusly constructed and radiated or emitted by the inside layer.

Hence the Energy is absorbed by the wood/paper and then transmitted layer by layer to the final layer of metal. Using such laminated structures, one is able to construct a box or cubicle within which an individual may sit or lie. The outer surface of the cubicle absorbs ambient q. Energy and transmits it to the interior space where it accumulates and concentrates.

108 AA ~ QUANTUM ENERGY AND SUNLIGHT

> "the only things we know about the quantum world are the results of experiments", (from Paradoxes And Possibilities).

> "We have to remember that what we observe is not nature herself, but nature exposed to our methods of questioning": (Werner Heisenberg).

Researchers in the United States and Europe have conducted experiments with plants and sunlight based on the findings of *Von Reichenbach* and others. (See 'Supernature', 'The Secret Life of Plants', & 'The Psychic Power of Plants; - *Colin Wilson,* and *Lyall Watson*).

A controlled experiment was conducted where plants were placed in a dark room with no sunlight permitted. Half of the plants were used as controls and all plants were placed in separate pots. A wire was run through an exterior facing wall of the room and was then connected to a large metal plate outside. Inside the room, the wire was connected to several of the plants through their soil. The metal plate outside was then allowed to receive direct sunlight.

After several weeks, the control plants were observed to wither and die - as would be expected in the absence of sunlight - whereas the connected plants *were observed to thrive vigorously*, - as if they had been receiving normal sunlight. This would seem to indicate that some kind of life-giving energy was being absorbed by or generated in the plate and conducted to the thriving plants by the wire. Similar experiments and results are also described in *Baron Von Reichenbach's*, 'The Odic Force'…

A researcher named *Hieronymus* (a Fellow in the Society of Electrical Engineers) was also an experimenter in the quasi-science of Radionics, and he claimed that matter radiates an unknown form of energy. He did many experiments with plants deprived of natural light: (he called his Energy "Eloptic Energy").

Reichenbach likewise discovered that such a conductor would display the generation and emission of Energy with even the simple application of a flame or a heat source. Similar results were obtained with an immersion of one end of the conductor into a beaker containing liquid wherein a chemical reaction was taking place. [See note 2].

Columbia University scientists *I.I. Rabi, P. Kusch,* and *S. Millikan* have proved that a form of "strange energy" passes from molecule to molecule, each molecule being a transmitter and a receiver of this energy.

It has often been claimed that sunlight is healthful, and the mother is wont to admonish her offspring to go out and play in the Sun and the fresh air. Why does the parent instinctively or intuitively understand that sunlight is healthful and the outdoor air so invigorating? The simple and one answer to these questions is that it is the abundant presence of quantum Energy which makes them so. (This is well known in the medical community and patients are often advised to absorb healing sunlight).

It may seem to some that it is claimed herein that E-energy is a magical and wonderful do-all and cure-all. In fact this is an improper, if accurate viewpoint to adopt. Instead, we should take a reversed point of view and realize that it is upon the ubiquitous quantum Energy that all life, and indeed, all existence are predicated. Therefore it is our right to expect an abundance of this Life Energy, - even to generate it in quantity.

For example, we are being virtually indoctrinated that we should stay healthy by being in the sunlight and fresh air, eating fresh fruit and vegetables, exercising, and so on. There is nothing wrong with this of course, but what is the purpose of it? It is simply and exclusively to obtain our quota of quantum Energy necessary for proper health. If instead we were able to obtain this quota of Energy from another - electronic - source, we would still be able to obtain and retain this health and joy of life, etc.. [Please see notes 1 & 2]. In this event, food would be redundant*!*

With Energy given to us by such an abundant source, it could be argued that we would be lazy, - but in such an environment we couldn't be lazy, you see - for q. Energy is the life force itself*!* And that is what makes Energy so wonderfully practical! Living with an abundance of E-energy would be like living with a natural "high", - and having excellent health into the bargain!

The universal benefits of this energy are ours by right. The knowledge of E-energy which has been hidden from Man for so many eons is rightfully ours, and it is here now within man's grasp to generate it and use it to build himself a new world and civilization, - where all can be truly happy…

From this and other articles, we can learn that Energy is generated easily and likewise in large quantities, - and that it may be conducted along and through materials - some better than others. Fortuitously, metal is a superior conductor of electricity, heat, and quantum Energy.

As is to be expected then, the velocity of E-energy in matter is also between that of heat energy flow and the instantaneous transmission of light in an optic fiber. In fact, as will be seen - the extended spectrum of E-energy parallels the entire electromagnetic spectrum and is therefore disguised. Since it is naturally hidden and disguised, it may therefore be said to be a genuine "Secret of the Universe" - just as electricity once was!

Where E-energy becomes most unusual in its manifestations is where its spectrum exists in the sub-atomic range between X-rays and the electron's wavelengths (and beyond). In this range of the Em spectrum, there is known to be a gap. Quantum Energy completes this gap, (please see diagram 1). This is quantum Energy proper, and when quantum Energy or 'Energy' is referred to in this book, we are referring to this particular portion of the spectrum.

NOTE 1: I am sure that many purists and naturists will decry this claim and state that nothing can take the place of natural purity. To say this is to completely miss the point, for it is *quantum Energy* which is the sole source of health, life, and the joy of living. In other words, quantum Energy - although we may perhaps regard its use as "cheating" nature, is indeed of nature and therefore is itself the quintessential source of the purity which is sought for!

To put it another way, one could spend his entire life in a dark cellar with just bread and water of the poorest quality, and yet if he were allowed a *sufficient* continuing dosage of quantum Energy, he would live a long and healthy life suffering no ailments whatsoever; - this is the power of q. Energy! To put it yet another way, one might bask in an Energy field for a few minutes and receive the same benefits which are derived from a basket of oranges and a day of sunshine!

The nuclear fusion plasma is the quintessential arena of quantum Energy activity and generation. The interactions of energetic electronic particles generates the entire spectrum of Electro-magnetic (Em) radiations, and with it, *the entire spectrum of quantum Energy.* Radio waves, microwaves, heat energy (although heat energy is largely retained in dynamic organized plasmas), light photons, ultra-violet, X-rays, electrons, protons themselves, - and also including **ee** particles (electron pairs), and alpha (a) particles (proton pairs).

Naturally since the quantum Energy is a part of this generated spectrum, it too is emitted and radiated. It is this essential *quantum Energy* ingredient which makes sunlight the health and life-giving radiation it is. In fact, if the entire electro-*magnetic* spectrum could be eliminated from sunlight, allowing this portion of Energy radiations to remain - these emissions would continue to provide the essentials for life. (See 108 A). Even the absence of electromagnetic light photons would not hinder visual abilities.

Ball-lightning, being identical in function to the Sun and stars (as will be explained) - also generates and emits E-energy - which is why witnesses of this phenomenon sometimes observe other unusual occurrences in its vicinity. It is quantum Energy radiations and emissions which precipitate these Poltergeist-like effects. These and *paranormal phenomena* all originate with the presence of quantum Energy. (> section 03). For example, if strong emotional energy of some type, such as anger, grief, and so on is stored in the environment, any unseen paraphysical activity associated with this emotion, such as knocks, sobbing, footsteps, or rapping, etc., is often replayed in the presence of ball-lightning. This sounds 'spooky' and 'creepy' but in fact is simply quantum physics at work, displaying "new" laws of physics.

NOTE 2: quantum Energy may be generated in various ways. One should remember that this energy form is everywhere and in all objects; it is the source of all other forms of energy and is created by other energies. In short, it is interchangeable with all energy forms, and this will be explained. However, q. Energy is the primary force in all phenomena. One primitive form of low quantity generator would be in the shape of a solid metal pyramid or cone. If we sit the cone on a heat source such as a flame, the bottom of the cone/pyramid will become very hot.

This heat energy will travel slowly up toward the apex of the cone. However the high point (apex) of the cone will not become hot since the heat energy is radiated away and dissipated before it reaches this point (if the cone has a reasonably good slope, - i.e. at least forty-five degrees).

The heat energy and consequent atomic/crystalline activity generates q. Energy within the metal and this Energy travels upward toward the apex and point. Hence one may receive a flow of Energy from the point of the cone. The cone/pyramid is an ideal shape because the wide base allows for a large input of heat energy while the apex point allows the Energy to be concentrated in a practical fashion. (The idea in Chinese acupuncture of heating the end of a needle, as in 'moxibustion' is a parallel technique to the one just described). (> *The Great Pyramid*).

The amount of Energy produced in this way will depend upon the temperature of the heat source and the area of heat input. In any event, this is an inefficient and primitive method to generate quantum Energy and relatively little Energy will be obtained. However, it may be enough to register an effect upon small chemical reactions and the resistance of fine electrical components. The technically minded may find this interesting.

108 C ~ SUNLIGHT AND RADIATED E-ENERGY

Sunlight consists of visible light photons, heat energy, X-rays, and indeed, the entire electromagnetic spectrum. The spectrum, being complete, necessarily includes quantum Energy as a parallel energy form. Since the Sun is a sphere of plasma undergoing nuclear fusion, vast quantities of q. Energy are generated, emitted, and radiated. Any electrically active plasma will generate and emit E-energy; this certainly also includes electric arc welding equipment.

In our everyday experience, almost all electrical plasmas (found in 'neon' signs and fluorescent lighting, etc.) are contained within a glass tube or envelope. Glass - unfortunately, does not permit most Energy radiation and blocks it, thereby depriving us of its benefits. If these glass tubes were replaced with those of quartz (which permits Energy) we could then receive these radiations. ('Glass' envelopes designed to permit "hard" ultra-violet radiations are actually made of quartz; found in 'sun-lamps').

Photographs taken using this emitted and radiated light will on occasion show unusual Energy phenomena on the film. In this case however, the camera used should have their glass lenses replaced with those of quartz (i.e. silicon oxide, SiO). Otherwise, a "pinhole lens" should be substituted. The Energy phenomena displayed will be for example, the Energy field remaining *in situ* where a biological limb has been amputated. This Energy field is an actual dynamic, electronic replica of the original limb which is constantly maintained by the body and mind/will.

When using this type of camera, it may be found on occasion (with a very sensitive film, or Uv film) that sunlight itself may be adequate to reveal this phenomenon. A nearby naked electric arc however will emit stronger radiations than we will find in sunlight, and a successful photograph may be expected. (refer to *Kirlian Photography*).

Many stars have greater powers of emission (possibly due to a greater concentration of hydrogen fuel) and consequently generate and emit greater quantities of quantum Energy. In such cases, life on their orbiting planets will exhibit greater development, both physically and mentally. Indeed, this may have been the situation in our own planet's history for we know very little about the conditions of our Sun in those days long ago.

(The Pituitary gland regulates the growth of an organism, hence a stimulated gland under certain conditions may well stimulate accelerated growth, and even growth beyond the geometric boundaries of what is acceptable today. Similar conditions are true of the Pineal gland which is associated with powers of the mind and will.

108 D ~ DOES IT REALLY EXIST?

One comment which the Reader is most likely to make is, "this quantum Energy sounds too good to be true". This will then quickly be followed by the important question, - which the Reader is most certainly entitled to ask, - "does this *quantum Energy* really exist ?".

Imagine an Indian living in the Amazonian rainforest. He has probably never seen a radio, a television, an airplane, or what-have-you. His own world may be awash with radio and television signal wave energies and yet he has not the slightest suspicion or indication that such things exist. He has no means to detect them even if he should conceive of their existence. But most would deny their existence having seen no evidence for them. One could try to explain such things to him but try as you might, he will not understand nor believe what you are talking about.

He may even condemn you for being mad and try to eat you! His world is concerned only with the everyday tasks of surviving in a primitive environment. Even we, with all our technological sophistication are in a somewhat similar situation, - and like the Indian, will demand 'proof' that such things really exist.

We now however are in the superior position of knowing that physical energies can indeed exist which are undetectable by non-technological means. This book is an attempt to present a logical and reasonable argument that not only do such energies exist, but also that new and as yet undiscovered Energies can likewise exist.

< first radiations >

The discovery of electricity goes back to the eighteenth century with *Galvani*, followed quickly by the discovery that it can be conducted through organic material, metal objects, and wires. Until *Hertz* made his discovery of electromagnetic radio-frequency radiations, i.e. Hertzian waves, invisible electric waves through space were not known to exist, - and remained unsuspected.

For many years, it was not known that electric flow in a conductor also generated other radiations, such as heat energy (*Joule Heating*), microwaves, and even light photons. These were discovered one by one so that at present, physicists claim to have discovered and classified *all* such radiations. [please see note]. Is it possible then that there could still be even more but subtler radiations as yet undiscovered? Certainly the possibility seems remote, - and yet the possibility remains!

> It has already been suggested by two physicists, (the Nobel Prize winning *Dr. Millikan* who discovered the electron, and *Dr. I.I. Rabi*), that atoms and molecules constantly radiate electric waves, - the frequencies depending on the atom's size or energy level or substance, i.e. the number of electrons and electron shells contained.

NOTE: Atoms in molecules will oscillate and thereby generate microwave and heat energy. These atoms must be in a state of electrical charge however, which means that they become ionized as they oscillate. Their electrical bonds with other atoms are stressed and broken. Hence, heat energy is generated by electrically charged oscillating atoms.

A "normal" atom/ion, being a particle of 'normal' matter, has its share of gravity component and inertia. In 'normal' matter that we are familiar with, an atom/ion usually oscillates at a frequency which generates electromagnetic radiation of microwave frequencies. If this oscillation frequency and velocity of the atom/ion were increased, the emitted radiation frequency would also naturally also increase.

108 DE ~ THE QUANTUM ENERGY EFFECT

The "Poltergeist Effect" can often be experienced on a common everyday basis at the quantum level. Years ago when clocks were mechanical and powered by springs, the owner would find that his new clock would run either a little slow or a little fast. It was necessary therefore to adjust the clock by means of a small lever.

 The owner of this clock would eventually be satisfied with its ability to keep the correct time. If he were to lend this clock or give it to someone, that individual would invariably find that this clock would not now keep the correct time, necessitating a new adjustment. The ticking *loudness* will also be affected. (the act of giving and the consequent change of ownership will actually have a more profound effect on the situation since the space-time condition of the clock is likewise affected; more on this later: section 03).

The reason for this necessity is quite simply that the new owner generates a slightly different quantum Energy field, and the fine components of the clock respond to this difference which becomes measurable over time. This field affects various aspects of the clockwork including the inertia of the flywheel, the friction of the bearings, expansion/contraction of parts, the torsion and tension of the springs (mainspring and flywheel, and so on).

Another effect is the so-called "cooling effect". If a heater is allowed to warm up completely then allowed to cool, one will invariably hear clicking sounds as the device cools down. These sounds are usually attributed to contraction, tension, and the sudden release of this tension in the materials in the heater. (This is of course a quantum threshold effect related to atomic vibrations, valence shell activity and friction, etc.. One can perceive the slow accumulation of tension in the cooling metal, and when a threshold of tension is reached, the metal contracts suddenly - thereby creating the clicking effect).

It will be found that this clicking sound occurs at different rates depending upon the emotional state of the observer, and the observer himself. Different observers will have different effects such as the rate of clicking and its loudness: [see note]. The cause of the variation is in effect the variation in the quantum Energy field generated by the observer.

The q. Energy field has effects in other situations such as in laboratories where quantum Energies are induced, - for example in chemical reactions and fine electronic measurements. However, delicate chemical behaviors are the most easily modified. This modification of chemical and mechanical behavior is done on a (usually) entirely subconscious or unconscious level. Strong emotions such as anger, grief, or joy, etc., will have profound and more easily perceived ("objective") effects. Even the effects of the Moon's phases have been observed. Just as sea creatures respond to tides and the movements of the Moon, so also do water and electro-atomic bonds.

In summary we can say then that this quantum effect is common and widespread in the natural world. The size of water drops and the rate at which a tap drips are likewise quantum effects.

Similarly, such phenomena as the apparent tightness of a bottle cap, the ease and manner in which paper tears, the smoothness of a surface to the touch, the dynamic patterns made by cigarette smoke, the flow of ink from pen to paper, the sharpness of a razor-blade, the movement of a raindrop across a window pane, the cracking of an egg, the rate at which a wet towel dries, the growth of bacteria and mould on a fruit, the quality of recording on a video-tape, the life-span of a light bulb, even the throw of dice and the shuffling of cards, - are all variable quantum effects. Beyond this, even biological growth, muscular strength, and indeed sickness and disease are included!

(One may perceive here a definite reference to the ancient Roman method of divination by observing the patterns formed by spilled lamb's blood. We find it ludicrous now but in those days, this method actually worked! The reasons why will become clear to the Reader as we proceed).

NOTE: It will be found on occasion that the ticking of a mechanical or electro-mechanical clock will also sound louder. This will depend upon subconscious Energy flows generated by the mind, and indeed will be thought of as, - and termed a "subjective" experience. As to the ticking of a clock - I too, have noticed on occasion that the ticking seems to be inordinately loud. I wondered if this was my own subjective perception, or if perhaps the clock was trying to communicate with me like a narcissistic lady, saying, "I'm here, look at me"!

The question posed by the Reader will no doubt be: "is it subjective or is it objective?" In this book, I have strived to bring the Reader to the conscious awareness that objectivity and subjectivity are merely viewpoints, as are the dichotomic pairs. In other words, they are not necessarily separate concepts, but only exist as such by the decision of the observer. It should also be evident, as also previously proposed, that all experience must in the final analysis be subjective, since the experience of the observer - whatever it is - is his alone.

As to this proposal, we can say that an experience shared with others is commonly called "objective" but even here upon close analysis it will be found that those involved in a common experience will have different versions if asked to retell it. Police investigators are well aware of this.

Further to this, Man makes a continual effort to organize his universe into the categories of "objective" and "subjective". Man is unique in that he has a need, not experienced by animals or children, to view the world 'objectively' in order to control it and to bring order, and to create.

In the worlds of children and animals, all experiences are subjective. They have no need to control or to organize since they intuitively know that eventually everything will work in their favor. This again is "Taoism" formulated by the great *Lao Tzu*.

Certain branches of Man still live according to this principle, notably the Australian aboriginals, and the Indian tribes of South America. It is highly probable also that earlier forms of man, e.g. the Neanderthal and the Cro-Magnon, also lived this way.

Note: This is probably a good time to introduce the notion that the brain is divided into two lobes, left and right. This has been well researched and documented, and evidently - the left and right sides are each one concerned with different functions relating to the personality and characteristics of the individual.

It is fairly common knowledge that the right side of the brain is involved in the control of the left side of the body (including the sensory organs), and vice versa. It has also been found that the left side of the brain is intimately concerned with characteristics and abilities such as logic, pattern recognition, details and analysis, mathematics, music (which is a mathematical construction, by the way), factual analysis, practicality, precision, map-reading, risk-taking, spatial orientation, ability to do repetitive tasks accurately, sequential recognition, organization, scientific reasoning, objectivity, and literal analysis.

The right brain on the other hand, is more concerned with such characteristics as symbol recognition, emotional experiences, subjectivity, philosophy, religion, creativity, imagination, intuition, concepts, empathy, multi-task execution, figurative perception, and recognition of irregular shapes and geometric forms. It is also recognized that the right brain displays those characteristics usually associated with the "irrationality" of the female, while the left-brain is associated with the right-handed logic of the male. (Irrationality here refers not to insanity, but to that side of nature which is not rationalized, - such as the ratio of a circle's diameter to its circumference, and so on).

What is not acceptable to the politically correct is the fact that it has been shown repeatedly that due to these differences, men are definitely superior at some tasks - such as piloting aircraft (which of course involves spatial orientation, mathematics, logic, objectivity, repetitive precision, and map-reading, etc.), while women excel in others, such as home-making (which involves time-sharing and multi-tasking), child-rearing, clothes designing, and patterning, etc.. [see note].

In the case of piloting an aircraft or the captaining of a ship particularly, these individuals have special talents and these are shown by their ability to optimize the use of both sides of their brains. In these cases, they are required to do not only mechanical, logical, and repetitive tasks and their organization, but are also required to be able to deal with people, their behaviour patterns and idiosyncrasies. They must have the ability to control their emotions, and the organization of staff and personnel at all times, - especially in times of emergency when self-control is essential to the survival of all.

This ability to optimize both sides holds true for scientists and physicists also. Consequently, we find that many such individuals - while involved in the quest for solutions to mysteries concerning the physical world, are also deeply interested in philosophy and religion.

NOTE a: the idea that men and women are equal is really a no-brainer and incontestable. Men and women have *always* been equal, albeit having complementary roles necessary for procreation and survival of the race. The media (and the education system) today has substituted the meaning of the word 'equality' with 'sameness'. It should be obvious that men and women are not the *same*. (An orange can equal an apple but it is *not the same thing!*).

People are being subtly taught that men and women are the 'same' - although it is never stated as such, since this would make the implication obvious. Women are given jobs corresponding to the traditional male role; they ride motorcycles, drive buses, and so on. It is the proven structure of tradition which is under attack. (Likewise, we find that all other socially destructive elements are being promoted).

No-one ever thinks however that women still expect men to be the aggressors in courting, to follow them, to open doors for them, and to pay the restaurant bill. It is the woman who dresses provocatively to lure the male. It is the woman who plays hard to get, and so on. The woman expects and encourages the chase, to be the prey, - and behaves accordingly. It is her natural inclination to want the male as the predator; ultimately she wants to be taken, - affectionately.

Men tend to be direct in their relationships with others while women tend to be indirect, both in their comments and actions. A women will often observe others indirectly by watching their reflections in glass, etc..

These natural differences will always exist, just as the other differences between men and women exist. To imply that men and women are 'the same' is ludicrous. It is an insane ideology created by the anti-social.

NOTE b: for those who are interested in gender differences, you will be aware that there are obvious physical variations. However there are also physical differences which you may not be aware of. Every chromosome of a female carries an XX gene while the male carries XY genes. This means that every cell in the body of a female is specifically of the female type.

The eyeball, the teeth, the fingernail, the flesh, the bone, the hair, the blood, the saliva, - yes, and the brain; - these are all specifically female, as any forensic scientist can tell you.

When a man looks at a girl, he is attracted not only to the shape, the hair, the colouring, etc., but to every square centimeter of her visible body. He sees an individual totally and completely different from himself, - this is interesting. Her bio-electronic field is likewise specifically female. Obviously, females find a reciprocal attraction to men for similar reasons.

< male and female Energy forms >

The two genders have Energy waveforms peculiar to their sex. The waveform of the male is essentially trapezoidal. If you have a doorway in which the two straight vertical sides slope outward toward the bottom, the opening thus formed is a trapezoid. It represents a cross between a square and a triangle. One could liken it to the outline of a pyramid with a flat top. The female waveform on the other hand is more sinoid (like waves on a pond) and represents a smoother, "softer" Energy flow. Other parameters which define Energy emissions can apply, such as varying wavelengths and amplitudes, etc.. (If you were to draw these two types of waveform on paper, you would recognize that the trapezoid form is simply a sine wave form with straight sides).

We may also note that the products of men, i.e. machines and electronic devices, etc., in their basic format tend to have square, trapezoidal, or rectangular shapes. Likewise those features of non-female technical devices which attract males tend to be squarish, while the features attractive to females are rounded and soft.

The question asked now is what causes this variation in Energy flows? Is it the biological cells of the body, is it the brain, is it the mind, or is it the Being itself? The answer can be found at all levels, however one must remember that the fundamental source of all phenomena - regardless of their apparent sources, is the Being. If the Being chooses to own and operate a male body, it will be responsible for the Energy flows originated and emitted. This must be true since a dead body emits no Energy whatsoever. Why are male and female Energy waveforms mutually different? Obviously, it is in order to attract the opposite sex.

These waveforms - which are peculiar to each gender, are designed to be attractive to its opposite. They are attractive (i.e. aesthetic) only because they are different, not because of any intrinsic quality. It is the Being who places significance upon this difference and then perceives that this significance is attractive. Hence, we see here then that maleness and femaleness are aesthetic forms, and the aesthetic is generated by the beholder. Ultimately, the aesthetic must be defined as a physical Energy form since it has a wavelength and frequency.

An example of subjectivity and objectivity is well illustrated in the wine and alcoholic beverage industries. One may sample a bottle of wine, etc., produced in one year and another wine, etc., of the same label bottled the following (or preceding) year. This is ostensibly the same wine produced by the same company using the same equipment and fruit, etc.. If these two wines are analyzed chemically they will be found to be identical, and yet the taste will be different. Now if they are chemically identical, how do we account for the difference in taste?

A chemical analysis is an *indirect* 'objective' process, while the taste of course is a *direct* subjective one. It is reasonable to say then that there exists a difference which is not subject to contemporary scientific scrutiny. The difference is evidently subtler than crude molecular structure. Hence we should look for differences which are finer: [see note 1].

The same phenomenon holds true when wine ages. The (subjective) taste changes, but the (objective) molecular structure is *apparently* the same. [See note 2]. We may suppose in the case of the wine that the difference lies in very minute traces of a slightly different molecule and its structure; but in the case of wine which has aged, why should this improve the taste (an entirely subjective experience)? (> 301 BB (2)).

It has been suggested by researchers that there is a slight shift in the molecular arrangement and geometry due to the presence of expanded (i.e. "Energized") atoms, and this is entirely feasible. In the case of water subjected to an influx of "Chi" (quantum Energy) generated by a healer or practitioner, it has been found that the water molecule has indeed been slightly altered in geometry, i.e. expanded in volume.

The normal arrangement of this molecule is two hydrogen atoms separated by an oxygen atom, and the angle thus formed is 103 degrees. In 'energized' water it has been found that this angle changes slightly, thereby in effect creating an entirely new substance, but which may still be called "water" – (and which indeed serves the same purposes). Hence we may expect to find similar differences in the two samples of wine. (The practice of blessing water and wine to make it "Holy" in various churches has this as a basis of fact. This technique of blessing water is very likely a holdover from the pre-Christian days which the church then adopted).

It has been found that so-called "Energized" water is preferred by both plants and animals, which then flourish with its use. It is likely also that children will generally prefer the "treated" water. It has also been found that plants given this water tend to grow more rapidly, to be healthier, more luxuriant, and larger. Again, this is quantum physics. [note 3].

In summary then, we may say that the difference between the samples of water and of wine lie in the differences of the energy levels of the atoms involved. These energy levels then directly affect the slight differences in molecular structure. Evidently, the atoms retain the increased energy levels which moreover do not decay over time, but which are apparently *permanent* changes. This again introduces the notion of non-existent time at the quantum level: (elsewhere explained).

In other words, the atoms themselves have changed. The character of this change is evidently an expansion of the electron shells in the atoms which now have increased energy levels and consequently higher rates of 'vibration': (to be explained). [see notes 2 & 3].

NOTE 1: there are some who would claim these differences are caused by less obvious factors such as the positions of the Sun and Moon, etc.. This may seem ludicrous at first and yet there are many serious books on this subject by respected authors, - for example: 'Astrology: The Space Age Science', by *Joseph Goodavage*. 'Cosmic Bonds', by *E.A.Lawrence*. and 'Cycles; The Mysterious Forces That Trigger Events', by *E.R.Dewy.*

NOTE 2: this is a somewhat similar problem to that found in Homeopathic medicine where a remedy is concocted from distilled water and another added substance. A very dilute solution is formed, and in fact this "solution" can be so dilute that not a trace of the additive agent may be found in it. Yet the medical efficacy *increases* as the solution becomes ever more dilute*!* From this we may deduce that some undetected and unknown properties of the included substance are passed on to the water molecules and/or atoms.

NOTE 3: Most people are probably not concerned with the colour of the eggs they eat but some prefer brown eggs over white. This may be for several reasons. Maybe they are racists. Some simply prefer the colour, or feel that a more natural colour may imply that the egg includes more nutrition. The so-called 'experts' tell us that there is no difference in the nutritional value of the egg, regardless of its colour. (Please see 104 B: *Structured Perception*).

Now there are several hidden factors in place here. In the first place of course, the "experts" don't always know what they're talking about, and this has been amply demonstrated time after time in various fields. As we shall see, the *way* in which something is observed and *perceived* has a very definite impact and effect upon that which is being observed and perceived. This fact is closely related to quantum physics, creativity, intuition, cause and effect, and also to subjectivity and objectivity.

The scientific 'experts' refuse to thoroughly investigate phenomena such as Homeopathy (or Poltergeist phenomena) since they do not know where to begin their investigations. It is something which apparently does not belong in the 'hallowed halls' of science - which in turn is clearly delineated by predictability, structure and function, cause and effect. Something is going on in homeopathy which apparently works, and yet there seems to be no explanation forthcoming.

In fact homeopathy *can* be explained, and it works on a quantum level which physicists are at present largely ignorant of. The same is true of the egg phenomenon. It may well be in fact that the intuitive belief that brown eggs are better really is correct. The scientists refuse to acknowledge the possibility simply because they are ignorant of the more subtle influences in nature and of the mind.

108 E ~ QUANTUM PHYSICS

> "the only things we know about the quantum world are the results of experiments", (from Paradoxes And Possibilities).

> "We have to remember that what we observe is not nature herself, but nature exposed to our methods of questioning": (Werner Heisenberg).

The new physics described herein may accurately be called Quantum Physics and/or Quantum Electro-dynamics. Now with quantum physics and Energies we will find that matter, energy, space, and time will behave in unexpected and bizarre ways. So-called "reality" can become confused with - and interchangeable with illusion, and vice – versa.

Likewise objectivity and subjectivity will become interchangeable and merged. It is a strange, surrealistic world where the image of an object may communicate with the object itself, where cause can become effect and effect becomes cause, where space and time can expand and contract, where space 'opens up' to reveal another space-time, where things appear and disappear - and be where they 'should not be', - or vice versa.

This is where time can stand still, where objects can have no weight or no solidity, where they may merge with other objects, and where things can occur spontaneously without apparent cause. These things and more will be presented, explored, discussed, and explained.

It matters little if physicists agree or disagree with the ideas presented herein. Physicists are always disagreeing with new ideas and those which they do not understand; - it is the result of their arduous and intensely focussed and narrow training. This is why a book such as this must reach the general population.

It is not unexpected for example that single objects may be perceived to be in two or more places at the same time, or conversely – several objects in the same place at the same time, even though there is no apparent agency causing their relocation. Likewise, we will find that while one cause may have several effects, we can also find that one effect may have several causes! In all of this strangeness however there is only one underlying essential factor, and that is *time* and its peculiarities.

> "Imagination rules the world": (Napoleon Bonaparte).

109 A ~ ACCUMULATION OF QUANTUM ENERGY AND CRITICAL MASS

In the context of accumulation, quantum Energy may be closely compared with heat energy. Energy may be generated and accumulated in matter. However while it may be generated, it is also being conducted away to other parts of the circuit and eventually to ground. It is also radiated and emitted into the surrounding environment and air at the surface interface of the conductor. If the generation of Energy in a unit of time is great enough, the Energy will accumulate in a suitable conductor or mass despite its rate of loss.

Alternatively with a larger amount of mass, the Energy will generally accumulate more rapidly since surface losses will be relatively slowed down. (As with heat energy, loss is at the same rate for each unit of area. However, a larger mass has more volume, which increases exponentially in ratio to the area according to the formula d^3/d^2, where d = the average dimension in the three dimensions of space: (in a sphere this would be the diameter). Therefore in a larger mass, the rate of loss is reduced in ratio to the rate of accumulation. This means that energy accumulation will proceed more rapidly than it is lost. (As with heat energy, the ideal geometry for a mass to accumulate Energy is a sphere or spheroid).

For example, if the Energy be made to accumulate in a small *cavisphere*, i.e. hollow sphere such as a tennis ball, the rate of loss may exceed its ability to accumulate Energy. However if the sphere be a very large and dense 'solid' mass (assuming the rate of Energy inflow to be the same), the Energy will accumulate more rapidly than it is lost. (This may be called "positive dynamic accumulation").

Heat behaves in a very similar fashion. The reason for this is of course that as a sphere physically expands (e.g. a balloon), its volume increases numerically and exponentially more rapidly than does its surface area. This means that a large massive sphere will accumulate Energy more rapidly than it loses at its surface, just as a large cannon ball will accumulate and retain heat energy more rapidly and efficiently than a much smaller musket ball.

Therefore if one would wish to create certain objective Energy effects with an object, it will be necessary to have sufficient mass, (and be ideally spherical), - or *alternatively*, the Energy produced and generated be of *sufficient intensity and quantity* to offset the losses.

< bucket analogy >

An analogy of the foregoing is as follows:

If one takes a bucket of water then drills a hole in its bottom, the water will of course drain away. If one now places the empty bucket under a slightly open water tap, the bucket will not accumulate any water because it is escaping through the hole and cannot be replenished by the feeble flow from the tap.

However if we increase the water flow rate to more than compensate for the loss by drainage, the water will then accumulate and fill the bucket. (We may then call this a "Positive Dynamic Accumulation" as opposed to a "Negative Dynamic Accumulation"). One can see then that it is possible to balance the rate of inflow against the rate of outflow and thereby dynamically maintain a full bucket, - or a partially full bucket at any desired level. (This could then be referred to as a "Static - Dynamic Accumulation"). The volume of water in the bucket under these conditions however is not in a "relaxed" or placid state and is necessarily maintained in a <u>dynamic</u> and "<u>energized</u>" (turbulent) state. (> *Dynamic Maintenance*, next article).

So it is with quantum Energy (or heat energy) in a conductor or circuit. The Energy, if not maintained at a desired level by its generation, will "drain away". This input flow of Energy and consequent outflow must therefore be maintained and balanced.

Normally we should expect to find the bucket water in a turbulent state under these conditions. However, it is possible to induce a rotation in the water while it is draining out of the hole. The amazing thing about this is that as long as there is an inflow and an outflow, this vortex will be maintained. It is the dynamic inflow and outflow which maintains this rotation of fluid. (More on vortices later).

There is another aspect to the 'bucket analogy' not yet mentioned. Let us now suppose that the bucket in the analogy is punctured with a row of holes in its side from top to bottom. If water is put into the bucket, it will begin to pour out of the first hole near the bottom. Hence, in order to maintain the level of the water it must inflow at a given rate. If now the input rate is increased, the water level will rise up to the second hole and pour out of it. Hence the input water is now required to sustain an outflow from *two* holes.

In this event, the inflow and outflow of the water has increased to maintain its dynamic level. The same argument applies for the third and fourth holes, and so on to the top of the bucket. In this case, it can be seen that in order to maintain an increased accumulation in the bucket, the input energy must likewise be increased. When the water level reaches the top of the bucket, both the input and output energies (water) have increased enormously. In simple language, if it goes in faster, it comes out faster. This system represents one wherein the Energy losses correspond with the amount of Energy put in and vice-versa.

In other words, the output and the input may both be considered to be either the *cause* or the *effect* - a typical characteristic of quantum mechanics. This is because these events are occurring <u>simultaneously</u> - and in a sense, we have eliminated time.

To illustrate this last point, let us refer to the example set by a river. If the land is slightly inclined, the river will move slowly and will consequently meander having a tendency to deviate from a straight line and wander from its intended course. If however the incline is steep, the water will travel more rapidly and its course will be correspondingly straight. In the first case, you see we have introduced a longer duration of travel, i.e. more time. It is this time factor which introduces the tendency for deviation, whereas if the velocity is high, the time factor has been reduced and the deviation is consequently less. This of course is obvious but it also a quantum principle. The time factor then is determined by velocity of flow.

109 B ~ DYNAMIC MAINTENANCE:
(Static – Dynamic Accumulation)

The phenomenon of 'dynamic accumulation' (109 A) is commonplace and is in fact one of the fundamental mechanisms of the physical universe and its maintenance. It is as commonplace and as important as the principle of Inflow and Outflow, - and vortices: (elsewhere explained).

For example, the body as a chemical engine is composed mostly of water (98% by weight) and this percentage must be maintained. As water is lost by surface evaporation, perspiration, urination, and secretions, etc., water must be inflowed. This inflow and outflow of water serves to maintain the essential balance of fluid composition.

Another simple example of dynamic maintenance is the magnetic field surrounding a conductor carrying an electric current. As long as the current is inflowing and out-flowing, the field is maintained. The field may be maintained for years without let, but in order for this to occur the dynamic current must always be present.

If you have ever taken a dip in a mountain pool, you will have realized that this pool is maintained by "Static ~ Dynamic Accumulation". In other words, the pool is supplied by an inflow of water which in turn continues on and drains away at another point. Hence the body of the pool is maintained by constantly moving water, i.e. a mountain stream: [see note 3]. When you get out of the pool, the water is not the same water you got into! The pool then can be likened in many ways to a living organism. (The public swimming pool is maintained by the same process).

The same principle may be applied to all biological systems, and in fact to every system which includes inflows and outflows in its structure and function. This includes virtually every dynamic entity in existence.

Stars must inflow hydrogen and outflow helium and energy to maintain their existence and *"raison d'etre"*. Even a mechanical engine must thusly be maintained. It must be fed air and fuel and allowed to exhaust the deadly waste. The oil required to lubricate the bearings must be continually supplied to the reservoir and maintained due to inevitable losses. This "static-dynamic-accumulation" therefore is another mechanism which defines a life-form: [see note]. Many of the parameters which define a life-form are therefore found in a mountain pool, - including the generation and presence of quantum Energy.

< dynamic maintenance >

Far beneath the seas and found in volcanically active regions there may be found so-called sea-vents, sometimes called "black smokers". From these vents issue forth super-heated water together with particles and minerals. This water may be heated to well above its boiling point at sea-level, and may be in the vicinity of three hundred degrees Celsius, or 750° F.. This water does not boil however, since the water pressure at these depths is so great. The water is thus super-heated to well above the boiling point.

Normally, the surrounding sea water at these depths is very cold, – the cold and the pressure are sufficient to maintain methane in a 'frozen' state. In fact, the water itself may be at the freezing temperature, but the pressure is too great to permit ice to form. (This is because in order to form ice, the water must expand). However, in the vicinity of these vents, the water is very hot with the temperature gradually cooling as the distance from the vent increases.

Hence is maintained 'islands' of heat in the frigid depths. Around these hot water vents exists an infra-structure of life forms, - flora and fauna. It is in fact an entire self-supporting eco-system. Now here we have a classic example of "dynamic maintenance". The hot water from the vents supplies heat and nutrients to the surrounding life which in turn enables small sightless animals to survive, - even in this tremendous heat and Stygian darkness. In fact these life-forms gain necessary energy not from sunlight, but from these hot-water vents! It is truly an alien landscape: [note 4].

No-one knows how long this situation has existed, but it is obvious if these 'smokers' were to cease, so too would the eco-system they support.

In the same way, life on Earth is dependent upon many factors - but primarily the Sun (and hence nuclear fusion) which supplies the necessary light and heat. It is a tenuous situation and one of dependency. Many architects use this principle of *dynamic maintenance* to construct domes over playing fields and Olympic arenas, etc..

The domes themselves are made of light flexible materials and are kept in position overhead only by the continual use of energy input in the form of compressed air flow. Without this dynamic maintenance by a continued energy input, the dome would collapse.

Similarly our entire modern way of life is based on, and reliant upon this principle of dynamic flow. All our electric lighting, heating, computers, communication, transportation, etc., are totally dependent upon the continual non-stop supply of electricity and energy input. An entire lifetime is now usually spent living under this dependency. In other words, our very way of modern, technological life is totally dependent upon a continuing supply of oil. Think about it!

Hence modern Man has placed himself in a position of dependency upon many controlling factors including oil, electricity, money, the clock maker, the car maker, the grocery shop, the doctor, the dentist, plumbing and water supply, and so on. All of these are required in order to obtain the bare essentials of life, - food, medical care, work, transport, and so on. Man is in a self-made utter dependency - and dependent upon those who control the supplies.

We should now see that everything which exists as a part of the physical universe depends for its existence upon flow and stasis and dynamic maintenance. In other words, all is temporal and passing. Everything which exists requires constant maintenance - and this in turn requires a constant input of energy. There is nothing which is really stable, even the sturdiest stone castle requires maintenance, it is not a perfect stasis. If this maintenance is eliminated, the castle will eventually decay and collapse (long after being abandoned by its resident life-forms)…

Thus is the universe maintained and sustained, - thereby is the universe continually created.

NOTE 1: much more on this will be given. > Also 9013 A.

NOTE 2: the super-heated water does not boil and the super-cold water does not freeze, - a peculiar state of affairs, and all in juxtaposition. Hence pressure alone creates a very interesting state of affairs. Ice has a lower density than liquid water and expands when it forms. Since the pressure is so great at these depths, ice therefore cannot form.

NOTE 3: such a pool of water may be said to be "alive" - like a living creature, since it is constantly dynamic. A stagnant pool of water by contrast may be said to be "dead". In fact we *intuitively* call it "dead water".

NOTE 4: in fact the "black smokers" are a geologic feature belonging to an entire chain or seam of volcanic crustal instability which wends its way around the planet so that it resembles the seam of a baseball. Part of this chain is the 'Mid-Atlantic Ridge'. It also forms part of the Olduvai Gorge and Rift Valley located along the great fault forming the bed of the Dead Sea, the river Jordan, and the Red Sea. This is one of the reasons why the Dead Sea is so full of minerals.

109 BA ~ INFLOW / OUTFLOW OF Q. ENERGY, & HEALING

As well as food, water, and air, etc., the inflows and outflows which maintain the dynamic electro-chemical organism, there is also the inflow and outflow of quantum Energy ('Chi'). The body itself maintains a certain level of Energy accumulation in order to function optimally (see 109 B). (This Energy surrounds the chemical body as a field and it is claimed by some that this Energy can be seen; - more on this later).

In children and animals, this Energy flow is relatively free and unhampered. As we grow older however we develop negative and unwanted attitudes including anxieties and bad emotions, and it is these Energy ridges which contribute to the blockage of Energy flows (*W. Reich*). Hence we may find that there are blockages of the Energy inflow which then affects the accumulation level in the body and the availability of Energy to the body. (Such a flow blockage may be called an Energy "ridge"; - more on this later).

The Energy sources for the body are usually the mind/will and the electro-chemical activity, although the source may also be exterior to and independent of the body. If the accumulation level is too low for the body to function properly, the body will get sick and less resistant to disease and psychosomatic disorders, etc.. Since the Energy is already associated with the endocrine system - including the Pineal and Pituitary glands, we find that the mind/brain combination is likewise also affected. [note 3].

In other words, we can say that negative attitudes and emotions can introduce further illness both to the body and the mind. In this case, we can observe a self-perpetuating negative cycle which can lead eventually to further degeneration and perhaps even eventual death. On the other hand, positive attitudes and emotions contribute to good physical, mental, and spiritual health. Indeed these two conditions are mutually supporting! One only has to observe children and animals to recognize this truth. [3b].

Acupuncture is an attempt to remedy an existing or potentially unhealthy situation and imbalance of Energy. However with the application of quantum Energy fields, a much more effective and efficient remedy can be affected. If we refer to the bucket analogy we can see that there is another way in which the body is depleted of Energy.

Simply stated, the body may outflow Energy more rapidly than it can be replenished. We can see then that the body's accumulation of Energy will be depleted. This can occur as a result of emotional and physical imbalances such as overwork, lack of adequate sleep, lack of nourishment, mental anguish and over-exertion, anxiety and stress, etc.. It can be seen then that an ailing organism exposed to a high value quantum Energy field can accumulate Energy which will then have a beneficial effect upon its physical and emotional well-being.

There is yet a third mechanism however by which the organism may suffer. In this case, the outflow of Energy is blocked while the inflow is maintained. In the water/bucket analogy this will create a spillage effect. In terms of Energy and the organism this means that the inflowing Energy cannot outflow in a normal and healthy manner and must therefore create a "spillage effect" and find un-healthy outlets.

This can result in various maladies such as hyper-activity, anger, anxiety, stress, sleeplessness, nymphomania, and Poltergeist activity. It may also be the cause of unwanted physical conditions such as cancer and so on. If quantum Energy is allowed to accumulate beyond its natural limit, it may degenerate into lower forms of electromagnetic energy such as light photons, ultra-violet light, and even X-rays and electrical charge! (These energies may be registered on photographic film in the proximity of such an organism. This will manifest sometimes as a fogging of the exposed film). One method of releasing this Energy surfeit is simple.

The patient can be *grounded* by having his body connected with an electrical conductor to a grounded object such as household metal plumbing pipe, a drainpipe, a hot-air duct, or a steel structural member. This remedy may be done regularly over a period of several days similar to the method employed by an acupuncturist. Another more creative method would be for the patient to engage in regular physical exercise or mental discipline such as archery or other martial art, etc,: [notes 1a, 1b, & 2].

The mind/will may be developed and trained to utilize this excess of Energy so that it can be projected from the body toward and into various objects in the environment. This will bring about the manifestations of unusual effects which are commonly referred to as Poltergeist phenomena (see *The Geller Effect*). This was the goal sought after by the ancient magicians. In martial arts, this Energy is directed toward a specific goal.

Natural unaided healing then becomes not so much a matter of bio-chemistry or bio-electro-chemistry initiated by the body itself but rather one of focussed attention and directed quantum Energy to the affected area. When we suffer traumatic damage to the organism, our natural and immediate response is to direct our full attention to the wound. This is almost an unconscious and intuitive realization that healing can be aided and accelerated by the mind/will directing attention (and Energy) to the affected locality: [see note 2].

In fact, when one visits a hospital with some type of ailment, a large factor in the healing process is set in motion by the *attention* given to the patient and his disorder. In addition to this is the psychological preparation of actually going to the huge imposing hospital building. All of the equipment used, the language used, and the soothing, reassuring words spoken (or brusque and business-like attitude) are parts of this magical process, (for indeed, Magic is just what it amounts to). And of course one is surrounded by a tremendous amount of agreement.

The equipment in the hands of professionals simply amounts to a via enabling the patient to heal himself - aided by the attention given by others, and his own belief system. It all constitutes a grand illusion of self-deception, i.e. a Placebo. This is why placebos work; - it is the patient's belief in the system which allows it to function as a healing agency.

In actuality, _the patient heals himself_. This should come as little surprise to those who have studied and practiced acupuncture, homeopathy, reflexology, Western medicine, and other healing arts. It has been a philosophical truism for thousands of years.

One might say the patient has been "brainwashed" or indoctrinated into investing faith in the lucrative medical system, - or indeed any such system where healing is purported to come from without and not from within. (This is especially true in today's technological Western world). The entire medical system therefore may be described as a placebo, i.e. an illusion! Hence it is easy now to understand why primitive peoples put so much faith in their shaman or witch-doctor; these methods work simply because they do nothing - except convince the patient!

NOTE 1: in fact it is advantageous wherever possible to be thusly grounded so that the balance of Energy flow in the body may be maintained at a proper level. (It is not recommended however during electrical storms or when working with electrical equipment). This is likely why people who live a very simple lifestyle and walk with bare feet are healthier. In a phrase, "Be well grounded!"

NOTE 1b: youth - especially young men, tend to have a surfeit of quantum Energy and this is why they should be encouraged to participate in physical sports, etc..

NOTE 2: as we shall see, q. Energy associated with the sentient organism is directed by the individual simply by the focusing of attention and direction of gaze: (more on this later).

NOTE 3a: quantum Energy behaves in a manner counter to what may be expected from the usual behaviour of energy. For example, if one lacks the ability to produce enough q. Energy to sustain good health, he will become ill. When he becomes ill, he will lack the ability to produce enough Energy to maintain health. Hence, we see a dwindling spiral having apparently no way out, and the organism finally dies.

If however the organism has the ability to produce enough Energy to maintain good health, he will maintain his ability to produce enough Energy to stay healthy. This is a natural process so that in nature, weak animals succumb and the strong survive.

Man, through emotional and/or material attachments tries to counter this natural process by attempting to keep the weak and dying alive (which may seem to some to be irrational). We see here a mechanism which operates according to the principle that nothing is static; either it progresses upward to survival, or it degenerates downward to death. In order to survive therefore, an organism must be able to create; it must be able to create Energy. This is the way of life. Those unable to sustain life do not belong in the physical universe. This is the harsh reality of material existence. If you want to be a part of it, you have to be creative!

NOTE 3b: we may refer to (109A) wherein the description of a meandering river is given. We see that the river has 'too much time' and consequently meanders (deviates). On a human level, this deviation may be considered to be a form of insanity since the main purpose of a flow is to get from one point to another. Deviation implies inefficiency and lack of good order; *sanity* implies the opposite. Any flow can be described with the equation (Flow = pv) where (p) is quantity of particles in motion and (v) is velocity. If the velocity is very low or zero, the flow is compromised.

In fact when a flow is severely compromised and deviated, this is the beginning of the end for it and its death will ensue.

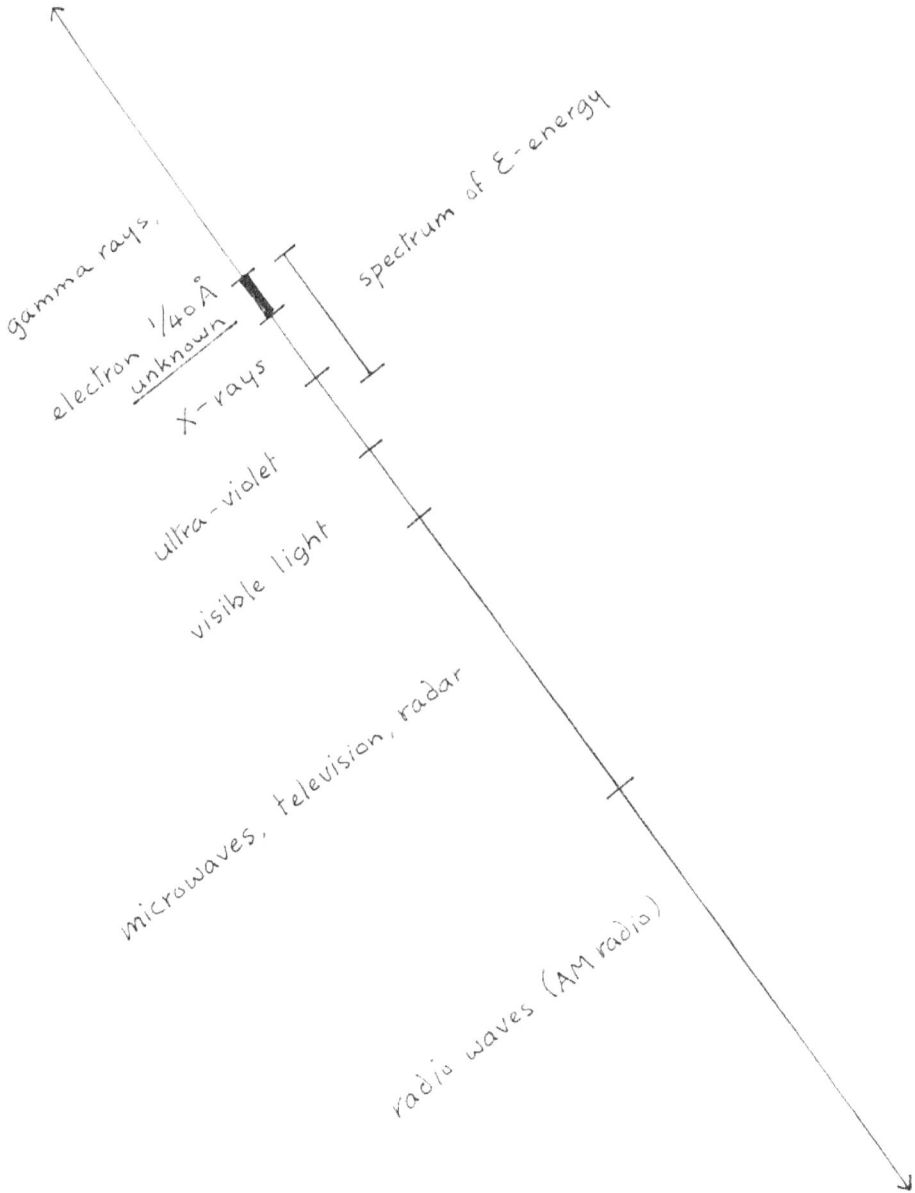

Electromagnetic Spectrum

gamma rays,

electron 1/40 Å
unknown

X-rays

ultra-violet

visible light

spectrum of E-energy

microwaves, television, radar

radio waves (AM radio)

Diagram 01

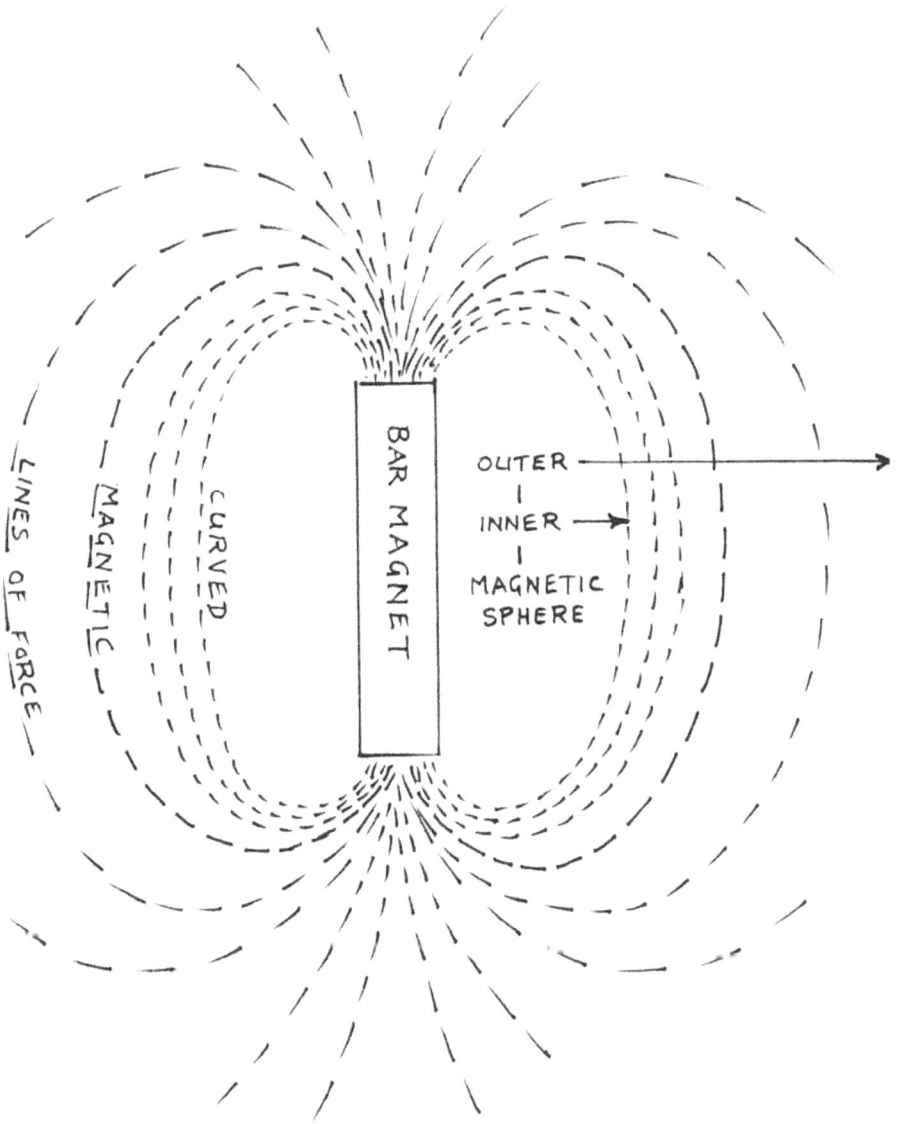

Diagram 03

electrostatically charged cloud

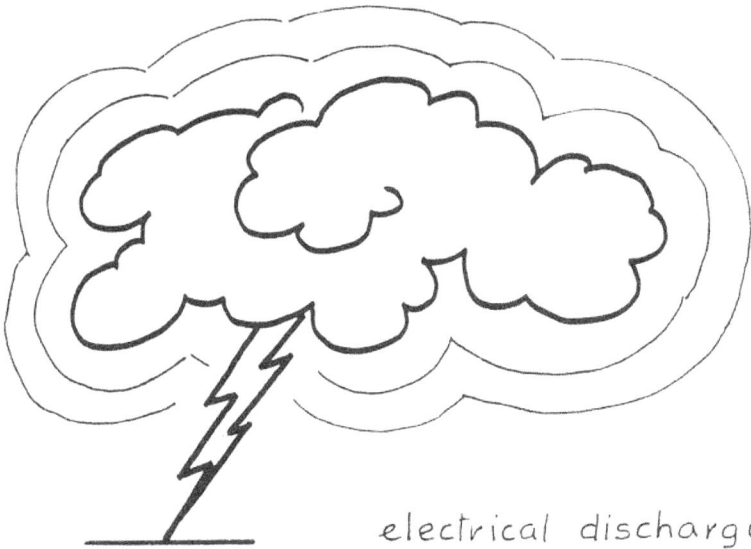

electrical discharge
to ground causes cloud to emit collapsing
field pulse wave of a tractor type, (～～～)

Diagram 06

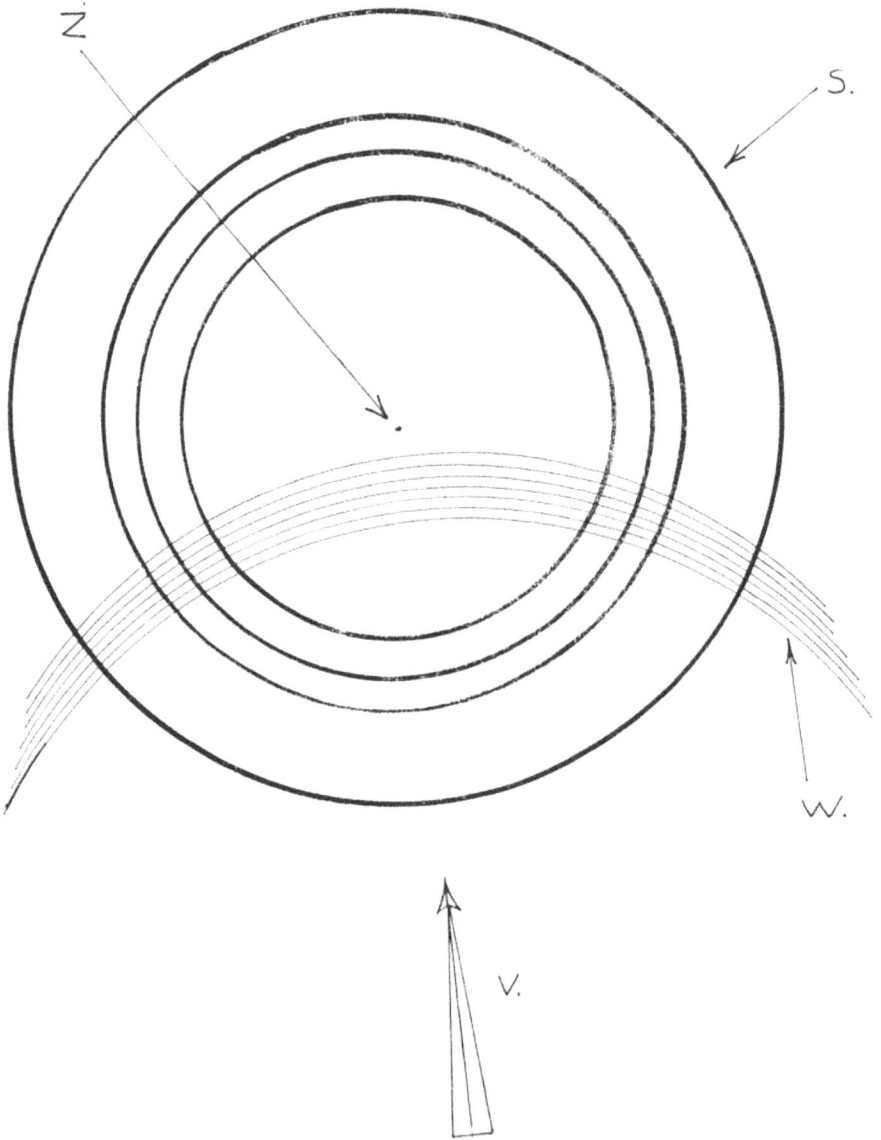

Diagram 08

CROSS SECTION OF A CONDUCTOR

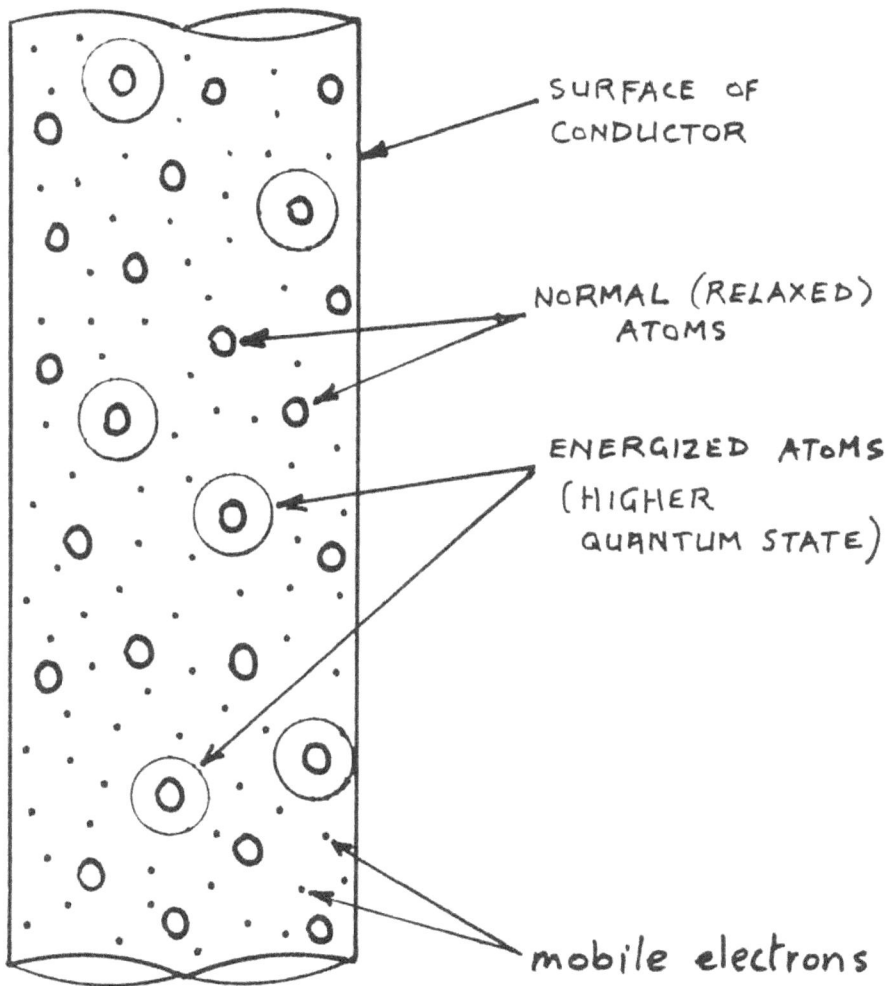

SURFACE OF
CONDUCTOR

NORMAL (RELAXED)
ATOMS

ENERGIZED ATOMS
(HIGHER
QUANTUM STATE)

mobile electrons

Diagram 10

'population inversion'

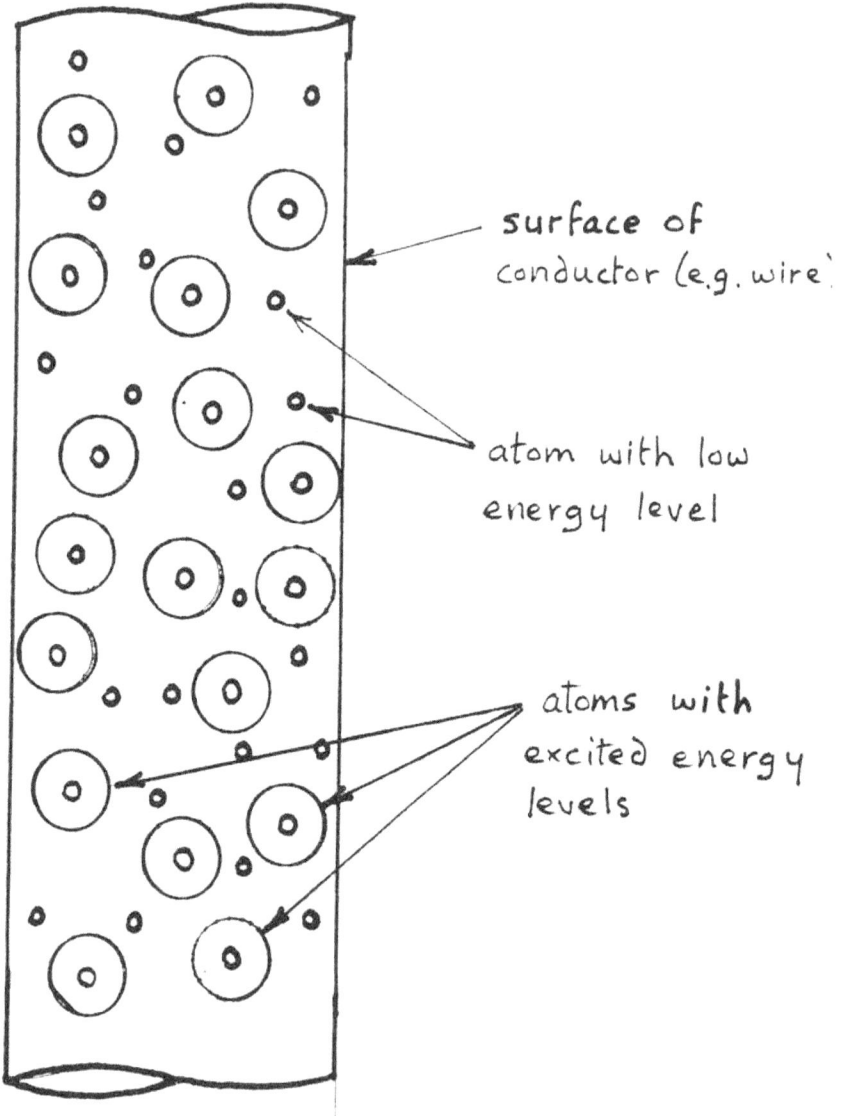

surface of
conductor (e.g. wire)

atom with low
energy level

atoms with
excited energy
levels

Diagram 11

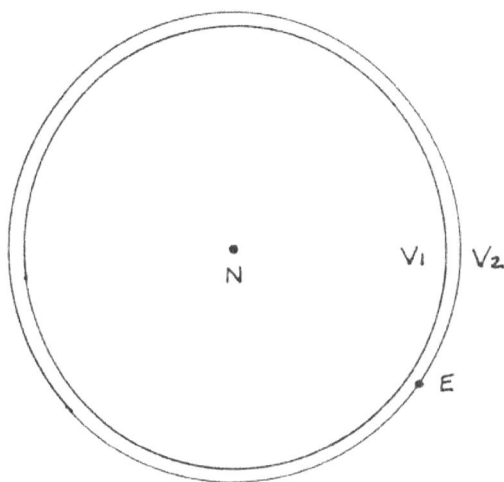

N = nucleus,   E = electron

V_1 = lower, or 'relaxed', energy level of valence shell

V_2 = higher, or energised, energy level of valence shell

　　　The diagram shows a Hydrogen atom, however the energised valence shell is representative of any atom.

Diagram 16

IRON ATOM

SECTION VIEW

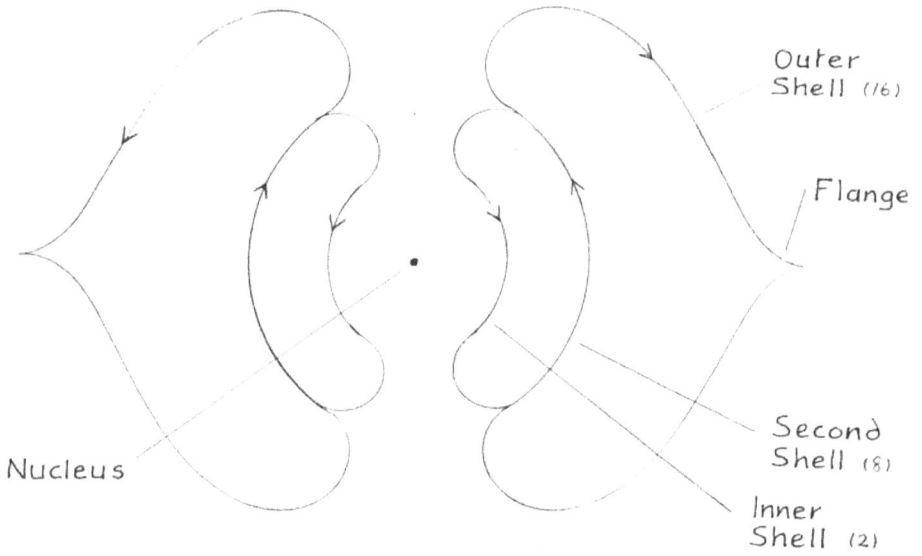

Outer
Shell (16)

Flange

Nucleus

Second
Shell (8)

Inner
Shell (2)

VECTORS SHOW ELECTRON FLOW

Diagram 17

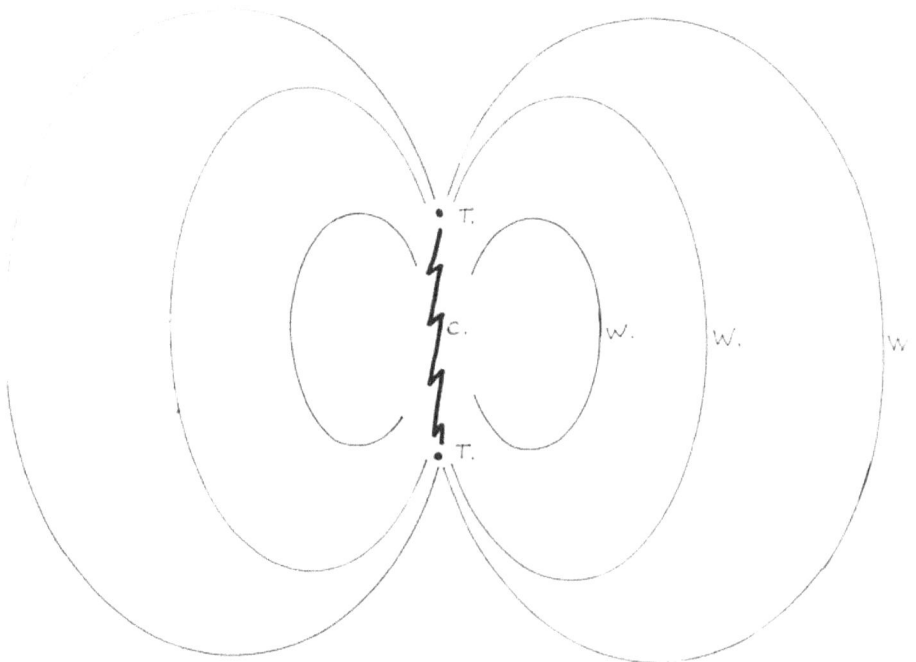

T.

C.

W.   W.   W

T.

Diagram 25

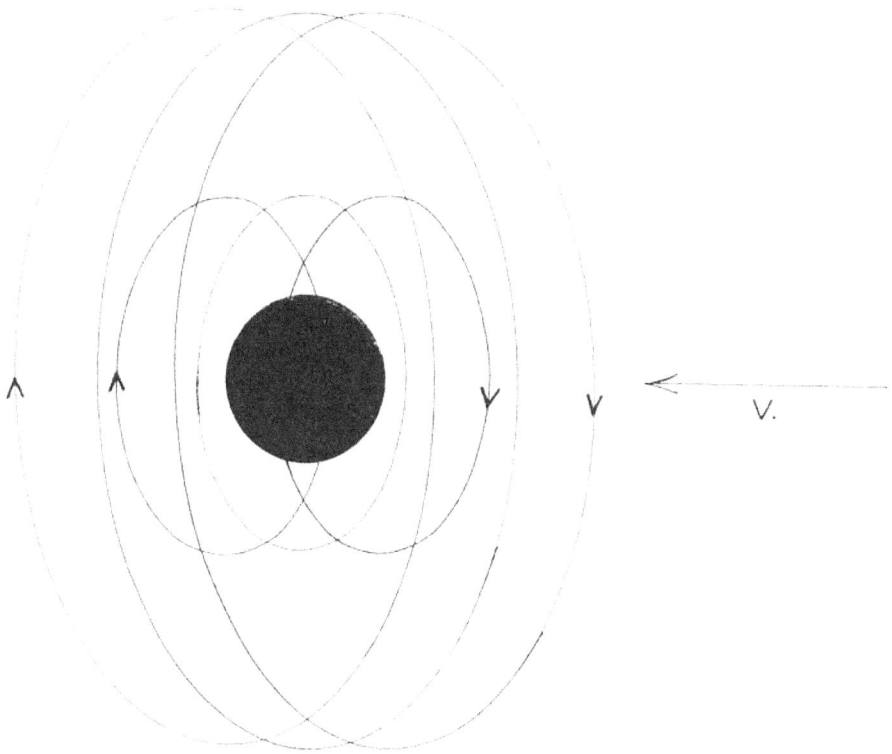

Diagram 26

V.

109 BB ~ QUANTUM ENERGY AND LIBIDO

Quantum Energy as explained is all pervasive, hence there is no substance which forms an isolative barrier against its flow. (There is however a special type of effective barrier which can be constructed; this is described later). Only a dynamic force can form a barrier against its pervasiveness. It cannot therefore be statically and passively stored - as electricity is stored in a battery, and in this respect it is very much like heat energy. Energy however can be accumulated dynamically (see previous) just as heat can be. (It may be said that heat energy is the electromagnetic cousin of quantum Energy).

This accumulation requires a constant input of Energy from a source which itself is dynamic. As is well known, sexual energy or libido is an Energy which, once discharged in a short period of time then requires a longer period to accumulate sufficiently to permit another adequate discharge. This is in good harmony with the established rules of the physical universe and cannot be changed. This accumulation is dynamic as described and so requires a constant inflow from the Energy source.

In a biological organism, this Energy source is the combined chemical and electrical activity within the organism. To some degree as explained, it is also generated by the mind/will, either consciously or sub-consciously.

Hence we find that those who eat good food and get plenty of exercise and oxygen, as well as having a good attitude and mental posture also have higher libidos. We can also find that those with highly disciplined minds capable of generating greater flows of Energy also have higher and more rapid accumulation.

109 BC ~ CLASSES OF FLOW

One may observe that there are essentially three classes of energy flow and activity. They are the vortex; dynamic balance; and static accumulation. (Actually no accumulation can be purely static since an accumulation represents an imbalance and this imbalance will always seek to re-adjust itself. However we may use the term for this immediate discussion). If we fill a bathtub with water, this fluid will find its own level so that its surface forms more or less a flat plane geometry. The same is generally true for any body of water such as a pond, a lake, an ocean, etc.. Unless there are unusual energy conditions prevailing, this is the normal behavior for fluids on the Earth's surface under the influence of gravity.

Hence we should never normally find a mountain of water in the middle of the sea, and so on. The same is of course true with a mountain pool which is being maintained by an inflowing and outflowing stream of water. We may call this manifestation of energy a "static-dynamic accumulation" or a "balanced flow". Another type of activity is the vortexial flow which is amply explored elsewhere in this book.

The third type of flow activity is accumulation - whereby energies, fluids, and solids gather to form discreet and recognizable accumulations. Examples of such activities may be found in space where solid masses accumulate due to mutual gravitational attraction.

As they accumulate, so does the gravitational force - which then exerts an even greater attraction, thereby pulling in ever more larger masses. Hence we have a self-perpetuating mechanism which feeds its own accumulation. This may be termed "acceleration of accumulation".

Another such example is the increasing growth of cities. A small town may grow slowly at first but as the population increases, so also does the economy, the flow of money, and manufacturing. This attracts more population growth and an expanding economy: [see note 1].

Again we can see a self-generating situation. Money itself is yet another example of this type of energy flow. If a man (or company, etc.) grows wealthy, his ability to generate more money and accumulate more increases in proportion to his wealth. Again, this is a self-generating phenomenon much like an increasing mass generating a gravitational field. It must be pointed out here that an accumulation of quantum Energy behaves in a similar fashion and follows a similar rule: [see note 3].

In fact it is the former examples which are following the fundamental rule set by q. Energy. We have seen three systems or classes of energy flow, and if we analyze the essential differences between these three, we will find that the fundamental and primary defining factor is the presence of quantum Energy, - whose activity is dictated by the collective will.

Money (a form of energy or a *symbol* of energy) is of course intimately associated with intelligent and conscious direction and manipulation. Without this guiding force of intention the concept of money becomes meaningless. It is an energy form generated by intention and agreement and must be constantly created. It is purely an intellectual construct not understood by children or animals.

Money is a symbol which only has significance when it is moved or transferred from one location to another. Hence money may be defined as a symbol in motion or a symbol undergoing a process of relocation. In other words money only exists as a dynamic symbol; it cannot truly exist as a static accumulation, - if we say it does, then it isn't money: [note 2]. Money can only truly exist as a dynamic force. In a similar way, a star can only exist as a dynamic entity. If we eliminate the dynamic flows of a star, it ceases to exist - both as a concept, and quite literally.

If the water in a bathtub would suddenly move to one end of the tub leaving the other side dry, this is only likely to occur either by mechanical or electronic effort, or by the direct application of will.

The analyses and technical explanations for all phenomena are merely attempts by the modern human mind and its "logic" to explain what is ultimately too simple for today's man to accept. He is so completely overwhelmed and embroiled in material and technical explanations for all existence that he can no longer accept the intervention of mind/will, i.e. so-called "spiritual forces".

NOTE 1: this accumulation is commonly referred to as growth, and indeed a parallel is found in the biological organism. However in all cases, we will find there are limiting factors (i.e. threshold limits) which always and inevitably manifest. The growth of a city cannot continue forever due to factors such as increasing distances, congestion of traffic, means of supply, state control of land, geography, and so on.

One way of combatting this is to create townships and municipalities which then become autonomous entities – each responsible for its own growth, geometry, geography, and functions. In effect this is precisely what happens when a single cell divides and multiplies, so that each cell-group then becomes an autonomous entity. In the case of a bio-organism, growth is normally limited by the programming in the DNA, etc..

In the case of solid and liquid masses accumulating in space, they too are eventually limited by their inevitable collisions with other such bodies and gravitational stresses caused by their proximity to stars and so on. In every case, it will be found that accumulating masses and energies will have some type of limiting factor or "threshold limit". This threshold limit once again re-affirms that inflow must be balanced by outflow and vice-versa. Stars on the other hand, are not limited to the same degree since they possess a highly organized and dynamic structure which is self-maintaining, (in effect much like a living organism!).

In the case of the accumulation of wealth, there are limiting factors such as death, dissipation, squandering, taxes, corporate corruption, failing markets, political situations, and so on. Even family dynasties that have survived for hundreds (perhaps even thousands) of years surrounded by great wealth must eventually fail and degenerate. Civilizations and social orders likewise will eventually fall since this is according to the ordained natural law which apparently has been in existence since the beginning of time: i.e. Birth – Existence – Decay.

Two other classes of flow are dispersal, and its opposite, aspersal - also discussed at length elsewhere. Dispersal for example is the flow pattern of energy we experience emitted by a light bulb, and includes heat and light photons. These energies radiate in all directions. Aspersal is the reverse of this and may be compared with implosion. Gravity is also a type of energy dispersal. Although it is a dispersal it creates the illusion of an *aspersal* when matter is pulled toward a gravitating body.

NOTE 2: to add clarity, let us imagine that instead of paper representing money we use cows, goats, and chickens, etc., as was done in ancient times. A goat or chicken, etc., is merely an animal and has no other significance except as a direct source of food. If we trade or barter animals in exchange for other goods, these animals then become real money with real value and gain significance. When they are grazing quietly in the fields they are not true money, - only animals.

Likewise a pocket full of money is useless and has no meaning unless it is spent and traded for goods. At that moment of exchange it becomes a real symbol of money. It is easy to see then that true money is really nothing other than energy which uses a symbol so that it can be measured and quantified. <please refer to 109 A: 'bucket analogy'>.

Money behaves in a very similar fashion. The bucket represents a business, an individual, or a household, etc.. money may be accumulated but it must be realized that unless it is active and dynamic it is useless - except for the feeling of security it provides the owner. As we can see, it follows the rule for all forms of energy so that in order for there to be an outflow, there must also be an inflow. Likewise, in order for there to be an inflow there must be an outflow. This latter may seem a little like magic and indeed one may consider it to be so. Money is nothing other than a particulate flow and hence, it will obey the laws of quantum behaviour.

As one can observe in the bucket analogy, the water pouring out of the row of holes may represent disbursements of money. Obviously, the more water coming into the dynamic pool, the more and greater (and inevitable) are the possible disbursements. Likewise the more and greater the disbursements, the greater will be the possibility for inflow. If one outflows he will surely inflow, - this is the quantum law of the universe.

One can see then that in order to have disbursements, the greater must be the reserve pool. However this pool must not be a static one - just like the water, it must be dynamic, - and this is one of the secrets of money. By outflowing while we are simultaneously inflowing, we have eliminated time and its consequent abberrative effects. Furthermore the more rapid the flow, the more we eliminate time and its effects.

NOTE 3: money behaves this way because it is a dynamic energy form; likewise the growth of an organism, an organization, and q. Energy behave the same way. Gravity has this same characteristic since it too is a dynamic energy form - and this will be explored in the appropriate section.

109 BD ~ ACCELERATION OF ACCUMULATION

Accumulation, like most other dynamic energy phenomena - progresses not with a simple unchanging rate of flow but rather by a decreasing or increasing, i.e. accelerating rate. In other words, it follows a dynamic curve. The simplest example of this is a falling object. As most of us know such an object will accelerate so that its rate of motion constantly increases, i.e. it experiences an accumulation of velocity and kinetic energy: [see note].

Another example of acceleration of accumulation is a flame or fire. A fire may be started by a single match and as long as there is fuel, the rate of the flame's growth will accumulate indefinitely in an accelerating fashion if the fuel remains as an unlimited supply. (In special cases such as with candles and oil lamps, the flame is controlled and therefore does not accelerate beyond a given predetermined point. The flame is controlled simply because the fuel supply is controlled).

We may give another example of this acceleration of accumulation. Imagine an iceberg floating and moving on a path which takes it to warmer latitudes. As it moves it melts - thereby shrinking in size. As it melts however several factors come into play. Let us put aside for the moment that the air and water become warmer as the 'berg travels, and imagine that these factors remain unchanged. As the iceberg melts (as ice will normally do when immersed in water) its geometry changes so that its surface area and volume both decrease. [see note 2].

However they do not decrease at the *same rate*. The volume decreases more rapidly in proportion than does its surface area. This proportion is quantified by the simple formula m^3/n^2, i.e. volume over surface area. This means that a large piece of ice will have a larger volume/area ratio than will a smaller piece. Since ice melts and suffers loss of mass *at its surface* this means likewise that a smaller piece will melt more quickly per unit of volume than will a larger one.

Hence we can see that a smaller piece of ice will diminish in relative size more rapidly than will a larger one. Similarly since it has a larger area/volume ratio it will absorb heat energy from its environment more quickly than will a larger one. Hence the ice mass accumulates heat energy ever more quickly as well as increasingly diminishing in volume.

Yet another example of dynamic acceleration of accumulation is demonstrated by a sinking ship. The hull of a ship may be breached by a small hole perhaps only 10 cm. in diameter. The ship will consequently accumulate water very slowly at first, (relatively speaking) – imperceptibly - but as it gets lower in the water more of the ship's openings and orifices are exposed to the water and immersed.

While the ship accumulates more water through time, it likewise loses buoyancy through accumulating loss of supporting air. Hence the forces pulling it down increase while the force holding it up are dwindling. This two-fold interplay of forces causes the acceleration of the ship's demise.

The chain-reaction of nuclear fission will likewise accelerate indefinitely as long as there is fuel (e.g. U_{235} & U_{238}) until a threshold point is reached where the chain-reaction cannot be contained, thus resulting in an atomic explosion. (in a nuclear fission reactor, this chain reaction is controlled and does not accelerate beyond a predetermined point (unless in the hands of buffoons and incompetants)). It may be noted that even the rates of acceleration can vary so that one form of acceleration is constant, such as with a falling object (i.e. 32 ft./sec./sec.) while another may increase exponentially, i.e. an "acceleration of acceleration"!

Again, some organizations owe much of their expansion to word of mouth. As the organization grows, there are more mouths disseminating news of the organisation and hence the organisation's growth accelerates.

In a similar manner, the positive accumulation of quantum Energy generated in a circuit will accelerate as long as the generating electronic condition is maintained.

In this situation also, there is a threshold limit where other phenomena will occur. This is explained at length elsewhere. In a situation where q. Energy is caused to accumulate by electronic means, we will find that this accumulation will accelerate exponentially in a fashion similar to nuclear fission.

NOTE 1: this rate is countered on Earth by air resistance so that the maximum rate of fall at sea level is about 120 mph or 192 K. p. h.. Hence we have two counter forces in this dynamic situation.

NOTE 2: even if the ice and water are the same *temperature*, the water normally still contains more heat energy than an equal weight of ice, i.e. 'latent heat', thus causing the ice to melt. However in unusual situations, the water may be "super-cooled" in which case it can actually be colder than the ice which it supports.

109 BE ~ DISSIPATION OF 'BAD ENERGY'

We are at all times experiencing emotion of one kind or another. In fact, all emotions are actually generated by the individual himself. Emotion is the result of quantum Energy. We sometimes generate bad or "unwanted" emotion such as apathy, grief, fear, anxiety, covert hostility, and anger, etc.. (Emotions are Energy flows of specific frequencies and wavelengths arranged in various dynamic patterns, as will be explained).

These created Energies will, over a prolonged period of time - have consequent side effects such as biological stress and illness, (i.e. they accumulate). Strong emotions create an overabundance of q. Energy which will quickly accumulate unless it finds a path of release. Stress and anxiety on the other hand, consumes Energy and makes one lethargic.

When we generate anger (or embarrassment; i.e. fear) for example, we find the body accumulating heat energy: [see note 1]. This heat energy is the result and secondary effect of flowing quantum Energy of a particular wavelength: [see note 2].

Hence in order to diminish the effects of the Energy, we should find a way to release this overabundance. After such an experience, people usually and intuitively engage in strenuous exercise to release this accumulated Energy. Laughter is also a method of release.

It is often acknowledged that children have a "lot of energy". The Energy referred to is of course q. Energy (Chi). Generally speaking this is true. In addition, children have small body masses which allow a greater expression of this Energy and physical activities. Hence we can see that there is a further parallel between quantum Energy and heat energy.

If we take two body masses - one having the volume of say a three year old child, the other of an adult human - we note that the adult has a mass approximately three or four times that of the child. If we also suppose that the two bodies generate and accumulate roughly the same amount of Energy, it is easily seen that the child has more Energy reserve to "play" with. Hence, we observe that Energy is a factor in growth (and healing).

An abundance of this Energy in children stimulates and is directed toward growth, and apparently when the body mass reaches a certain level relative to the available Energy (i.e. a threshold point), this growth rate diminishes until it ceases. Growth however does not cease completely for we observe that it continues as a healing force and as a continuing regeneration of cells throughout the body's life-time. As old age encroaches, even the Energy required to heal diminishes and also the ability to maintain health.

Naturally, a greater body mass will require more Energy to continue normal functioning. Likewise in the adult, Energy is channeled into various other activities such as physical or mental work, sexual activities, sports, anxiety and stress, ill-nesses, etc.. We know that normally, q. Energy flows freely in the body. Therefore an over-accumulation may be released from the organism by grounding it. This may be achieved by connecting the body to a grounded object. > see previous article.

For many thousands of years around the world, Man has very often included water in his various religious rituals. The purpose is most often to "cleanse" or "wash away" so-called sins or evil influences. In fact these influences referred to are none other than bad or "negative" Energy – that is, an over-accumulation of quantum Energy. Furthermore this ritual cleansing or washing has a basis in truth. A wet stream-bed is an excellent grounding for Energy. It is also a good supply for those in need of it.

Water both absorbs, conducts, and generates Energy so that another method is to immerse one's body in a pool of water, a stream, the sea, or a bath-tub, etc.. A full bath-tub is more effective if connected to a metal drain-pipe. Moving water is best such as in the sea, a stream, or a shower, etc.. This where the idea of baptizing originates.

NOTE 1. why does the body accumulate heat energy when anger (or embarrass-ment/fear) is generated? It is due to the free flow of q. Energy. There is both an inflow and an outflow thereof. The Energy generated produces heat energy as a by-product when it flows freely in the body. In some cases such as the expression of anger, Energy is generated by the Being which is then emitted or broadcasted to a point or points exterior to the body. It can even cause Poltergeist occurrences. As noted elsewhere, q. Energy and Em energy are interchangeable through atomic matter.

Other states or activities where heat energy is produced in the body are during moments of anxiety (e.g. embarrassment), during the sexual act, during the process of telekinesis or psycho kinesis, and during the healing act termed the "laying on of hands". During this last, heat is more localized and is generated in the vicinity of the body contact where the Energy flows freely from the hands to the second body.

NOTE 2: with the emotion of anger, the generated Energy accumulates in the body but it also has a path of release since this particular emotion has the attention and intention to be broadcast and radiated out - often to a particular target, - usually an individual, and sometimes an inanimate object. The individual who receives the emotion/Energy of anger will himself experience an accumulation of heat in his own body.

110 ~ VELOCITY OF QUANTUM ENERGY IN MATTER

E-energy, like heat, is capable of penetrating all matter, since all matter is constructed in fundamentally the same way. This construction consists of atoms joined together by electronic bonding forces. However as may be expected, q. Energy of different frequencies will travel at different velocities in different types of matter. [see note 4].

In substances such as glass (which is actually a very stiff liquid ceramic), concrete and stone (ceramics), and rubber, etc., Energy travels relatively slowly. (The velocity through such substances is on the order of a few feet per second or less. In a silk cord, the velocity was perceived to be about a meter a second. ('The Odic Force': *Reichenbach*)).

In these substances, the arrangements of the molecules and atoms are rather chaotic and random. Remember also that E-energy, being of such extremely fine wavelengths and frequencies, will interact readily with particles of the same order of dimension (such as atoms, atomic nuclei, protons, electrons, and so on) due to simple frequency resonances.

As an analogy, one may visualize a stream-bed strewn with pebbles and a trickle of water wending its way slowly along the jumbled path. It can be perceived that the water will make its way quite slowly from one point to another. This slowness of flow is caused by the random arrangement of the pebbles and the consequent friction.

If however the streambed were a regular arrangement of pebbles, all of more or less the same size (like atoms in a crystal lattice or the gravestones in a military cemetery), it would be observed that the water would then be able to proceed more rapidly and would select easy and straight paths to follow between parallel rows of pebbles. Therefore as one may expect, q. Energy will travel faster through a substance having a regular array of atoms such as with metal, rock crystal, or ice, etc..

Quantum Energy will travel more slowly also through a liquid (water or oil, etc.) for the reason that the constituent atoms and molecules are in a chaotic and dynamic jumble. Energy is propagated in solid matter by first being generated by atoms. The Energy is then emitted and radiated. These radiations impinge on neighboring atoms which then reflect them or absorb them. They are consequently energized by the principle of resonance. Then these absorbed emissions are again re-emitted by the atoms as spherical waves which are then again propagated. (*Huy-*

gen's Principle). (see diagram 08).

This process continues in a way similar to the electron particles as they jump from atom to atom. Electrons being particles, usually travel much more slowly however, - (it has been estimated that in solid metals, their velocity normally is on the order of a centimeter per second, more or less, depending on the conducting material and the driving voltage which determines their velocity).

Being particles with electric fields and hence obliged to interact with the atoms, electrons behave in a way similar to a hiker who is slowed down by having to walk through bushes and underbrush! The thorny bushes act as a drag on the progress of the hiker in the same way that the atom's valence shell fields hamper the progress of the electrons: [note].

The Energy also travels from atom to atom, but being essentially electrical waves and not particles, their velocity is greater (Reichenbach's 'The Odic Force'). The propagation of q. Energy through a long conductor is very similar to the conduction of smoke through a long chimney.

Because of this method of propagation, E-wave travel to successive neighboring atoms in a crystal lattice is a simple "house-to-house" affair. However in the case of liquids, the situation is very different since liquids have greater dynamic activity and have no static or "fixed" structure like solids. Their molecules and atoms are always moving relative to each other. Therefore any energy radiated by an atom will usually affect only its immediate mobile neighbors.

When energy (e.g. heat energy or q. Energy) is absorbed and re-emitted, any close neighboring atom in another molecule has already departed to another area in the liquid mass. Because of this, the Energy is not propagated atom-to-atom in direct sequence and hence is propagated slowly. In time however, liquids can absorb all the Energy given to them.

For this reason also, the Energy is absorbed by a liquid more rapidly if it is at a higher temperature where the molecules consequently move more rapidly through the liquid mass. Liquid water therefore "<u>absorbs</u>" the Energy whereas ice, - being a crystalline solid, will <u>conduct</u> it.

Also for the same reasons, crystalline solids such as ice will conduct heat energy more rapidly. Paradoxically then we can see that water and similar liquids will conduct Energy more rapidly at both below and above certain 'critical' temperature ranges.

This incidentally is quite possibly the reason why we find icebergs so beautiful to look at. They are conducting huge amounts of quantum Energy up from the depths of water - which has already absorbed much, as well as the additional Energy being constantly generated by the action of friction of ice against the water - and water against water in its turbulence. Since ninety percent of an iceberg is submerged, it has the opportunity to collect vast amounts of Energy from the depths. In a sense then, an iceberg in the water is something like an acupuncture needle in flesh, - both generating and releasing stored Energy from below the surface. [note 2].

The Energy is then radiated out into the air by reflected sunlight from the exposed portion of ice. It is also carried away and to the observer by the reflected sunlight. This Energy combination enters our eyes and in turn activates the endocrine system, the pineal and pituitary glands, (both associated with psychic awareness and abilities, - see Glossary). The resulting beauty then becomes almost a "spiritual" experience.

In this example, we can see then the interplay of subjective and objective experience, - each one a part of the other and inseparable. Gazing at the full Moon, Niagara Falls, or indeed any strong source of q. Energy gives us similar feelings. [see note 3].

NOTE 1: likewise in the same way that a hiker will cause the brambles to snag then suddenly release to bounce and jiggle, - so the passing electrons will "snag" and cause the electron shells of the atoms to release, bounce and jiggle - like quantum sized plates of jelly!

NOTE 2: We also know that quantum Energy is the 'Formative Energy' (*Reich*) and its great abundance in the ice creates the wonderful sculptured forms that we associate with these "castles of the sea".

NOTE 3: the fact that icebergs are so beautiful is a direct result of the interaction of the observer's perception and the great abundance of Energy communicating between the viewer and the ice. It is a situation where the physical and the non-physical are in communication on the most fundamental level.

NOTE 4: In fact ordinary light passes through or is conducted through our bodies to some extent depending upon its intensity. Try the experiment of putting your thumb over the lens of a flashlight. You will observe that your thumb glows red - indicating that the light is passing through the flesh. It is due to this phenomenon that light has therapeutic values for the body since it stimulates electronic activity. Red light is better since it has more penetrating qualities.

Likewise, you can shine a light through a page of paper. This of course implies that the light will pass through two sheets, four sheets, eight sheets, and so on. In other words, a light with sufficient intensity will 'shine' through a stack of books! Furthermore, light photons will 'shine' through stone, concrete, even steel, etc.. This is exactly what takes place when light radiation is generated by an atomic explosion. In fact, just as much damage to property is done by light photons as by heat photons.

So why does the increase of intensity also increase the penetrating power? This is simple. We can refer to the equation $(P = pv)$ where (P) is penetration, (p) is particle density, and (v) is particle velocity. Obviously, the velocity of light is not increased - and if anything, it is retarded. It is the increase of particle *density* or *volume* that does the trick.

The first atomic 'layer' of a substance will stop a percentage of the photons. Subsequent 'layers' will block more photons. However if the density of radiations is high enough, a percentage of photons will penetrate all the way through say - a block of steel. This means that if you shine a weak flashlight beam at a block of steel, eventually a few photons will actually penetrate all the way through it! This sounds amazing but when analyzed logically like this it becomes a very reasonable occurrence.

In other words, atomic matter acts as a kind of filter for photons. The smaller photons (e.g. X-rays) get through more easily than the larger ones (e.g. microwaves). In fact, the wave or particulate nature of radiations is partially determined by how the radiation interacts with atomic matter.

Furthermore, the degree of penetration of a type of radiation in a given substance is directly related to its wavelength. We now see how this is true for quantum Energy, being of such a fine wavelength.

111 A ~ SOME PROPERTIES OF QUANTUM ENERGY

Because of the lack of interest on the part of mainstream physics in "Chi", 'Orgone', and subjects such as meta-physics, Yoga, Feng Shui, dowsing, Shiatsu, and para-physics generally (including Poltergeist phenomena), no large scale or serious investigation of this Energy form has proceeded, - none that we are informed of, anyway! (There is evidence however that the Russian scientific community is especially fond of borderline research, paraphysics, parapsychology, and so on).

Therefore the full scale of the properties of this Energy is unknown. There are a few hints given to us by private researchers and their limited investigations - and the strange reports of such phenomena as homeopathy, miraculous healings, Poltergeist activity, levitation, ball-lightning, etc., which are only some of the more salient phenomena witnessed. Such subjects however are usually never touched upon by the broad spectrum of the general media, and are often relegated to the fringe "eccentric" area. There are probably many unknown and un-guessed at properties connected with Energy that at present we may not yet imagine.

The range of q. Energy wavelengths that interests us is quite large, extending from about 1000 Ångstroms ('hard' ultra-violet) and beyond to $\frac{1}{40}$ Å., and beyond. Every different wavelength may have differing attributes – somewhat as does the more familiar spectrum of electro-magnetic energy. (Diagram 01). [See note 1].

(Note: An Ångstrom is a unit of length used in physics. It is equal to one ten billionth of a meter or one tenth of a Nanometer. It is also the official diameter of the smallest atom, i.e. Hydrogen,).

< distance unity rule >

We are all familiar with the distance squared (d^2) rule which regulates how electromagnetic radiation is distributed from a point source through surrounding space. As the radiation propagates, its intensity diminishes therefore according to this rule.

Pure, un-modified quantum Energy however does not follow this d^2 rule. As it pervades the surrounding space, its intensity <u>does not diminish</u> in the slightest. It therefore follows its own distance unity rule (d^1). Its range of propagation however is limited by the intensity of its 'radiation'. Its range moreover is demarked by a well defined boundary and the intensity of the radiation is <u>constant</u> within this boundary. In some ways then it has similarities to laser light, i.e. its energy is of constant intensity throughout its range, and it too follows the d^1 rule. (In other ways, its emission resembles the smoke patterns as it curls up from a cigarette).

As an analogy of this, imagine that the Energy extends outward from its source a given distance and forms an invisible spherical limit of demarcation - like the skin of a balloon. With the application of more input energy, i.e. gas, the balloon can be expanded to a greater diameter with the resulting increased tension and stress in the latex membrane. The gas inside is of course of a higher density than the outside air and likewise its density is constant throughout.

Also as the balloon gets bigger, we will note that the skin must be tougher and stronger. We could say that this balloon skin then represents a kind of "force field" enclosing the compressed gas within. Indeed on a quantum level, this is precisely what it is! In any event, the larger the area or volume to be covered by this energy field, the greater will the input Energy necessarily be. In the case of the balloon, the strength and thickness of the membrane will need to be determined before inflation. In the case of the self-constructing force-field however, it is the input Energy which itself decides all parameters.

NOTE 1: actually the spectrum of quantum Energy is as wide as the electro-magnetic spectrum but since we are dealing in this work with the sub-atomic aspects of this Energy, we are only interested in the 'quantum' frequencies, i.e. from 1 Angstrom to $^1/_{40}$ Å., (and very likely beyond) - that is, those frequencies which can interact with the atomic and electron structures and - it will be seen - with space-time itself.

111 AB ~ SPECIAL PROPERTIES OF QUANTUM ENERGY

As explained, Energy extends parallel to the entire electro-magnetic spectrum from radio-waves to the electron particle, and beyond. The Em spectrum can be considered to be divided into two major characteristics with visible light as the approximate dividing line. Longer wavelengths of energy are *wave*-like in nature and effect, while the shorter wavelengths are *particulate*. Likewise we may consider that q. Energy also may be divided into two major portions of longer and shorter wavelengths.

This is important to know since the longer and shorter waves have essential differences just as does the Em spectrum. However one point of division here is apparently in the approximate area of Ultra-violet of one thousand Angstroms wavelength. The entire range of E-energy may be called by various names. However the shorter waves should be referred to as 'quantum Energy' since these frequencies are associated with sub-atomic behaviors and interactions, i.e. at the quantum level.

At this point of division, the nature of E-energy shifts from a *radiating* one of the longer wavelengths to one of *'emission'* or *non-radiating*. 'Radiating' in this context means behaving like electro-magnetic radiation which travels outward from a source point at Relativistic velocities. The shorter q. Energy wavelengths are rather emitted or *emanated* in a way similar to the smoke which curls upward from a lighted cigarette. This is an important distinction and should be kept in mind. This means that at longer wavelengths, the Energy follows the d^2 rule for radiation while at the shorter wavelengths it follows the d^1 rule. (to be explained).

This finer Energy form will behave in a variety of ways depending upon the parameters and conditions the Energy is exposed to. For example, it can flow along a conductor in much the same way as an electric current in a wire. It can also exist as a 'field' effect, something like a static electric field or gravity. [see 'Energy wavelengths': Glossary].

At the intermediate frequencies, i.e. at visible light wavelengths, Energy appears to be *target selective* so that generated Energy at these frequencies will be self-guided to the nearest large object within effective range, or another quantum Energy source. It also has the peculiar ability and propensity to seek out and enter caves, crevices, tubular openings, pipes, orifices, and places like rabbit holes, etc..

111 B ~ CHARACTERISTICS OF Q. ENERGY

Quantum Energy may be found most abundantly in places where there is also found dynamic activity and other forms of energy such as vibrating objects (bells, etc.), sunlight, reflected sunlight - such as Moonlight, etc., mountain peaks, fire, flowing water, icebergs, living biological structures, electrical and chemical activity, etc..

It is influenced by emotion and/or emotional thought, (as in Theta or Alpha brain waves), - or the will. Several minds working in concert have a more pronounced influence, as many great orators and "charismatic healers" have found to their gratification.

In conjunction with matter (e.g. gas, electro-plasma, bio-organisms, etc.), it is attracted to orifices and enclosed and tubular spaces, especially if the orifice is lined with electrically conductive material. Such spaces may be metal drain-pipes, soot-lined chimneys, wet caves, tunnels, and so on. It is also attracted to spaces surrounded by structures such as trees, and also spaces between the leaves of vegetation, etc..

An example of a wonderful accumulation area for this Energy would be a sunlit clearing in a woods where is also found a stream and a pond, and biological activity. Bonfires are also a strong source of Energy, which is no doubt why they were so popular with ancient peoples. When 'flowing' over surfaces, or along other interfaces of matter, its natural tendency is to move in spiral or helical paths: (more on this later).

It is attracted to metal and shiny surfaces such as mirrors, pots and pans, etc.. (In this context, it has been claimed by researchers that shiny bits of metal attached to trees and shrubs, etc., will stimulate their growth. There is no doubt that shiny metallic surfaces are attractive to life-energy and this is exemplified in nature by the naughty behavior of Magpies (beautiful black & white birds) who love to collect such trinkets).

The reason q. Energy is attracted to shiny surfaces is no doubt due to the electronic activity caused by impinging light photons and their reflection. It is absorbed by standing water, liquids, and insulators of heat and electricity. Light and electromagnetic radiations (both particulate and wave-type) are active carriers of Energy, - and wires or string, etc., are passive conductors.

Q. Energy is the primary source of all forms of electromagnetic energy, *and is interchangeable with them through the agency of matter*. It is the subtle link between matter, electromagnetic energy, gravity, inertia, and the mind/Being/non-physical agency. It flows from an area of low E-potential to one having a higher one, and thus in many respects operates contrary to the established rules of electromagnetic energy behavior and parallel to the characteristics of gravity (and money!).

It flows along metallic wire conductors more quickly than heat energy, but very slowly compared to electricity (not electrons - they travel slowly), which is practically instantaneous. Its rate of flow along a non-metallic conductor (i.e. silk) is roughly three feet (one meter) per second, more or less, (*Von Reichenbach*).

('Electricity' *per se* is actually a field effect impinging on the electrons in the conductor. In this respect, electricity is similar to the water obtained from a tap. The water/electrons you receive at your terminal is delivered instantaneously but it is not the same water/electrons being pumped into the system miles away). [see note].

It has a cohesive quality and if influenced by the will, can condense in free space into a fine type of "etheric" substance, (which then quickly dissipates into its immediate environment). This substance is not ordinary matter in that it is not composed of atoms or particles, it is then pure condensed Energy. It has been referred to as "etheric matter", "angel hair", or "ectoplasm", etc., by those engaged in research and associated matters. Energy is generated by fire, electronic plasmas, mechanical vibration, and chemical action, and with electronic technology. When flowing in matter, it will be attracted to and repelled by sharp points.

E - energy interacts with atoms, thus giving rise to electro-magnetic (and atomic) phenomena, quantum phenomena, gravity, and inertia. Energy may be carried by electro-magnetic waves, and similarly reflected, refracted, diffracted, and focused. It can permeate virtually every substance, albeit at different rates of flow. It tends to accumulate in spherical objects, such as crystal balls, pebbles, balloons, planets, etc..

The Energy doesn't particularly "prefer" spheres, but rather the spherical configuration is a more efficient accumulator than any other form. (This may be why children and animals intuitively delight in ball-like objects, including marbles).

It is attracted to other sources of "Life-energy". It is absorbed by wood, water, bio-organisms, leather, and other substances, especially concrete in which it moves very slowly. It is affected by and attracted to the presence of electric fields, magnetic fields, and electro-magnetic radiation, just as are other forms of electrical energy.

It will kill harmful bacteria and viruses, in the same way that all electromagnetic radiations of very high frequency will likewise do, - and it improves the taste of food. (Subjective taste, of course, along with smell, vision, hearing and touch are good examples of how the mind and matter interact, with quantum Energy as an intermediary).

In generated quantities, it will decrease electrical resistance in conductors, regardless of their material composition (similar to the way in which all very high frequency Em radiations will do). E-energy suffusing an object will decrease its inertia and gravitational moment. It also has other effects on the space-time continuum, - (to be explained).

E-energy will stimulate the glandular system, including of course the Pineal and Pituitary glands, hence affecting also organic growth and regeneration. In a natural setting, the strength of this Energy is affected by weather phenomena, geography and landscape, geological activity, - and, interestingly, as will be described, *astronomical alignments*. ('Earth Magic': Francis Hitching).

E-energy may be produced by the mind or will, - or produced electronically. It is strongly attracted to hydrogen and electrical flow.

NOTE: this apparent slow characteristic of Energy to flow in a conductor is due to the fact that q. Energy interacts with atoms as it flows due to its very fine wavelengths. This implies that the Energy is first absorbed then emitted by the atoms. The result of this is that every atom thus affected retains a measure of Energy and may be said to be 'Energized'. A crude analogy may be used to explain this.

Each atom receives a quantity of Energy before it is allowed to continue flowing. We could imagine a series of locks in a canal where the water will spill over from one lock to the next. The water cannot proceed further until each lock in turn has been filled. This of course takes time. This is not exactly like the *Huygens Principle* although for our purposes we may refer to it as such.

112 ~ ENERGY AND MATTER INTERACTIONS

Interesting effects on matter can be made by the application of various forms of energy. Examples in nature are the glow of the fire-fly, the flight of birds and insects, the column raised in water by a dropped stone and the resulting expansion of wavelets, the vortex produced by water (or other liquids) when draining through a narrow constriction such as a bathtub drain hole, - the tornado and its cousin the water-spout, etc..

We take all these things for-granted intellectually and yet they are truly fascinating and even exciting phenomena in their own right. Matter without energy would be truly "life-less" and "dead". Indeed, without energy (including especially E-energy), matter could not exist*!*

Another way of expressing this is to say that if matter exists, then it cannot be "energy-less", "lifeless", or "dead"*!* (More articles on the conditions of existence are forthcoming). By extension then it becomes apparent that 'life' (and "life-energy") is everywhere, and that "death" is really only the point of view of a specialized interest*!* (What physicians call 'death' is actually a diminishing accumulation of Energy in a body, - to a degree where the resident mind/will is no longer able to animate the body. An axiom to this is that:

> Wherever there are found sufficient concentrations of Life-energy, there will be found also <u>animation</u> and life itself, - including sentient life. Quantum Energy is a completely new energy concept for modern Man generally, and much more so especially if generated in objectively large quantities by means of electronic technology.

> "I know of nothing sublime which is not some modification of power", (Edmund Burke).

113 B ~ WATER AND QUANTUM ENERGY

< see The Wealth of Sea-water: and 303 AE (2) > While q. Energy may be considered to be "quasi-energy" or "etheric energy", water may likewise be considered to be its material counterpart. Like Energy, water too is ubiquitous – it is everywhere on this planet; in the biosphere, the seas, the rivers, the lakes, the soil, the ice-caps, the clouds, the trees, the mountains, and in the air as vapor. Biological life as we know it would be impossible without it.

The same is of course true of q. Energy. A bio-organism may contain at least 98% water as a constituent by weight. The human brain is composed of over 99% water by volume. In other words, your head is full of water.

The body is suffused with quantum Energy and it forms a dynamic field extending several feet from the physical body itself thereby constituting the so-called "aura" of tradition. Likewise, the body is maintained and sustained by a continual inflow and outflow of both water and Energy.

It is water and Energy together which maintain the structural integrity and functions of the body. The brain itself is virtually all water and may be thought of as the main repository and accumulation area of Energy and functions with this generated Energy.

As will also be explored in depth, we shall see that in the biological organism, we find two essential ingredients associated with the nuclear fusion process. These are hydrogen (bound up in the H_2O molecule) and an energy form which operates at the subatomic, i.e. quantum level.

Why do we all feel lethargic and listless on a cloudy, rainy day? It is simply because of the diminished levels of Energy in the atmosphere and environment. The clouds and the surrounding air are full of water and water vapor and as we have seen, water is a strong absorber of q. Energy, especially the vapor. Contrarily, if we spend a sunny day at the sea-shore, we feel elated and cheerful in spite of the abundance of water nearby. Hence we find that the healing process is accelerated in the sunny weather while retarded in the gloomy and grey weather.

This is due to the fact that the Sun is supplying a greater quantity of Energy than is being absorbed by the water, and further, the waves themselves are also vigorously generating Energy. We also know how gloomy and 'lifeless' the sea and the shore can be when there are dark clouds abounding.

Quantum Energy has a strong affinity for water in all its forms - and anything which contains water such as bio-organisms. The reason for this affinity is that water itself is two-thirds hydrogen and the real affinity exists between the Energy and *hydrogen*. Hydrogen is the simplest element – consisting of a dynamic electron surrounding a single proton at the atom's nucleus. (Hydrogen may be called "packaged electricity").

E-energy is an electric field wave which, due to its frequencies, interacts strongly with electrically charged sub-atomic particles. Since the hydrogen atom has only one electron, the electron shell is tenuous - that is, it is "open" enough to permit Energy of almost any quantum frequency and wavelength to penetrate the shell and interact with the particles.

We can imagine that an electron shell is like a spherical open-weave basket surrounding the proton. The more electrons the shell possesses, the closer and tighter is this "basket weave", - likewise the fewer the electrons, the more "open" it is. Hence the hydrogen atom is highly susceptible to influence and modification by impinging quantum Energy.

Another and perhaps better analogy would be that of a soap bubble representing an atom. It is observed that virtually all types of electromagnetic radiation are able to penetrate and pass through the soap bubble, including of course light photons. In the same way - because of the sub-atomic wavelengths, Energy is able to penetrate and "shine through" atomic shells. Hence, Energy is able to interact physically on a quantum level with sub-atomic particles such as electrons, protons, etc., in a way similar to that in which light photons interact with electrons.

On the scale of a soap bubble, incident ultra-violet light on its surface will cause electrons to be ejected therefrom - as if indeed it were an actual atom. This is simply because the surface is made up of countless atoms, and it is these atoms which are being affected. (To paraphrase Pythagoras, "as below, - so above"!)

It therefore resonates more readily to q. Energy frequencies and wavelengths, and is thus easily dismantled by incident Energy. Hence we find that water molecules are likewise quickly dismantled, and so water in the presence of Energy is highly disposed to rapid evaporation.

Bio-organisms are therefore subject to loss of water, i.e. dehydration, and dead organisms to desiccation, i.e. "mummification". Mummification is simply the drying out of bio-matter before the organisms of decay (which need water) can do their work.

Evidently - according to much research, q. Energy modifies and re-organizes the water molecule (H_2O) structure more or less permanently. This results in the molecular structure of water expanding slightly like an accordion. The overall effect of this is a "new" kind of water which contains more inter-atomic space, - like a stretched sponge. [notes 1 & 2].

Minute particles of impurity such as minerals and metals are normally suspended in water. However they will not remain so in this new Energized and expanded water, and will consequently be released to precipitate out and become sediment. [see note 3].

In sea-water, such particles include valuable metals such as gold, silver, magnesium, etc... Hydrogen then is highly susceptible to ionization by quantum Energy. Hydrogen and q. Energy are therefore highly compatible as a "dynamic duo"; - the one, passive matter - the other, active Energy. [see note 4].

Dynamic hydrogen plasma - especially when undergoing nuclear fusion, becomes a prodigious generator of quantum Energy. Plasmas are intimately interactive with quantum Energy, the one begetting the other. Constant generation therefore assures further ionization of the hydrogen.

NOTE 1: This expansion is accomplished by changing slightly the angle (103 degrees) between the two hydrogen atoms attached to the oxygen atom. This would be affected by the valence shells of the oxygen atom and of the hydrogen atoms becoming Energized, thus causing the shells to expand slightly. When this occurs, the attraction between the outer electrons and the protons at the nucleus is weakened. The electrostatic repulsion between the hydrogen electron shells consequently has a greater effect, thereby moving the hydrogen atoms further apart and increasing the angle formed by the three atoms. Evaporation of the water occurs at an accelerated rate in the presence of sunlight - not so much from the heat radiation as from the quantum Energy which is absorbed.

NOTE 2: as elsewhere described, the two hydrogen atoms of the expanded water molecule may be likened to two pith balls suspended on strands of cotton. If they are electro-statically charged they will move apart - thereby increasing the angle between the two cotton strands.

NOTE 3: incidentally, quantum Energy will improve wine enormously. Wine improves with age. 'Ageing' is a slow process but need not be. By interacting with the wine on a quantum-atomic level, the time required for this "ageing" process will to a large measure be eliminated. This effect of Energy on *time* will be discussed at great length later on. In effect, the *time flow rate* is increased.

NOTE 4: another interesting thing about hydrogen is the fact that it exists only in two states, one as ordinary matter, and the other as non-matter, i.e. "quasi-matter". There is no in-between state as with a larger atom. When the atom is ionized it becomes - not an ion, but a completely dismantled atom consisting of two sub-atomic electrical particles. Furthermore hydrogen is easily ionized, requiring the least energy.

115 ~ QUANTUM ENERGY AND THE SPACE-TIME CONTINUUM

There is some evidence that q. Energy, which also constitutes the dynamic electronic field associated with the matter and form of an object, <u>will</u> <u>interact with the space-time continuum</u>. Experiments by Russian scientists working with '*Kirlian Photography*' in the 1970's discovered that any object, when moved to another location, leaves behind an "imprint" or electronic image in its former location. ('Psychic Discoveries Behind The Iron Curtain': Ostrander & Shroeder). (>Aether Trails - 127 AB).

This 'imprint' apparently affects not only the material environment but also the very space itself, and remains in a local time field. (It is thereby suggestive of the concept of an "aether", a nineteenth century concept of space which is just now being reconsidered anew by physicists: > 207 AA – part 4; & 229 C – section 4.2).

With Kirlian Photography, this imprinted image becomes visible (something like a "ghost"), and as what the Hindu sages would call a recording in the 'aether' or "Akashic Records" (after *Edgar Cayce*). It is such energy fields that are perceived by visually oriented animals such as cats. (> *Akashic Records*: - & - *Peripheral Vision*).

It is naturally quite interesting to speculate on the effects of such an object imbued with an unusually high level of q. Energy. Such an Energized object would leave a more permanent and powerful influence in the area, which would in time perhaps become to be regarded as a "holy" or "sacred" place by the initiated and the uninitiated alike. In fact as we shall see, this is precisely what occurs.

It is even possible that unusual events might occur there from time to time, - the timing being influenced by the confluence of cyclical and synchronous occurrences, i.e. the weather, the Sun, the Moon, and yes - even the planets, - and so on. (> Synchronicity, section 03).

Just such a place may be the so-called '*Oregon Vortex*'. This is a circular area of land approximately 165 feet in diameter located on the banks of Sardine Creek near Grant's Pass in Oregon, USA. The size of the zone seems to vary at ninety-day intervals, i.e. it "pulsates". This 'pulsation' is a clear indication that Energy is exerting its influence. (*Wilhelm Reich* on '*Orgone*'). Inside this zone, there is reported to be a sensation of increased or intensified gravity towards its center.

Small pieces of paper or a puff of smoke released in still air within the area begins to spiral upward with a helical motion, ever faster. (Another clue to the presence of q. Energy. This phenomenon is apparently electronic in nature since a photographer's light meter registers different readings inside and outside the area. ('Stranger Than Science': Frank Edwards, > also *Unusual Radiation Characteristics of E – energy*).

It seems possible from the above description that the Oregon Vortex is "powered" by an unusually strong source of q. Energy. This might be an underground river, possibly combined with a deposit of metallic ore, etc..   This example of the Oregon Vortex is doubly interesting because it brings together several aspects related to electrodynamics and Energy - namely, the effects on electromagnetic radiation (light), gravity, dynamic helical motion, the ebb and flow of q. Energy, and even space-time itself.

If such a field remains *in situ* after the removal of an object which generated the original field, it will be seen further on in this book that it must be a quantum-electronic phenomenon. Further this phenomenon gives us a clue as to the nature of space and time, since the field must obviously exist in time (and space) in order for it to exist at all*!* [note 2].

NOTE 1: many questions may be answered and much data is given in later sections about the interrelationships between time, energy, space, matter, velocity, space-time continua/geometry, - and how they may be altered, modified, and so on. It is apparent for example, that the increase of quantum Energy into any situation will decrease the time factor.

NOTE 2: If you have an older television set with a glass tube, there is a trick which can be done as an analogy to illustrate this. While the set is turned off at night and while the room lights are on, place your hand - or any object next to the screen for a minute. Now turn off the lights and you will see the shadow image of your hand, etc., remaining on the screen. Older B/W sets of the 1950's will display this better.

116 A ~ REFRACTION

A glass lens is able to focus light due to the nature of refraction. When light passes from the medium of air into glass, it actually slows in velocity while travelling within the glass (with the consequent change of direction), and it is this phenomenon which is utilized in the lens. This means that it takes more time for a light photon to travel through an inch of glass than it does through an inch of air or space. (The same is true to varying degrees in the case of all transparent substances).

This further means that the velocity of light is <u>not a constant</u> (K) from medium to medium. Furthermore, one could argue that the transparent medium in effect constitutes a kind of space-time "warp'" or "anomaly".

In other words, the space-time continuum is altered within the substance of the glass. There is nothing magical or special about this event, the glass, or any such medium, since this space-time warping is caused by the electronic fields and bonding energies associated with the atoms. This same space-time warping can be induced simply in the air itself and even in "empty" space, (as is demonstrated by light passing through a powerful gravitational (i.e. electronic) field, and electromagnetic fields associated with stars, etc.).

The shimmering "heat waves" in the air that one sees on the highway or desert on a hot summer's day is another example of this modification of the refractive index of the air by electromagnetic energy, i.e. infra-red light (heat), microwave energy, - and importantly also - q. Energy.

Light passing through the 'Oregon Vortex' (see previous article) slows down because the <u>time</u> <u>flow</u> <u>rate</u> has been altered within the local field of the Vortex. Hence this Vortex behaves somewhat like a lens and refracts the light passing through it.

One may also be interested to learn that reflected photons from a smooth surface are shifted slightly to a slower time flow rate. In other words, this reflected light exists in a slightly removed "parallel universe". This is why sensitives and psychics are sometimes able to see things in mirrors which others do not, - and this is largely the principle involved with scrying and "crystal gazing". (more on this later).

116 B ~ REFLECTIVITY & LUMINESCENCE

Vision for the most part is the ability to perceive ambient light photons reflected from the surface of objects. Reflection, like vision, is an electronic phenomenon. When a photon (an electro-magnetic field wave ~ particle, or "wavicle") impinges upon a surface, it is absorbed by the valence electron shell of a surface atom and the energy level of the atom is thereby raised.

If the frequency of the photon corresponds to the resonance frequency of an electron in the valence shell, this photon or one like it is then spontaneously emitted by the excited atom. This in essence is reflection. Reflection then may be thought of as the "re-emission" of photons.

For example in the case of a gold surface, if white light is shone upon the surface, the re-emitted (reflected) photons are predominantly of a yellow-orange color, - i.e. gold color! Any photon or energy capable of raising the energy level of a gold atom will cause it to emit a 'gold colored' photon. The spectrum of pure white light contains all colors, including of course the gold color. This is the frequency of light photons which are absorbed and re-emitted by surface gold atoms. A photon of another color is usually absorbed by the gold atom but not re-emitted since the other color does not resonate with the surface atomic shell.

This excess of unused, absorbed light energy is then passed on to other atoms via atomic bonds - which are then caused to oscillate in the crystal lattice (or amorphous structure in the case of non-crystalline materials). This oscillation, accompanied by vibration, is manifested as *phonons* of infra-red and microwave frequencies, i.e. heat energy, and also q. Energy.

Any object of any color will generally reflect and emit its own characteristic color. If a piece of gold for example is observed indoors, the reflected light will be dim since the ambient light intensity is generally low and the photons impinging on the surface are therefore few in number.

If this piece of gold is observed in bright daylight however it will naturally appear brighter. This is because the direct sunlight is more intense and hence the quantity of photons impinging on the surface is greater. In this case, more photons are being absorbed and re-emitted, i.e. reflected, and we therefore see these extra emitted photons.

Let us now imagine that the piece of gold - while emitting light photons of "outdoor intensity", is taken indoors. We would now say - upon observing this emission of increased intensity, that the gold appears to be actually "glowing". Any light emitted by an object - even if the emitted light is only reflected, is *de facto* a light *emitted* by the surface atoms of the object. In other parlance we might say it is "reshined" light.

This 'reflected' light is therefore "second hand" light emitted by the constituent atoms. Actually if a piece of gold (or any substance) were left in the direct rays of sunlight, it would of course absorb the Energy carried by the sunlight. It would also "generate" heat energy and would as a result be suffused to a small degree with an excess of q. Energy.

If now this piece of gold, etc., were taken indoors, this extra absorbed content of Energy (and heat energy) is released into the environment. Some of this Energy is converted into microwave radiation, infra-red (heat) radiation, as well as photons of "visible" light. Any 'sensitive' with an increased ability to perceive subtle energies would be able to perceive and recognize this increased photon emission. In other words, the 'sensitive' might say that the gold is slightly phosphorescent, i.e. it glows.

In fact using this example, we could put on night vision equipment allowing us to see infra-red light. Then the gold or whatever would appear to luminesce or "glow" for some time after until the heat energy within the material has equalized itself to the surrounding environment.

If we were now to Energize this mass of gold electronically so that its content of Energy were increased, by means of an Energy "field effect", we would be able to cause the gold to visibly glow with its characteristic color. (We may refer to this phenomenon as the 'Moses Effect'). This glow is of course composed of generated and emitted photons. (Similarly, by allowing a current of electricity to pass through the gold, thereby generating heat energy, a small number of light photons are also generated and emitted). [see note 2].

Many of the incident photons impinging on the object's surface are "lost" by absorption so that only a percentage of these, - including 'gold photons' - are actually re-emitted, i.e. "reflected". If the object were quantum-Energized, its surface atoms would absorb the incident photons as usual but a higher percentage of these atoms would emit and re-emit these photons. Hence would the overall reflectivity and luminescence of the surface atoms increase. (This percentage may be called a "Dynamic Percentage", or "Momentary Population"; see Note 1b).

A relatively smooth surface like that of glass, liquid, or polished metal would thereby appear to have increased reflectivity and/or luminosity. It would also appear to "glow" in a dim environment since it is now emitting more photons than it is reflecting. (It is also reflecting more photons than it would under normal conditions).

NOTE 1a: a Phonon is an oscillation in the crystal lattice of a substance. It has the characteristic of a wave induced in a skipping rope except that the "rope" is a long row of atoms strung together. It is actually a 'sound' wave of extremely high frequency.

NOTE 1b: to give you a clear idea of what a 'momentary population' is, imagine that you have a thousand people in a large room. These individuals are breathing regularly, so that some are breathing in, some are breathing out, and others are somewhere in between. A certain percentage of these will be breathing in *at the same time*. Since people all breathe at different rates, it will never be the same individuals who breathe in during the following cycles. We can say however that the *number* of people breathing in at the same moment will be a constant. This constant may be termed the 'momentary population'. Another example may be given. Imagine a film running through a projector which shows a picture of a tree. The frames of film move rapidly through the projector yet the picture of the tree remains - although it is is never the same film frame from moment to moment.

NOTE 2: in fact any material may be made to glow thusly by energizing its atoms. This can be done simply by subjecting the mass to a powerful irradiation of Hertzian waves as from an arc-welder.

117 ~ MORE NOTES ON QUANTUM ENERGY

This Energy form is of course very unique and demonstrates properties quite "unusual" when compared to other forms of energy found in the electro-magnetic spectrum. It is non-the-less constituted of particular electronic wave characteristics. It has an extremely short wavelength, comparable to short ultra-violet light, and shorter. If permeating an object in sufficient quantities, it can cause the object itself (being an electro-atomic structure) to glow with a visible light - even if the object is a biological organism. (> *Ageing and Health*: - and also 'The Moses Effect').

Because of its unique electronic nature and frequency, it has the ability to permeate virtually all matter on a sub-atomic level. No technical instrument has been devised by mainstream science which can objectively isolate and measure it. Indeed, because of its very elusiveness and apparent subjectivity (and largely un-guessed at existence), it has never been acknowledged to exist by mainstream science. Hence no initiative to devise such instrumentation has heretofore existed: [see note].

Under "normal" conditions, this Energy exists in matter and living organisms in very subtle quantities and hence has only been experienced subjectively by gifted people who are sufficiently sensitive and perceptive, (*Von Reichenbach*). If generated in larger quantities, it can become a more objective experience and can be sensed then by any observer.

When a flow of electricity is initiated in a metal conductor, e.g. copper wire, many forms of energy are generated which are subtle and not objectively experienced by the observer. For example when an electric appliance is turned on, such as a light bulb - the initial surge of electricity in the tungsten filament generates a radio wave pulse. As the current continues to flow, it encounters increasing resistance, - which is normal. This resistance generates heat (Joule heating) in the filament, and the final result is heat and visible light. (This is actually a very crude way of generating light photons).

Light and heat are actually electro-magnetic quasi-particulate radiations emitted by oscillations created in the electrons of atomic valence shells and the crystal lattice of the metal. Furthermore, if this initial flow pulse were intense enough, the emitted radio-wave radiations would become objectively apparent. (Actually heat radiations are not very particulate, - they are more like microwaves and radio-waves: - explained later). Microwaves are also generated by the same current flow and are normally too weak and subtle to be observed or experienced.

(Another good analogy of subjective and objective experience would be if one were to pass a small electric current from a one and a half-volt battery through his body. There would be no objective sensation. However, if this current were enormously increased in both amperage and voltage values, it is obvious that this experience would become very objective to the individual, - and to bystanders also).

(Oscillations in the crystal lattice are called 'phonons', which are simply mechanical ultra-sonic waves in the crystal lattice of the conducting material. Also generated are photons (light wave particles or so-called "wavicles"). Because this flow of current in the wires is engineered to be efficient, all of these energies are not readily or objectively experienced by the average observer, – they are too subtle to be perceived).

Another form of energy is also generated which we are similarly unaware of and that is the quantum Energy. This fine electronic wave energy is capable of "flowing" through or traversing the interstitial spaces between the atoms of matter (and through the atoms themselves) much like the waves at sea will pass through a flotilla of fishing boats and their nets. They are able therefore to radiate away like heat from between the surface atoms of the conduit and into the surrounding space.

It is useful therefore to think of Energy as a fluid flowing through the interstices of the atomic structure of matter in much the same way that water might flow through a sponge, or indeed, as heat energy itself.

This is why there is no known effective insulator or barrier for this peculiar type of energy. Bakelite, concrete, water, and leather however have been found to have superior retarding and absorptive effects: (*Reich*). It has also been found that glass (a ceramic) will block some frequencies of radiated q. Energy, just as it will block and reflect some frequencies of ultra-violet light: (> *Geller's Frequencies*). Stacks of newspapers or books will constitute excellent temporary insulators.

Mirrors - like crystal balls and other paraphernalia, are subtly fascinating and "mysterious" to occult-minded people, i.e. psychics and 'sensitives', on an intuitive level. The reason for this is that - being made of glass (and backed by metal), mirrors reflect Energy - and change its frequency slightly while doing so; (> *The Crystal Ball*). (All electronic wave radiations go through a slight frequency change when being reflected or refracted. This Energy can be transported along conduits of metal or other material, just as light waves can be conducted by a glass rod or optic fiber, (*Reichenbach*). In some respects, this Energy can be manipulated like other forms of electro-magnetic energy.

It can be refracted, reflected, and focussed by lenses and prisms. However like ultra-violet light, certain frequencies of Energy are not passed by ordinary glass, although glass will absorb the Energy slowly (as does concrete). Useful lenses and prisms would necessarily be of quartz, diamond, or ice, etc.. Quantum Energy is also able - like other forms of electronic energy and radiations, to impinge actively upon electrical conductors to render them more conductive by decreasing resistance.

This Energy can also modulate longer frequencies of Em radiation such as light "wavicles" (i.e. light photons) similar to the way in which radio waves are modulated in FM radio waves to carry message frequencies. In this respect then Energy may be transmitted by other forms of Em radiation including sunlight, which includes all known Em radiations.

E-energy may therefore be transmitted by virtually any type of emitted, reflected, refracted, radiated or beamed electronic energy form, including Laser and Maser 'light'. This means that if any such radiation is passed through the energy field of a powerful source of q. Energy, this radiation will become modified to carry quantities of Energy from its source to another point. (Technically speaking, Laser light is not 'radiated', since radiation implies the d^2 rule. More properly, the light is 'emitted' since laser light follows the d^1 rule).

The intensity of Energy will diminish with distance in the same way that the electromagnetic carrier wave diminishes, i.e. according to the d^2 rule. This mechanism of conveyance is exactly why sunlight is able to convey quantum Energy to our hills, countryside, rooftops, clotheslines, windows, trees, cows, and wheat.

Ideally then the carrier wave or emitted energy will be as a focussed beam such as a Laser beam, since this is a very efficient way of transferring light energy from one point to another. The overall resulting effects upon the 'target' may then be referred to as the "Jesus Effect".

NOTE: In fact, private independent researchers in the field have devised such an instrument, and it consists of a fluorescent screen similar to the one constituting your television tube. Energy impinging upon this screen will stimulate atoms in the coating substance and these in turn will fluoresce slightly, enough to register on a sensitive photographic film. Indeed, a photographic film will also serve as a register, as with Kirlian Photography and as in the *Thomson Experiment*.

Another important point should be made regarding the flow characteristic. When electricity flows in a conductor, it tends to create a resistance to itself. This resistance is expressed as Joule Heating and magneto-resistance, etc.. Quantum Energy on the other hand when flowing in a conductor will *decrease* electrical resistance.

118 ~ QUANTIFICATION & MEASUREMENT

The physicist tomorrow will find himself in a quandary when he attempts to measure and quantify the various unusual phenomena associated with quantum Energy. The problem will present itself to be as difficult as measuring and quantifying emotions or thought. For example, how is one to measure and quantify the loss or reduction of weight in an object when this measurement relies on the object's physical connection with the planet? (This problem is similar to an attempt to weigh a dirigible while it is in flight, or trying to determine the original positions of dice after they are shaken. Again, how does one balance an object against a set of weights whose gravitational value is continually shifting? How does one measure the q. Energy content of an object when this Energy cannot be isolated or stored in any way?).

In fact if one analyses the situation carefully, he will recognize the fact that indeed all dynamic phenomena are measured *indirectly* – whether it be the flow of water in a pipe or electricity in a wire. When we measure the temperature of bathwater, we use a thermometer. This method of measurement – although we call it "direct" measurement - is actually the effect of the heat energy (i.e. molecular motion) of the water upon another liquid enclosed in a glass cylinder or upon a coiled bi-metal strip connected to a pointer and a dial. In other words, this is an *indirect* measuring method.

A more direct method would be to actually place one's hand or foot into the water. However we are then obliged to admit that a more direct method is also a more subjective one. Likewise objective methods are more indirect! "Objective" observation then requires an indirect via or method of approach.

In any event, the simple action of merely reading a thermometer (or any measuring device) is in the final analysis still an objective process, as will be explained. This must certainly beg the question, "how can something be an objective experience if it is experienced indirectly?"

119 ~ OBJECTIVITY, SUBJECTIVITY, AND ENERGY

As described previously, objective observations must be indirect while direct observations are necessarily subjective. Apparently, there are two kinds of subjectivity, and likewise two kinds of objectivity. The one form is predicated upon perceptions in the physical aspect, the other upon *agreement* with others. A very few people are able to see manifestations of light and shadow which others do not. For example, an electrically active conductor generates light photons which are so few in number that only a small percentage of the population is able to perceive them. On the other hand, the appreciation of an iceberg is a very subjective thing and yet many people will agree (objectively) upon its aesthetic properties.

In scientific or statistical terms, one may say that "objective" experience of subtle phenomena is that state which is limited to one or a few individuals (i.e. 'sensitives') within a population. (One study firmly suggests it to be 4.99 %, or one in twenty of the general population). In other words, the agreement of the existence of the experience is not universal. When any particular experience becomes accepted as a commonly shared experience by the majority of the population, we can say then that it is no longer subjective but is now objective. We may call it "objective" even though it is obviously still a subjective one - as in the case of the appreciation of beauty.

It is claimed that beauty (that found for example in a rainbow, the full Moon, a snow covered mountain-top, or an iceberg, etc.) is a subjective, personal, and individual experience. Yet almost everyone will admire a particular phenomenon and even agree that this phenomenon is beautiful. So this raises the question of whether this is a subjective, or an objective experience: [please see note 6].

If indeed it is a subjective experience, then it is a subjective one which everyone objectively agrees upon*!* In view of the fact that any objective experience must ultimately be a subjective one in any event, - it is then fair to say that beauty observed and agreed upon by many must perforce be both objective <u>and</u> subjective. If then an experience is both subjective and objective, we are obliged to consider that either subjectivity or objectivity ultimately do not exist at all as discreet concepts - or that they are both one and the same, – depending upon your point of view and perception*!* [see note 0a & 0b). (Refer to the *Dichotomies*, 104 BP).

Life-energy, i.e. quantum Energy, normally is a very subtle energy, - which is to say it is extremely difficult usually for a human being to detect with the senses. It has always been described as so. The subtlety of this Energy makes it a subjective phenomenon (like subliminal advertising), - being consciously sensed by a few and not by the majority. Therefore its existence is denied by this majority. The few who know the truth are frustrated, not being able to inform the ignorant masses that they are indeed ignorant! ("it is the prerogative of the fool to speak truths that others will not utter"… William Shakespeare).

The subtlety of any energy is simply a matter of degree. That is, if any energy is produced in larger quantities - it becomes less subtle, less subjective - and more "*objective*", - so that it can be then more easily sensed, perceived, and agreed upon by a larger population. [see note 3].

(It is probably true that there is an entire spectrum of such sensitivity running through a population - just as there is with for example, intelligence levels, color perception, tactile, auditory, olfactory sensations, and so on. Furthermore, the degrees of this sensitivity vary from culture to culture, age to age, background to background, and the various levels of education. In addition, it is fairly well acknowledged that children (and animals) have a more acute awareness of subtle changes and influences around them, so that things which are readily experienced by them go all but unnoticed by adults.

A family cat for example will enter a familiar room and upon detecting a slight change from the expected therein - such as the small change in the position of a chair, a small rug, or a book placed on the coffee table, will immediately investigate the vicinity of change. To an adult human family member however, such small changes generally go unnoticed, partially because such changes are expected. Other such small changes for example might be a spider on a doorhinge, or a clock which has stopped ticking. [See note 1].

Therefore it is seen that subtlety, subjectivity, and objectivity are intimately associated with an individual's sensitivity, perception, and <u>awareness</u> of his environment: (> 42-141 A: Pure Frequencies; > please see footnote 2).

Many kinds of energy that we are familiar with are in fact "subtle". For example, radio waves, television Micro-waves, infra-red light, Cosmic rays, mild X-Rays, ultra-violet rays, and even electrons (the latter three are emitted by fluorescent lighting and television sets) - are all forms of radiation that we do not sense generally under everyday conditions.

Any radiation which is not sensed or which has no observable effect to the majority is normally termed "subtle" and "subjective". Visible and infra-red light (heat) are the only two forms of subtle electro-magnetic radiation we can normally detect with our senses - since the human body is especially sensitive to these two energies. However even here, if these radiations are in turn reduced in intensity, some of us may fail to detect them: [note 4].

All other forms of radiation are normally too subtle to sense and therefore fall into the "subjective" category. If any one of these forms of radiation were increased to sufficient intensity, they too would be sensed 'objectively' and would cease to be entirely subjective: [see note 5].

For example, even though we don't normally sense radio waves or microwaves, our bodies certainly do. Since the body is largely composed of water and ionic salts, it is an electrical conductor and behaves like an antenna. When radio waves or microwaves impinge upon an antenna of suitable dimensions, an oscillation of electrons occurs and hence an oscillating electric current is affected in the antenna.

This happens in one's body constantly on a daily basis. Microwaves (including television broadcasts), radio waves, (and Hertzian waves generated by lightning storms around the globe) are everywhere about us constantly. Since most people are between five and six feet tall, their bodies respond best to wavelengths and harmonics in this range, i.e. microwave emissions. Children and small animals respond accordingly.

(The body, having a complex geometry - is of course composed of a wide variety of dimensions from the length of an eyelash, - so that it will respond to many wavelengths of energy). By simply increasing the power of the Em wave emission, the oscillating currents in the body will be increased to the same degree and may actually be felt or experienced as an electric shock or 'tingle' by the individual.

It can be perceived therefore that in many cases, the difference between subjective and objective energy behavior is simply one of a degree of power setting, wave intensity, or amplitude. By increasing the power output even more, it would eventually be possible to actually cook the body by this method, (e.g. by the use of high-energy radio wave or microwave emissions). We may then say that these effects are objective!

Other effects obtained by different settings would be a complete overwhelming of the body's electrical system, - with the result of total paralysis and possible permanent nerve damage. (Cases are on record where workmen standing too close to microwave towers and dishes have been internally burned and cooked, i.e. 'microwaved': > 'Mysterious Fires and Lights': V. Gaddis).

Quantum Energy is unknown to present-day orthodox physics because of four factors. Firstly, the intensity of environmental emission is normally very low, hence the effects are entirely subjective if noticed at all. Secondly, it is an electric wave - not an electro-*magnetic* wave. Thirdly, with any object exposed to light and which also emits q. Energy, the Energy will be modified and 'absorbed' by the reflected light photons. Hence, the Energy will be carried by and bound up with the reflected light and 'disguised'. Fourthly, since the Energy is literally everywhere, it is very difficult to distinguish. It a case of "not seeing the trees for the woods". (Um, - or is it "not seeing the woods for the trees"?).

All electronic instrumentation is designed to respond to or operate with only electro-magnetic (Em) radiation, be it wave or particle. It is for these reasons that the electric E–wave energy was never suspected to exist, and hence was never looked for.

Furthermore, many types of electro-magnetic energy transmissions are only utilized for their electric wave component, while the magnetic component is of relatively little value. (It is of *some* value as we shall see).

As discovered by independent researchers in the field, Energy radiations can be detected by a fluorescent screen - and thus subsequently recorded on photographic film. (They can also be recorded *directly* onto photographic film. The recorded radiations in these cases are in fact photons generated by materials which have been affected and energized by q. Energy). [see note 8].

These emissions are normally too weak for an average individual to perceive, but when they are recorded and accumulated on such a screen, their effects become visible. (> *Thomson Experiment*).

To summarize the above then we may say that in everyday events, subjectivity and objectivity are - by convention - normally well separated and relatively easy to distinguish. One is a personal and private experience while the other is a shared and agreed upon experience. (These two types of experience form a dichotomy). [see note 7].

> One may see a tree or a pond and this - like all else - for the individual is ultimately a subjective experience. However we usually refer to it as an objective phenomenon since we can <u>agree</u> with everyone that this phenomenon actually exists as a shared experience.

To say that a tree or pond is beautiful however becomes a more subjective and less objective experience since, although almost everyone will agree that the tree is there, - not everyone will agree that it is beautiful. Some will view a beautiful tree as nothing more than an obstruction to profit or Feng Shui, etc..

Aesthetic energy is a very high level energy and has a sub-atomic wavelength, and is therefore a quantum Energy form. At the q. Energy level then, subjective and objective experiences become integrated or "merged" and indistinguishable. They are ambivalent, like the light photon posing as either or both a wave-train or a particle.

In the final analysis however, the single and powerful element which separates subjectivity from objectivity is, - <u>awareness</u>. If the awareness of those around you is increased, the subjective experiences which you enjoy will become more "objective" and will be shared by them also.

> Note: Man has been able to devise a complete technological civilization based on the intelligent and knowledgeable use of matter, energy, space, and time. This civilization may be said to be a culture of 'objectivity'. The products of all this technology and objectivity however, such as automobiles, computers, television, and so on are designed for *subjective* use. The pleasure one derives from watching TV for example or from going for a ride in the country is subjective, and this is what these 'objective' items are ultimately designed for.

Other types of items or systems, etc., are designed to expedite this ultimate goal. Apparently then, although we strive to obtain objectivity in all our endeavors, the ultimate purpose of this striving is actually to obtain subjective experiences. This must mean that we are intrinsically subjective creatures who use or create the illusions inherent in objective perception to further become subjective creatures!

> "Science does not know its debt to imagination", (Ralph W. Emerson).

NOTE 0a: in a subjective universe, nothing can truly be measured. This is abundantly true in the arena of Quantum Mechanics wherein the physical universe displays its inherent subject-ivity and eludes object-ivity. The attempt made by scientists to measure and quantify everything perceived is really an attempt to render the sub-jective as ob-jective.

In other words, it is an attempt to "solidify" that which of its true and fundamental nature is <u>neither 'solid' nor objective</u>! One example of this is the effort to quantify time by the use of clocks. Another is the attempt to rationalize the circle by finding the ultimate value of Pi.

We may say that a subjective experience becomes objective through the mechanism of *<u>agreement</u>*. The more agreement there is the more "objective" this subjectivity becomes.

NOTE 0b: we may introduce another factor, which is that beauty doesn't actually exist objectively - it is purely subjective. Hence, to agree that something is beautiful, is to agree on an abstract thing which doesn't really exist outside of your own perception. It is an agreement on a subjective experience and nothing else. Yet we acknowledge and agree that beauty exists. This is precisely how the physical universe is maintained. It only exists because we agree on the idea that it exists. It has no more substance than that.

NOTE 1: actually they are not unnoticed. The human mind has a mechanism which catalogues such subtle changes as being relevant or   irrelevant at the subconscious level. So although they are ignored and placed below the awareness level of the conscious mind, they <u>are</u> perceived by the subconscious and the unconscious mind (which incidentally never sleeps), - and is totally aware of <u>all</u> things at <u>all</u> times. In fact, it is this hidden mind that the Magicians of ancient times strived to achieve communication with, - as exemplified in the Tarot cards, etc..

NOTE 2: the degrees which separate the accepted notions of subjectivity and objectivity are actually degrees of awareness and perception. There are people in society (and animals) who have an unusually high ability to perceive their environment and can therefore perceive things which the majority cannot. Such people are sometimes referred to as "crackpots" simply because they do not follow the 'normal' trend. Examples are often cited concerning the ability of a mother to hear the cry of her child while others cannot hear anything.

Conversely, there are people who are unable to perceive things which the majority *does* perceive. An example of this can be the ticking of a clock. Some people, strain as they might, are unable to hear this sound. Indeed you yourself have likely had this experience: [see note 4]. This is innocuous enough, although such people are largely ignored while some are thought of as mentally ill or some such, - depending upon the experience in question.

In a population then we can find a wide range of individual abilities which do not conform to the '*status quo*'. This does not mean that these people are ill, but simply indicates that there are different levels of ability in different areas of perception. Some people may look up into the sky and see unusual objects there. This does not necessarily mean they are loopy, it may simply mean that their visual abilities extend into a wider range of perception and frequency, - like cats for example.

It is commonly known that children occasionally will see extra-ordinary things which adults do not. The same is more than likely true for many animals and insects, etc.. It is well known for example that children may entertain a friend who lives in the garden but which of course is regarded as "childish fantasy" by adults.

NOTE 3: apparently, there are two criteria which are available to determine both objectivity and subjectivity - if this determination is desired. An experience is subjective if it is very subtle and hence available only to a few having adequate perception. Likewise, it is subjective if it cannot be quantified and expressed in physical terms, - such as beauty.

Alternatively, an experience is "objective" if it can be quantified, and is likewise objective if a large number of Beings can agree upon its existence. Once again however, it must be noted that in the final analysis, all experiences are *subjective* and become "objective" only through <u>agreement</u>. Quantification and measurement are alike methods to establish agreement. (Standardization of measurement is of course a form of coerced agreement and serves again to further perpetuate the physical universe). It should be recalled moreover that all 'objective' measurements and quantification are managed through *indirect* means.

NOTE 4: there is another form of subtlety, however, such as the ticking of a clock. Even if the clock ticks loudly, anyone familiar with it will after a while cease to hear it. This failure to perceive the sound is not due to any inability of the physical aural mechanism; - it is strictly an "inability" of the mind/Being; - a 'mis-perception'*!*

NOTE 5: generally speaking, the greater the percentage of the population which perceives something, the more 'objective' it is, and the lower the percentage, the more subjective it is. In a group of artists for example, things which are discussed very objectively by the artists may seem very subjective to the general population (or vice-versa). This perhaps is why artists are often referred to as eccentric.

It is a peculiarity of the nature of things (again, this is paralleled in quantum mechanics) that a group or population may have a few percent who can agree upon some physical aspect - for example seeing disembodied Beings. The majority percentage will ignore this fact and ridicule the minority. Now let us say that twenty percent of the population develops this ability. Here is serious contention to the remaining majority, - no longer do they laugh and ridicule, - they start to wonder.

If the percentage is now raised to fifty, we find that the former majority actually begins to accept a possibility which they previously could not conceive of, even though they still cannot see. This process represents a gradual conversion of the entire population. At some threshold point however (perhaps around the thirty to fifty percent mark, we will find that this conversion process begins to accelerate. When the conversion reaches ever higher percentages, this acceleration will likewise increase since the former skeptics will be going into agreement with the new majority.

Not only this, but in fact we would find that an ever increasing percentage of people beginning to actually experience seeing things, - even the former sceptics. Hence we find that subjective possibilities are now becoming objective realities for the majority.

This type of situation is paralleled in the world of quantum activity. Let us suppose that you have a large sheet of pristine iron or steel. The surface has no oil or protective covering on it so that it is subject to rusting. If the sheet is left outdoors exposed to the elements, it will soon begin to show the signs of rusting. You will see a spot of rust here, another there, and so on. These spots will grow in size and in number so that the iron now appears to represent a map depicting an ocean having scattered islands in it.

At this point, one may see that the islands of rust represent perhaps ten percent of the total area of the sheet surface. Gradually these rust spots grow and cover more area. While these spots grow we observe that the area of pristine iron is diminishing. We may also observe that the rusting process is accelerating as the iron is on a 'learning curve' and has 'gone into agreement with' the rusting process.

We may imagine that the iron surface represents the population of a society and that the rust spots represent groups having a specific single ideology. Eventually the islands of rust will join to form larger areas and these areas will eventually form fifty percent of the entire surface. There will occur the ubiquitous threshold point. When this occurs, we will observe that the areas of pristine iron have now become islands in a sea of rust, so that the entire situation has reversed itself. This is precisely how a new ideology in a society can grow and accelerate, (such as a religion).

As we can see, this is another analogy of *'population inversion'*.

The former majority population has now become a minority while the former minority has become the majority. We can observe from this that in the physical universe and in most social situations also that the *tendency* is usually toward decay. We will never see a rusted piece of iron convert itself to shiny newness, with the rust disappearing.

The same is generally true in societies, unfortunately, unless a powerful counter force appears to reverse this trend. As long as societies are based upon physical principles while completely ignoring the mind's potential, these societies will travel the same downward path of all material things.

NOTE 6: let us suppose you were a regular patron of a restaurant and every week you visited it to obtain a bowl of your favorite soup. One day you find that your soup does not taste quite the same as it usually does. You may mention this to the proprietor who in turn insists that it was prepared in exactly the same way as before. Other patrons may likewise insist that it tastes the same as usual.

Here then is a problem; the first possibility is that the patrons and the cook are all mistaken, in which case you may declare that your situation is an objective one. Alternatively, the soup really is normal but your own taste has changed - in which case it has become a subjective matter. There is no way to know for certain which it is.

Now here we have entered the realm of <u>private universes</u>. For others, the soup tastes the same - while in your universe the soup tastes different. Is it different or is it not? Yes, it is - <u>*but only for you*</u>. The fact that others cannot perceive this change is not important; what is important is that it is real for you, in your private universe, (which you share with others' universes from time to time, sometimes, or frequently).

NOTE 7a: they may form a dichotomy but they can hardly be acknowledged to be a pair of true opposites for the following reasons. A subjective experience is a private and personal one exclusively. An objective experience must be said to be an experience shared and agreed upon by all. Nevertheless, each and every one sharing the experience is having a private and personal experience and this is the common denominator of the two. Hence all experiences, regardless of whether they are shared, agreed upon or not, are ultimately subjective.

NOTE 7b: this matter of subjectivity and objectivity can lead us into the quantum arena. Subjectivity deals with the experiences of individuals, (i.e. particles), whereas objectivity deals with masses of individuals. The mundane world is the experience of masses while the quantum universe is one of particles. (more on this later).

NOTE 8: experiments by *Reich* and others indicate that freshly cut garlic, ginger, or onion will emanate q. Energy. Hence, we can construct an experiment using a photographic screen to detect these emanations. Between the garlic, etc., and the screen, place a barrier having a small hole of any desired geometry. The emanations will affect the film.

120 A ~ HIERONYMUS, AND THE LIGHT SPECTRUM

The rainbow is an example of prismatic refraction of sunlight by cloud droplets. The outer curve of the bow displays the red end of the light spectrum - the inner, the violet. Beyond the violet color we will find invisible ultra-violet, and beyond that still higher electric frequencies.

In the electromagnetic spectrum, radiations beyond ultra-violet are called X-rays. The nature of X-rays is particulate. However the energy beyond shorter wavelength ultra-violet refracted by a rainbow is not particulate as one might perhaps expect, - it is essentially wave-like and composed of electric field wave Energy (i.e. quantum Energy).

Most kinds of prism, reflector, or diffraction devices - by their very nature cannot influence X-rays. (One would not expect X-rays to be refracted by any solid, liquid, or gaseous medium due to their particulate and atom-penetrating qualities. Furthermore, X-rays generated by the Sun are incapable of passing through the barrier of the Earth's ionosphere and layered upper atmosphere).

This range of the spectrum is quantum Energy "carried" by modified ultra-violet rays. This means in effect that if one were to look at the inner, concave area of the rainbow immediately adjacent to the violet band of color (where apparently there is no 'color', i.e. where there is an area of invisible Uv radiation), he would be looking at an area rich in E-Energy radiations. (One might be tempted to say then that staring at a rainbow could be therapeutic - in a very physical sense - and in the same way that staring at an iceberg, a mountain-top, or the full Moon, etc., would also be therapeutic; see footnote 1). (More on this effect later). [see note 5].

The healing Energy emissions enter the eyes from these strong sources and thence affect the endocrine glandular system including the pituitary and pineal glands (and also including the thyroid, the thymus, the solaroid, the gonoidal glands, and so on - which are all interconnected: - see note 4].

If one desired to use a prism to refract sunlight and thereby capture this particular ray, he is reminded that glass does not permit most Energy or Uv ray frequencies, - and that therefore a quartz prism should be obtained. However this experiment may be more simply done by using a diffraction grating such as the surface of a CD disc. In this case, the Energy would be refracted to the eye (beyond the Uv portion of the spectrum). An indication of this is the fact that most people prefer the blue/violet colors and find them more aesthetic and pleasing.

If one were to survey a large group of people to determine what their color preferences were, the largest percentage of choices would be in the indigo to violet range of colors (*Von Reichenbach*). This end of the spectrum carries a larger portion of aesthetic electronic q. Energy.

If we now go beyond the red end of the spectrum, we will encounter colors which are generally agreed upon to be unpleasant and lacking in aesthetic appeal. These colors include shades of "dirty" yellow or amber. These are the kind of "ugly" colors used by *Hieronymous Bosch* in his paintings of death, decay, demons, and spiritual filth. This color may be represented by the water of a stagnant pond containing debris and decay.

Beyond this part of the spectrum, we venture into very low heat levels and microwaves. If we were obliged to assign colors to these parts of the spectrum, the majority would probably agree that the most suitable colors would likely and intuitively be shades of "mud", "dirty amber", or perhaps the color of rust (i.e. degenerating or "decaying"iron).

Interestingly, *Reichenbach* in his researches found that his 'sensitives' were able to detect that the Energy 'Odyl', was "polarized" by crystals and magnets. He discovered that a quartz crystal for example would emit a "dirty-yellowish-orange" color at its base end where the crystal growth takes place and where the crystal structure is rather chaotic and opaque.

Further he found that the crystal would emit a violet "flame-like" emanation from its pointed end where the crystal structure was orderly and well formed: [see note 3]. Magnets in their turn would emit a violet 'flame' at their North-seeking pole. This apparent polarization is evidently the result of the Energy being separated into different frequencies.

> If we observe the colors of nature, we would soon recognize that colors other than "mud shades" are usually exhibited by living organisms whereas the muddy colors and dark ambers are generally associated with death, decay, dissolution, chaos, and destruction. Those animals who do display a 'muddy' appearance do so for reasons of camouflage.

Examples of the earlier mentioned "mud-colors" are soil (which is decayed matter), dead leaves and grass, fecal matter, any matter which is burnt or exposed to heat, - and including rusted iron, etc.. [see note 2]. It may be said that next to the color green in nature, the commonest colors are the earth-colors (if we ignore the blues of the sky and the oceans).

(Actually mud does not have a true 'color' *per se* since it is the result of a chaotic mixing of many colors - as any child who has played with colored paints knows). The "dirty" browns and ambers are the color representations of chaos, hate, avarice, neglect, confusion, filth, "the easy way" - and death: (the kind of environment where one might expect to find demonic activity, - which is probably why *Bosch* intuitively chose these colors for his paintings). Primary colors, on the other hand represent order, organization, discretion, separation, simplicity, cleanliness, etc..

We may say then that the 'spectrum' of earth-colours or dirty ambers is the extension of the light spectrum beyond the visible red. If we construct a dichotomy from the extended light spectrum, we may also say that the complementaries of the 'Hieronymus' colors (i.e. 'H-colours') are violet, ultra-violet, and q. Energy "colours".

If we now allow these "H-colours" to represent chaos, death, filth, and destruction, etc., - it follows therefore that its dichotomic complement, i.e. quantum Energy, - represents order, life, and creation. Indeed *Reich* claimed that this Energy was, or is in fact the formative and creative force in nature, - and that the universe as a result could never "run down" to entropic oblivion: (See article 102).

We have already seen that the "H-colours" are easily associated with the low end of the light spectrum, i.e. infra-red and microwaves. It would not be far-fetched then to say - judging from our Earthly experiences, that we are living in a low frequency "microwave world". In fact this is actually quite literally true since all matter in its 'normal' state vibrates and oscillates with, and radiates energy predominantly in this range of the Em spectrum. Since solid matter (composed of atoms) generally vibrates and oscillates at these frequencies, we are obliged in our Earthbound biological world to live with this frequency range of matter, - as is all biological life. To put it another way, - or perhaps as the Hindu Yogi would say, - we are living in a world of "low vibration".

Interestingly, the exterior dimensions of the human body cover the full range of the microwave spectrum. Virtually all biological life forms with definite structure fall within that same general range. (As to viruses, they are not now regarded as 'biological' organisms, although one is at present obliged to refer to them as "life-forms"). The largest land animal by contrast, i.e. the elephant, falls just at the limit of long microwaves and short radio waves.

> Note: The very largest animals such as the sperm whales have a body length extending outside of the microwave range and well into the short-wave radio wavelengths. (Their lives are consequently of "low vibration" - and slow movement, and interestingly are associated with heaviness, massiveness, inertia, and so on).

Whales however do not live in an electromagnetic environment. Sea water is a fairly good conductor of electricity and is therefore a reflector and deflector, not a conductor - of Em radiation. Living in the sea is akin to being constantly surrounded by a local and personal Faraday Cage. This fact is precisely why ships and submarines must use *sound* detection, i.e. Sonar. Radar or other electrical radiations simply cannot work under water, especially seawater.

The kind of vibrations common to their environment are sound waves which are of a low order of vibratory energy or frequency. It would seem therefore that whales and other sea animals live in a much more alien world than we could imagine. Since these creatures live in salt water - and since salt water does not carry electromagnetic wave radiation, their environment is virtually completely devoid of electronic 'noise'.

It is interesting to note also that the Triassic saurians (i.e. giant lizards) were compatible with longer wavelengths and therefore were living an experience with lower frequencies. Their thoughts, emotions, and so on were probably lower and concerned almost exclusively with food - hunger for food, hunting for food, chasing food, catching food, ripping food apart, eating food, defecating, procreating, and sleeping.

These are certainly not the high frequencies sought after by the Hindu Yogis or other such enlightened ones. (It still not known whether the Saurians were cold- or warm-blooded, or indeed why they perished). Some speculate there was an atomic war!

NOTE 1: The efficacy of this latter exercise however may be arguable since Moonlight is reflected sunlight and is therefore polarized light. It is claimed by some that polarized Moonlight is detrimental to one's psychic and emotional well being. According to experiments done by *Drbal* and others, it has been found that Moonlight actually dulls new razor blades which have been left exposed to the Moon's rays).

NOTE 2: Rust is the product of oxidization and this is merely a very slow form of combustion. Rust of course is a combination of iron and oxygen and therefore may be considered to be the "ash" of "burnt" iron. Iron is the commonest heavy metallic element in the universe and oxygen the commonest reactive gas; (followed by chlorine, bromine, and fluorine). (Aluminium is the commonest light metal on the planet's surface and interestingly has an atomic number exactly half that of iron, i.e. 13 to 26).

NOTE 3: this can easily be explained if we understand that the lower end of the crystal has an amorphous structure and hence the atoms are randomly interconnected. The atoms at the pointed end however belong to a more organized crystalline structure, thus they are more efficiently packed and require less space. Hence their inter-atomic bonds are tighter and when their shells are caused to vibrate, they naturally vibrate at a higher frequency, - thereby emitting a higher frequency Energy and thus stimulating a higher frequency Energy emission.

NOTE 4: In passing, it may be fortuitous to say that the entire sky is to some degree filled with this refracted therapeutic Energy. It would be to this degree beneficial to the upturned eyes.

Modern society accepts as a normal part of custom the wearing of sunglasses. In fact sunglasses are *detrimental* to the eyes and health since nature herself supplies the atmospheric healing Energy. Sunglasses manufacturers tell us that ultraviolet light is dangerous to the eyes. This promoted "wisdom" serves their profit margins well. In fact, the eyes *need* Uv light just as they need all other forms of light. Did you ever see animals of the wild wearing sunglasses?

UV light interacts with the eyes and the endocrine system. This light form stimulates this system beneficially along with all the connected glands and organs. In a sense, this form of therapeutic technique is somewhat akin to acupuncture and the technology of "Chi". Eyes then may be regarded as the organs of both psychological and physical health. In animals, the size of this organ is an indicator of the intelligence level.

Additional: I am often amused by the tobacco industry. They have created a tobacco *mystique* in order to sell their products. What exactly is tobacco? It is a leaf which grows in the field, just like any other leafy plant. Why have we chosen tobacco to smoke? Only because the American Indians used it in sacred rituals. It's a mystery to me why it is sacred since it doesn't give you a high like marijuana or any other such psychedelic plant. In any event, tobacco now is nothing more than a tradition and there is no good reason why we shouldn't smoke any other plant, such as cabbage, lettuce, spinach, corn-leaves, carrot-tops, or what-not. In fact, it is probable that many other plants are healthier to smoke*!*

NOTE 5: another direct source of q. Energy is mentioned elsewhere. Where two pure colors of radiant light are impinging on each other, a third energy of higher frequency is generated. This third energy form has no magnetic component and is purely electric. This is quantum Energy and is generated in quantity by the rainbow.

120 AA ~ THE RAINBOW EFFECT, Q. ENERGY, AND LIGHT

The closer a region of space is to a gravitating body, the more it is modified and 'curved' (*Einstein*), and the 'denser' it becomes: (explained elsewhere). Hence, in the vicinity of such a body, the space has a varying 'density' depending upon its proximity to the body. It is this varying density which causes light to 'bend'/'refract' as it passes near the object.

An analogy of this can be constructed in a simple experiment. Let us imagine a large glass filled with warm water. A piece of ice is now immersed into the water and held in place at the bottom of the glass. There will then be a temperature variation in the water, - colder and denser at the bottom and warmer/less dense toward the surface. If light, such as a laser beam is now passed through the water, it will be observed that the beam is bent, i.e. refracted, toward the colder/denser region.

In the case of a gravitating body in space, the light beam is not only 'refracted', - its velocity is decreased. (In fact, refraction generally is synomymous with the light's change of velocity and time flow rate, - and vice-versa). It does not change color however since it has lost no energy.

Space is actually a space-*time* continuum, hence we are obliged to say that time also becomes 'denser', - which is to say it 'slows down': (*Einstein*). In several respects, as it relates to light - bent space is similar to a glass lens.

Because of this retardation of the time flow rate, any activity within this space-time is also slowed down – relative to 'free space'. When the light photons again reach 'free space', they regain their former velocity since they have retained the same energy content. They have not lost any energy. Here we find an interesting comparison where energy is modified differently under differing conditions. When a light source is moving through space toward an observer (or away from him) the light perceived by the observer changes its *frequency*; its velocity however is unaffected. This is predicted by Relativity and is called the 'Doppler Shift'.

In the example given here however, we find the light *velocity* changes - but *not* its frequency. In both cases therefore we find a different mechanism employed to conserve energy; energy is not lost – simply converted: [see note].

If however a light beam should pass through a powerful quantum Energy field surrounding an object, the light photons will *gain energy* which is 'sequestered' from the field. This occurs because the energy composing a photon is in fact quantum Energy; (This will be explained in due course). Hence the photon's velocity may increase, or its frequency, - or perhaps both will increase.

Hence we should expect to see the light refracted to form a rainbow of colors where the color (frequency) depends upon its duration of travel through the quantum field (and hence its proximity to the object/field generator). The overall effect of this would be a changing rainbow - like a halo surrounding the object.

NOTE: As stated, the gravitational field of a body is equivalent to a lens when light passes through it. This means that when you are in space observing a planet, the planetary light received by you is actually of a lower velocity at close quarters and increases the further away you travel from the planet. (The frequency is not affected).

Such differences are subtle and will not be noticed at relatively small distances such as the Moon's orbit. When powerful gravitational fields and great distances are considered however, such differences will become measurable. They become vastly important at galactic distances and masses so that not only space and light are affected, but also *time* as well.

120 B ~ MORE ABOUT QUANTUM ENERGY & IT'S PRODUCTION

Electric waves may be generated and emitted by any surface which is electrically charged. This would include wire conductor surfaces as well as wider surfaces. For example, an instrument such as a harp has metal 'strings' which vibrate - thereby emitting musical sounds. If these strings are a part of an electric flow circuit or are electro-statically charged, they are then surrounded by an electromagnetic or electric field.

When the strings vibrate mechanically, the electric field component is translated into electric field waves (at the sound frequency) which are then emitted and radiated into the surrounding space: [note 2]. This type of wave is called a 'compressional' or compression/expansion wave – and behaves in a similar fashion to sound waves, except of course that it travels much faster and the medium is space-time.

(If a magnetic field could be "radiated" from a source such as a vibrating magnet, the field waves would propagate according to the $\underline{d^3}$ rule (distance cubed). However the intrinsic nature of magnetic fields is to condense and centralize rather than to expand and radiate. In other words, independent, purely magnetic field waves cannot exist. In many ways, the two fields (electric and magnetic) constitute a dichotomic pair. The magnetic field itself however can be made to vibrate as with a vibrating magnet).

Another harp - being similarly electrically charged and some distance removed (and perhaps located in another room), will receive these radiated electric waves and the strings will resonate electro-mechanically in response, - emitting sound 'magically'. (This would constitute a neat "parlor trick". However one is cautioned that playing around with electricity can be dangerous and should be supervised by an expert.

A harp has many strings and hence a wide range of possible emitted frequencies of sound. Any of these strings being electrically charged and plucked or made to vibrate, will emit both sound waves and electric waves of corresponding frequencies. [See Note 1]. It is obvious that the shortest string will vibrate at a higher frequency than the longer ones. Theoretically, strings could be made short enough (and thin enough) which would then emit ultra-sonic frequencies, i.e. beyond 20,000 cps. (In fact the shorter harp strings do exactly that - as harmonics!).

It may be seen therefore that any object capable of mechanical vibration will emit energy of a characteristic frequency. (Large solid steel objects such as anvils will emit ultrasonic frequencies when struck, as will small bells). Further, if any object is electrically charged so that it is surrounded by an electric field and made to vibrate, - it will generate, emit, and radiate electric waves corresponding in frequency to the mechanical vibrations.

When a steel ball bearing is dropped onto a hard, horizontal steel plate in a vacuum, it will bounce (for a time period almost indefinite, due to minimal friction). This bouncing behavior indicates that the steel bearing deforms in geometry while striking the plate. While the bearing is in vertical motion, this sudden deformation causes the bearing to vibrate at extremely high frequencies in the ultra-sonic range. (Similar to the way in which a struck golf ball vibrates while in flight). If this bearing is electro-statically charged, this mechanical vibration will be translated into electric waves of corresponding ultra-sonic frequencies. (With this information it should be possible to devise an instrument to measure this frequency electronically. The same would of course hold true for golf balls!).

It naturally follows then that the smaller (and harder) the steel bearing, the higher will be these vibrations and consequent emitted frequencies. Hypothetically if one could take a single atom of any substance (being a particle of matter) and bounce it in the same way, it would vibrate at its characteristic frequency.

This frequency would of course be extremely high, i.e. in the frequency ranges of Uv light and beyond. The atomic valence shell "surface" is electro-negatively charged and therefore the atom would generate, emit, and radiate electric waves of corresponding quantum frequencies. This frequency range would then constitute q. Energy.

It may be seen therefore that vibrating atoms may communicate with each other by means of this emitted energy in the same way that electrically charged and vibrating harps, tuning forks, or bells, etc., may communicate. This would be done by means of radiated and emitted electric waves. In fact this is precisely what occurs on an atomic level so that there is established a dynamic connection between all atoms in a mass or object.

NOTE 1: in fact, an electric-field harp is a superb analogy of an atom in that its strings, resonating as they do to various frequencies, - may represent the multiple energy levels of the various atomic shells. Each string or shell in turn responds to precise wavelengths of received wave energy, then re-emits it or converts it into other forms of electric or electromagnetic energy. Atoms in normal "dead" matter do not oscillate faster than infra-red frequencies because of the inter-atomic bonds; however their electron *shells* can vibrate at much higher frequencies.

Therefore atoms (and consequently molecules) are capable of generating and emitting electric quantum Energy. It follows then that any substance containing atoms is capable of generating quantum Energy, - whether it be metal, stone, bone, flesh, or what-have-you. They are capable of resonating to these same frequencies.

According to our primary definition, all of these wavelengths constitute quantum Energy. This range of Energy therefore constitutes an entire electric wave spectrum *parallel to* the existing electro-magnetic spectrum. If the Energy then exists in the frequency range of visible light, we must consider that it can *emulate* visible light and indeed is capable of emulating all the manifested characteristics of the entire Em spectrum.

This of course means that it can emulate radio waves, microwaves, (Radar), heat, light (colors), ultra-violet light, and X-rays, etc., all depending upon its source of generation, emission, and radiation. In a powerfully active medium such as an electrical conductor or plasma, etc., q. Energy of all frequencies should be found. Oscillating and vibrating electrified particles can then emit q. Energy of characteristic wavelengths.

NOTE 2: the metal string of a harp will carry an electric current flow and thereby be surrounded by an electromagnetic field. If the string is caused to vibrate, this Em field will vibrate too. However an electromagnetic field is composed of both a magnetic field and an electric field. In this event, the electric field will separate from the wire and radiate as an electric wave while the magnetic component will not.

120 C ~ E-ENERGY AND MATTER

Evaporation is the process whereby liquids will convert to gases at a rate determined by their associations with other substances, the surface area, the contained heat energy, the prevailing atmospheric pressure, temperature, and so on. This evaporation occurs at the liquid's surface. Any liquid exposed to the air remains a liquid largely because it has surface tension. It is surface tension which helps to prevent or slow down the evaporation process. [please see note].

If a water molecule tries to escape from the surface, this tension will pull it back into the body of the liquid. However if this tension does not exist - then the water molecules are free to escape and consequently the water quickly evaporates. This why some liquids such as gasoline evaporate quickly; they have low surface tension.

This tension exists due to the fact that molecules and atoms are bonded together by electrical forces known as the Van Der Waals' forces, metallic bonds, ionic bonds, chemical bonds, Coulomb Interactions, and so on. Within the body of the liquid, atoms and molecules are bonded in all directions, so that these bonds resemble the spikes of a Sea-urchin. At the liquid surface however, these bonding forces are arranged around the molecules sideways, parallel to the surface, and in all directions beneath the surface; (like a spiny Sea-urchin which has been cut in half).

This means that the available bonding forces of each atom or molecule at the surface are concentrated into half as much volume and are therefore twice as strong. As a result, this molecular layer itself has a stronger bonding force than the molecules within the body of the liquid. Hence the surface molecules form a kind of 'skin' which covers the liquid. This bonding force is called "surface tension".

If we put a drop of water on a smooth oiled surface, we will observe how this water 'skin' contains the liquid. The spherical drop flattens out under the influence of gravity but as we can see, the skin is strong enough to contain the water in a flattened sphere. An excellent analogy of this is a water filled balloon. If we place this item on a flat surface, we can observe how the skin of the balloon will contain the water even as the balloon flattens out to form an oblate spheroid.

Quantum Energy predictably has a desiccating effect on any matter which contains water, and particularly hydrogen as part of its chemical and particulate structure. This would naturally apply to biological matter, wet cloth, or wet sand or soil, etc., - and of course quantities of free water.

Any such object in the increased presence of quantum Energy will become dehydrated and dessicated in short order. This is due to the effect of Energy on the electronic forces which bind atoms and molecules together, and particularly its effect on hydrogen itself.

As elsewhere explained, the fine electro-dynamic nature of Energy interacts with these forces and "loosens" their electronic grip so that electrons, atoms, and molecules have greater freedom to remove themselves from bondage. (See *'Friction'*, and *'Superconductivity'*).

NOTE: there is another process by which solids may convert *directly* into gases, and this is called 'sublimation'. Solid frozen carbon dioxide for example will sublimate into gas with no intermediate liquid stage.

121 A ~ MATERIALS STRUCTURE & MEMORY

According to many researchers into so-called "pyramid energy" - including *Drbal, Kirlian, Pavlita*, etc., the Energy generated and accumulated within the cavity of a hollow pyramid form was capable of "mummifying", i.e. desiccating and preventing rot (in meat and fruit, etc.) - and also of sharpening razor blades, (especially older steel type blades).

Steel razor blades have a metallic crystalline structure. This crystal arrangement of atoms has a primitive type of "living" memory so that if the crystal is distorted - as razor blades are while being used, the crystal structure of the sharp edge will actually attempt to re-form itself over time to regain the former structural geometry it once possessed. If the blade is left in sunlight, the received radiations will assist and accelerate the metal's _self-healing_ process. A bath in warm water will also speed up the 'healing', as will sound, music, or vibrating electric field.

A piece of metal may be struck hard with a hammer. This will cause the metal to deform "permanently". The crystal structure now has a new "memory field" conforming to this new enforced memory. (In biology this enforced memory is called an 'engram'). The metal will now retain this shape since this is now a part of its enforced structure.

But wait, - if this metal is now strongly imbued with a powerful inflow of Energy, _the original shape will return and the metal will regenerate itself and be "healed"_!

If however the metal is deformed while also being subjected to q. Energy, high heat, and/or melting, this crystalline memory will become a new and permanent memory structure, thereby changing the Local Space-Time Continuum (STC) of the metal. This memory structure and space-time continuum is an electronic field: [see note below]. (> section 03).

Quantum Energy, because of its powerful interactions with materials on an atomic-sub-atomic-electronic-crystalline level will have a strong effect also on this structural memory reorganization in matter. The Energy will speed up the healing process to a phenomenal degree. (More later).

NOTE: it is said that time heals all wounds. This may be more than a mere platitude. If the bent metal for example is allowed to remain bent for a sufficiently long period of time, its former memory will be lost and its new shape will become a permanent record in its local space-time continuum. In effect, it will have suffered an 'engram' which now has become a newly instilled part of its geometric crystal-memory "program". (However, this cannot be said to be a 'healed wound' since the 'wound' has become a permanent structural memory).

121 AA ~ THE QUANTUM MEMORY FIELD

> "the only things we know about the quantum world are the results of experiments", (from Paradoxes And Possibilities).

> "We have to remember that what we observe is not nature herself, but nature exposed to our methods of questioning": (Werner Heisenberg).

< please refer to 301 BB > Every object is composed of atoms and in a solid

object, most of these atoms are connected so that together they constitute a rigid mass. If one atom oscillates, this oscillation is passed on to the surrounding atomic structure. In a crystalline structure, it is passed on to atoms forming a 'chain', - like the waves passed along a rope when it is jiggled. This interatomic communication occurs for example in a ceramic or metal object. (This atomic 'chain' resembles a string of 'beads').

The atomic structure of the object, e.g. a cup, generates a surrounding quantum electronic field. This field is the "memory field" – or local space-time continuum. Within this memory field every atom is in communication with every other atom in the network.

One may use the analogy of a fishing-net wherein every node or knot is in communication through vibration with every other node. (A more accurate analogy would be a three-dimensional net having nodes and strings in all three directions). If the cup is damaged in some way, one or more atoms are displaced and this is registered in the overall memory field. If the trauma occurs in a biological organism, this injury is called an 'engram'. This engram/injury is transmitted throughout the entire atomic/molecular structure and becomes a memory impinged upon the structure and the local space-time field. This dynamic field may also be called "the Aura" - and is called so by researchers: [see note].

A living biological structure itself has such a field and if an injury/engram occurs, the organism - having a memory and a genetic "blueprint" - is able to repair itself with the life-Energy as the dynamic rebuilding force. This is ordinarily called "healing".

Now an inanimate object cannot ordinarily repair itself since it normally has no access to sufficient quantities of quantum Energy. However if an ample supply of Energy were made available to the injured cup, - the cup will actually repair itself!! Examples of such damage may include cracking or even worse. If the cup, etc., is damaged by intense heat or electrical discharge however, this will result in a permanent destruction of the original memory field and atomic structure. This would mean that the affected portion of the cup may not be repaired. Such damage will result in the creation of a <u>new</u> memory field which then becomes the new pattern or local STC.

NOTE: the so-called Aura is a dynamic quantum-electronic field surrounding all objects, and is perceptibly stronger around living organisms and structures. Those people who have especially sensitive sight and who have the unusual ability to perceive other parts of the light spectrum are able to actually see this aura.

(It is interesting to note the effects of a word upon the mind. To many people the word 'Aura' rings of 'New Age', 'paranormal' hokum. Yet if we give it a more scientific sounding terminology such as 'quantum-electronic bio-field', it gains a new-found respectability. Nothing has changed, - just the name!).

121 AB ~ ENERGY, MOLECULES, & MATTER MODIFICATION

< please see 108 E >. What do the following items have in common; – vision, mind, thought, emotion, friction, evaporation, rainbows, the shape and form of trees and mountains, gravitation, inertia, transparency, electricity, electric and magnetic fields, superconductivity, the sun, Poltergeist activity, ball-lightning, Black Holes, galactic spirals, and the time flow rate, etc.. The answer is that these and more are all the result of *quantum mechanics*. Indeed the truth is that <u>all</u> phenomena are the results of quantum mechanics*!*

Hence if we can understand this fundamental and underlying force of all phys-ical activity, we shall then gain new insight into the hidden workings of the universe. These include the four constituents: Matter, Energy, Space, and Time, which in turn include gravity, inertia, re-location, and so on. Quantum Energy operating at the sub-atomic level can alter and dismantle atoms, and consequently modify molecules and crystal structures. Crystal structures may be considered to be specialized forms of molecule in that they are composed of atoms bonded together in various ways. (As explained elsewhere, water is constructed so that any body of water at or below 35 degrees Celsius (95 degrees Fahrenheit) may be considered to be a single mol-ecule; likewise it may also be considered to be a single crystal*!*).

All of these types of bonding however are essentially quantum-electronic, in-volving the outer electron shells of the atoms: [see note 1]. Any 'normal' mass is composed of a large number of atoms and we may use the analogy of colored ping-pong balls stuck together, or a bunch of metal rings all interlocking and forming a matrix. The rings represent the outer electron shells of the atoms.

If we now modify any of these rings, the group is likewise modified. For exam-ple, if we were to increase or decrease the size of one or more rings, the entire group will change in size, geometry, and structure. Likewise if we remove or insert rings, the size and structure of the group will again be changed. We may also split a mol-ecule in half, etc., using the same techniques - thereby rendering two or more new molecules of different substances, - perhaps completely new substances*!*

Hence it can be seen that it will be possible to create new substances and to modify the behavior of existing substances simply by modifying the constituent atoms at the quantum level.

< friction >

What does friction have to do with Poltergeist phenomena? Quite a lot actually. One well-known phenomenon in Poltergeist activities is the startling ability of quan-tum Energy to cause objects to levitate. Another even more bizarre phenomenon is the passing of such objects through apparently solid barriers such as stone walls or glass windows, etc..

Obviously there is an Energy connection between these two occurrences. Normally when we put a coffee mug on the table it stays where it is put. There is only one reason why this occurs and that it is the presence of friction - a quantum force. [see note 2].

If we now take this mug and try to push it through a window glass or wall, it refuses to go - why is this? Fundamentally, it is the same force of friction which prevents us from doing so. More specifically, it is the interatomic bonds which prevent us. This is what constitutes friction, whether it is a mechanical friction or the friction of resistance to an electric flow in a conductor. Hence, we can deduce that it is friction (the electronic interatomic forces) which are modified in the Poltergeist activities. In fact it is more than this; the atoms themselves are "converted" into their electronic wave characteristics. (> *'Bose-Einstein Condensate'*: 430 A).

NOTE 1: there is no need to know whether the bonds are ionic bonds, co-valent bonds, metallic bonds, or what-have-you. In the world of quantum mechanics all such considerations - useful in chemical technology, will become obsolete. In this new physics of quantum-electronics, all things are reduced to their utmost simplicity so that much of today's complexity will become unnecessary. In other words, a bond is a bond, and is fundamentally electronic.

NOTE 2: you may ask, "but what about gravity, - surely gravity holds the cup on the table?" the answer is yes and no - but in this case, friction is by far the more important force. The oceans contain many currents which go this way and that. The cause of these currents is the *Coriolis Effect* and is caused by the rotation of the planet. Hence, we find that this Force overcomes both gravity and friction in order to move great masses of water. If there were no friction binding the cup to the table, it would go drifting away due to the Coriolis forces, irrespective of gravity. Hence in its effect, friction is a far greater force than gravity.

121 B ~ GELLER'S FREQUENCIES

"... psychic healers have displayed the ability to mummify meat within a very short time with some sort of "X" force that seemed to flow from their hands. The phenomenon has attracted the attention of many scientists throughout the world, such as *Dr. Peter Kapitsa* in Russia, *Dr. James Errera* of Belgium, *Douglas Dean* of Newark College of Engineering". ('Psychic Discoveries Behind The Iron Curtain': Shroeder and Ostrander).

> These findings were also repeated by Soviet scientist *Inyushin*. His discovery confirmed that the _eyes_ *emit radiation* of a type which could penetrate a thin sheet of metal such as aluminium foil and which would register on a photographic emulsion sensitive only to *Ultra-violet light*. (My italics). (This would of course imply that the eyes emit a radiation or energy which has this range of frequencies).

Likewise, the famous Russian animal researcher *Vladimir Leonidovich Durov* became convinced after performing ten thousand experiments from 1923, that the eyes emit a type of ray or energy. (See the works of *Von Reichenbach. Pythagoras* also believed this). [see note 1 below].

"Kirlian photography" has shown that people like *Uri Geller* and "faith-healers" can project an "electrified" energy from their bodies and eyes. This technique also shows that the intake of <u>oxygen</u> causes the 'aura' to brighten. ('Psychic Discoveries Behind The Iron Curtain',...).

Geller made fame in the 1970's with his ability to bend spoons, keys, and other pieces of metal by the power of his mind or will. He was able to do this feat not only in close proximity to these objects but also hundreds of miles away from them, and even when they were hermetically sealed in glass bottles. (Some bottles were even sealed by melting the glass). This indicates that this Energy is also manifested in wavelengths which are able to penetrate glass.

Since glass is opaque to electronic frequencies in some ranges of ultra-violet light, it is evident that the "spoon bending" Energy is not in this range of frequencies. However glass is not opaque to other frequencies, which include visible light, some frequencies of Uv rays, and certain frequencies of infra-red, X - rays, and beyond.

It is therefore likely, based on what we know already, that the Energy he produces is in the higher frequency ranges, i.e. ultra-violet and beyond.

In the case of spoon and metal bending, researchers have concluded that metal samples having crystal structure dislocations are more easily bent. This in turn would indicate that the Energy frequency is working on an atomic or electronic level, and specifically on the atom's valence (outer electron) shells, - and the bonding forces uniting them.

(Metals such as brass (keys), or bronze, etc., would lend themselves more readily to this kind of manipulation since they are both alloys of different metals (copper and zinc, and copper and tin respectively) and therefore have many such atomic-crystalline dislocations). It is possible that alloys, being composed of a variety of atoms - respond more readily to a range of Energy frequencies.

Of even greater interest was the ability of Geller to cause or to be the agent of Poltergeist - type phenomena. These occurrences would be causing objects to <u>disappear</u> and relocate in places far removed from his own location, or causing objects to float in the air and <u>pass through walls</u>, seemingly unimpeded.

(Most amazingly also is the ability of Geller to actually cause the apparent reversal of growth. While holding a bean sprout for thirty seconds in his hand and concentrating on it, the bean sprout reversed its growth process and became once more a complete whole bean*!*

Now if this is true, and it is a documented factual occurrence, this then opens up a whole new field of speculation as to the nature of life processes, the flow direction of time, and of course the fundamental nature of reality itself. (Explored in detail in section 9000+). [note 2].

His watch-mending and metal bending, etc., never happen unless other people are present. He believes *he draws on their energies*. This is a belief shared by many orators, faith healers, magicians, and illusionists. The Life Energy generated and focussed by the attention of an audience of hundreds can and does actually work miracles.

> Some, who spend much time with Uri report a sudden weight loss … It is known that certain psychics *lose weight* during trance. Certainly, in both cases, energy is consumed… Uri 'repaired' a portable electronic calculator for *Dr. Werner Von Braun* and broke the scientist's wedding band *simply by staring at it*. (My italics). (taken from 'Uri Geller', by Pat Silver and Jesse Lasky Jr. : - 'Men of Mystery'; Colin Wilson).

Just as astounding is the report of hundreds of incidents where the mere broadcast of a Geller live demonstration on television caused the silverware of viewers to bend. Hence, we see that quantum Energy may be projected and broadcast via television transmission (i.e. microwave radiation). (This will be further explained elsewhere). This radiation is similar to Radar. (In fact television is a direct descendent of Radar).

Upon analysis of this scenario, we can perceive that there are two possible mechanisms for the duplication of the quantum effects at the secondary terminals (i.e. the television viewers). Initially we may suppose that it is a form of combined radiation which follows the d^2 rule since it has a single source point and multiple effect points (the viewers).

Quantum Energy is combined with the broadcast Em radiation which diminishes in intensity with the d^2 rule. Hence the effective range of the Energy would be similar to that of the TV signal. However we may suppose that there is another mechanism at work. It is likely that Geller himself is both generating and broadcasting the Energy he creates.

This broadcast Energy field finds a resonance with the television viewers and if they are suitably responsive, i.e. in agreement with Geller's abilities, they are stimulated to *themselves* generate unknowingly (subconsciously) and direct the Energy to obtain the resulting effects. This can be referred to as "sympathetic resonance and rapport".

Here then we have another case of where a single effect can have several causes, and where this effect will not accrue without all the causes in operation. This is a common quantum mechanics activity.

This secondary "re-emission" of q. Energy would be very similar to the *Huygens Principle*. Indeed we may well assume that it is in fact the same principle. Hence we have postulated two possible mechanisms for the cause and effect of this phenomenon.

We may not assume that quantum Energy situations behave predictably like those we are familiar with. In fact in the world of quantum physics, the cause of observed effects is just as likely to be multiple and collective as singular. In other words, both mechanisms are together likely to be the source of the observed effects. That is to say, that one cause is not enough, there must be multiple causes!

One may draw a simple parallel here. If you go into a coffee shop for a cup of coffee, the coffee exists and you can see it. You have the money to pay for it, and you can talk to the waitress. However before you can get your trembling hands on your Dolce Vita Ultissimo Full-bodied Arabian Mocha Latte with nutmeg, Swiss chocolato shavings, and whipped cream, - six criteria (causes) must be in place. You must be in the coffee shop, there must be a 'barista' in attendance, you must ask for and describe the desired item, you must pay for the coffee, the coffee must be available, and the waitress must bring it to you. All six criteria (causes) must be in place before the desired effect transpires. In fact there are many more criteria, such as the availability of Swiss Chocolato shavings: [see note 3].

NOTE 1: it is a well-known phenomenon - especially to school children, that if one stares at the back of an individual's neck, the subject will shortly become uncomfortable and will eventually turn around. You can do it too. This is a concrete example of the power exerted by a projected beam of Energy from one's eyes.

NOTE 2: in the physical universe, time moves only in one direction, albeit *at varying rates*. This is an immutable and inviolable law, - this is the way the universe exists and functions. If something appears to go back in time, this is a created illusion. It is quite possible that Geller believed he was turning back time and was duping himself into believing it. It is far more likely however that he was simply modifying the geometry of the plant to make it <u>appear</u> to reverse time. In other words, he was causing a shape-shifting phenomenon: (to be discussed).

Furthermore, it is unlikely that an isolated volume of space-time could be made to reverse while he and the onlookers were passing through time in the normal fashion. As will be seen throughout the book, time can only move in one direction – when it moves at all! By the exertion of electronic technology, it can be made to stand still in local vicinities, but *time cannot be reversed*. This is an immutable law of the universe, and is apparently so at all levels. (to be explained).

NOTE 3: actually, this quantum behaviour concept of having many causes to produce one effect is really very common. For example, if you want to boil water with an electric kettle, several causes must be in place to produce this one effect. For example, you must first have a kettle, the kettle must be functional, there must be water in the kettle, the kettle must be plugged in, there must be electricity, the switch must be on, etc..

Quantum forces are associated with quantum Energy, i.e. 'Life Energy'. when Life interacts with matter, it is inevitably a creative process.

> 'Disbelief and ignorance is the result of the suppression of knowledge' (author).

> "it is against reason", said Filby. "What reason?" said the Time Traveller. (H. G. Wells).

Most of us I believe have read at one time or another of cases where people have had strange and unusual - even "spooky", "creepy", or weird experiences. Perhaps you yourself have had one but were afraid to mention it to your friends for fear of ridicule and ostracism. Nevertheless, as you may know, such strange occurrences do indeed happen.

Examples are the unusual phenomena associated with *Poltergeist* activities, *Spontaneous Human Combustion*, *Ball-lightning*, lightning from a clear blue sky (the fabled "Bolt from the Blue"), people who have had cancer go into remission or have re-grown hair and regenerated lost teeth, or even limbs, or who claim to have had 'out-of-body' experiences, etc..

It is acknowledged by some - and in certain fields of research, that a few people are able to make small objects move without being touched by hand or influenced by any physical force. Let us suppose for the moment that such things do belong to the rational world of our mundane experience, and as such, constitute a legitimate part of our own so-called reality. Very well, - starting with this hypothesis, let us admit that someone is able to move a small object with the power of his mind.

Now we must consider a number of other things. If such is the case, we must accept without question that he is able to generate some form of energy with his mind or will which interacts with a physical object, - and by extension with the physical universe in general.

We may not immediately know what form of energy this may be but we can suppose that it is perhaps related in some way to an energy form that we may be familiar with, such as *"Chi"* or *"Ki"*. We do know that *Chi* is related to the mind and is generated by it. We also know that it can interact with matter, - in this case the biological body - to bring about its healing and even the regeneration of failed organs and limbs, etc.. [note].

Since the biological body is ultimately a part of the physical universe (being composed of atoms and molecules, etc.), we must then suppose that the mind can interact with the physical universe generally, using the created Energy as the via of manipulation.

In regards to Poltergeist phenomena, researchers have demonstrated that objects can be caused to levitate, to move through the air, and even to pass un-hindered through walls and other barriers to emerge unscathed on the other side. Objects have even been caused to disappear completely and re-appear at another location, perhaps Kilometers away or even in another country: (the ultimate Fax*!*).

Ball-lightning may also be classified as a Poltergeist phenomenon since it has been observed to appear where a human being was the likeliest source. Now if a mind or will is able to create an example of ball-lightning (i.e. a collection of energized particles, that is - a *plasma*, etc.), we must admit that it is capable of creating matter - for basically that is what it is.

We are now faced with examples of minds/wills creating energy forms and objects. Since objects are made of matter, and since matter itself is composed of pure energy, it is now reasonable to say that the mind or will can indeed create matter and energy. (In fact, the mind *per se* is not directly involved with the creation of this Energy as will be shown). Since matter is composed of energy and (mostly) *space*, we must therefore concede that the mind is capable of creating space also. Furthermore, since space is actually a composite of space and time, we are moreover obliged to admit that the mind (or whatever it is) can also create time*!*

Man has for thousands of years questioned the origin of the physical universe, the creation of matter, energy, space, and time. We have before us an intriguing possibility as to the origin of at least some matter and energy. Matter itself is composed of both energy and space-time. As we already know, *time is an illusion created by motion*, - i.e. energy, or matter plus energy. Hence we can say for the moment that evidently, the possibility exists for the mind/will to create matter, energy, space, and time. (Much more will be said on these intriguing possibilities later).

NOTE: it has been demonstrated - even on national television - that this 'mentally' produced energy can interact with material objects, as in setting fire to a piece of newspaper, etc.. [See also 'The Amazing Kreskin'].

122 B ~ STRANGE ENERGY

> "Occurrences which according to received theories ought not to happen, are the facts which serve as clues to new discoveries", (Sir John Herschel).

> "the only things we know about the quantum world are the results of experiments", (from 'Paradoxes And Possibilities').

Energy manifestations considered in this work are some of the following: psycho-kinesis (moving objects or influencing energy flows such as electricity by unintentional thought or mind power), telekinesis (moving objects at a distance intentionally), dowsing, levitation, Poltergeist activity, merging, so-called "spoonbending", "pyramid energy", Orgone, Odic force, ball -lightning, etc..

Today we have many books and articles given to us by researchers into these various subjects and in many cases, these researchers have lived in the dim corridors of history. All of the above listed subjects are relegated to the fringe areas of science.

Only through the hard and often unrewarded (and sometimes harassed and otherwise punished) work of private individual researchers are we permitted to read about them and enrich our knowledge of the underlying strangeness of our universe.

You will notice that when reading the aforementioned list, that all of the listed subjects are concerned with unusual forms of energy - in fact, forms of energy which are generally unfamiliar to us. The question to ask therefore is; - "can it be possible that the strange energies involved in these phenomena are related, - and furthermore could it be that they are all one and the same class of energy form. Are they related to the elusive 'Chi', or 'Ki', 'Orgone', 'Kudalini', 'Odyll', 'N-rays', 'Formative Energy', 'Serpent power', 'Vril', and so on?"

The world was astonished to learn of Hertz's discovery that there exists an unseen influence which can travel through space to create a spark in a wire coil some distance away from a sending apparatus; i.e. the so-called 'Hertzian waves' which today we call radio-waves.

The fact that there may be undiscovered or unusual energies should not seem so fantastic in light of the fact that physicists of today will admit that even ordinary fire and electricity are not entirely understood, and that the electro-magnetic spectrum itself <u>includes an unexplained gap.</u>

Likewise, there are aspects of electricity which are truly unusual. In 'Fate' magazine of Feb. 1955, there is a case reported by *Mr. H. M. Cantrell* and confirmed by *Dr. J. E. Taylor*. In the case of *Mr. Thomas Young* of Dukedom, Tennessee, who was dying of cancer. One day while lying in his hammock under a tree and a cloudless sky, one of these trees was suddenly struck by a bolt of lightning. A few weeks later it was discovered that *his cancer was completely healed* !

Other cases are on record where people have experienced lightning strikes or near misses and lived to report amazing results - such as the blind regaining their sight, or the deaf their hearing. There are other tales of missing hair and teeth being regenerated, and so on.

Another case in point is that of *Edwin E. Robinson* of Falmouth, Maine, as reported in the Japan Times of July 06, 1980, and confirmed by *Dr. William F. Taylor*. After being struck by lightning, the man regained his sight, partial hearing, and the once bald man has now a regenerated growth of hair. The libraries are replete with such casebooks.

The connection between health and electricity is enormously well documented. *Dr. Cesar Romero-Sierra* who is or was a professor of anatomy at Queen's University in Kingston, Canada, has developed a band-aid having two thin metal strips which act as electrodes. A small electric DC current is passed through the wound. He claims to have produced the same healing results in only twenty minutes using this device that would normally have taken up to two days.

Daniel Dunglass Home (1833 ~ 1886) became famous for his unusual abilities and is listed in the British 'Dictionary of National Biography'. As reported by the *Earl of Crawford* in the 'Report of the Dialectical Society of England', "Home was able to handle red hot coals and could pass on this immunity to others".

His other unusual abilities such as levitation (for which he became doubly famous), was reported by the also famous *Sir William Crookes* (scientist) in 'Proceedings, Society for Psychical Research', Vol. VI.. Crookes himself stated that Home was also able to make his body <u>luminous</u>, as reported by *Lord Adare* in the 'Encyclopedia of Psychic Science'. (See *the Moses Effect*). (As will be explained, the two phenomena - luminosity and levitation, are often associated).

Home's fame spread far and wide and permitted him to meet with other such greats as *Alexander Dumas, Count Alexis Tolstoy, Sir David Brewster, Napoleon III*, the *King of Bavaria*, the *King of Naples*, the *German Emperor Wilhelm I, Empress Eugenie, Alexander II - Czar of Russia*, and *Queen Sophia of Holland*.

E-energy is simply a higher form of ultimately physical energy which interacts with physical matter and energy as we know it. It also interacts with the <u>mind</u>, as explained. Emotional thought is that simpler type of thought process exhibited by children, insects, animals, etc., and includes such characteristics as curiosity, affection, disgust, aversion, desire, anger, and playfulness, etc.. These are the results of emotional thought.

Generally speaking, children, animals, insects, spiders, and what-not, apparently do not have a developed analytical mind such as do adult humans. We are a specialty. This can be determined scientifically by measuring their emitted 'brain-waves'. Adult humans have a well developed ability (through education) to generate Beta frequency brain waves. This is the frequency which is indicative of, and is associated with precise analytical thought. Children and animals on the other hand, generate predominantly Delta and Theta brain-wave activity. Theta waves especially are associated with dreaming, "hallucination", and creativity.

122 BB ~ BIOLOGICAL MASS AND LEVITATION

One is usually inclined to think of a biological body as a special kind of matter since it is 'alive'. The only thing which makes it different from so-called 'dead' matter is the rate of inflow and outflow of quantum Energy (and of course being handled and manipulated by an operator). Being in the state of 'aliveness' makes it habitable and operable as a useful machine. (One can have fun with it too!).

Largely this book discusses levitation of "dead" objects such as coffee cups, etc., but what about the levitation of living biological bodies? We have touched briefly on examples such as *D. D. Home* and the bounding lamas of the Himalayas. Apparently, the human will has the power to generate sufficient quantum Energy to permit a heavy body to be lifted into the air in the same manner as we find in cases of Poltergeist activity.

Essentially and in a physical sense, there is no difference between the various kinds of atomic matter except our perception of it. Whether we are discussing a stone, a tree, a jelly-fish, a clam, a piece of metal, a glowing coal, - whatever it happens to be, - we are talking about atomic matter, i.e. a group of atoms. Any atomic matter is subject to levitation, etc..

For the most part, all vegetation and animal life is connected to the ground, i.e. the planet. The Earth generates q. Energy and this Energy is therefore shared between the ground and the body (and lakes and seas, etc.). This means that the Earth gives us Energy through this connection but it also takes it away so that any surplus which we may have shortly is absorbed by the ground. In other words, we are "grounded". In this way we share a common bond with the planet and all of our fellow men (and women - which goes without saying, obviously).

If we should break this connection, we would as life units be independent generators of Energy. Hence if we feel good while in this condition, we should feel better, - and if we should feel gloomy, our spirits should be raised. Furthermore, we should experience an improvement in health over a period of time.

This improvement in feeling is generally experienced by air travellers. Furthermore the aircraft, as it generates friction while moving through the air at high speed, is also a powerful generator of Energy: [see note]. If the living body could be suspended in the air with minimal or no support, it would be free to generate and accumulate its own charge of Energy.

This does occur in the case of sky-divers and balloonists, etc.. As elsewhere described, this results in a feeling of euphoria for the individual and furthermore, it has been found that for a short time afterwards he is actually a few kilograms lighter!

It is suggested here (but not proven) that if one were to suspend himself for say a quarter or a half-hour every day from a long nylon rope, he would feel an improvement in mood and in health. In the absence of a rope, one should at least try to rest in a hammock suspended between two trees (another source of Energy), and preferably outdoors.

Another good way to generate Energy for the body is sex, especially where the emotions are involved. I am definitely not advocating random sex with anyone since this can generate bad Energy and can actually be bad for the health, both mental and physical - but sex with true feelings of affection is a marvelous therapeutic mechanism. (*W. Reich*).

NOTE: the same is true of ship-board travellers for different reasons. In the first place, the ship is packed with organisms full of life Energy. The travellers on board are generally in a good and relaxed mood, and this mood and Energy is shared by all. Secondly, there is the friction of the metal hull against the water. The third reason is that the ocean or sea presents a very flat geography for many miles. Quantum Energy - just like static electricity, seeks to travel upwards to high points such as mountain peaks and so on. The ship represents a high point on the flat ocean and therefore attracts both static electricity and q. Energy. Hence, the ship is charged with a greater than normal accumulation of Energy. For these reasons ship-board life is generally healthier and happier.

122 C ~ QUANTUM ENERGY AND MODIFICATION OF THE MATERIAL

There are actually four general methods by which matter, energy, space, and time may be modified. They are progressively mechanical, chemical, electronic (electromagnetic) - and now as we shall see, Quantum-electronic.   It is fairly obvious that matter may be modified mechanically, chemically, and electronically: - this is being done on a daily basis in factories around the world. Energy can be created and modified mechanically and chemically as with explosives, fuel cells, batteries, etc., and of course is also modified with electromagnetic techniques.

Space can be modified and altered mechanically simply by re-arranging the material components which define and delineate the geometry of the space in question. A housewife (oops, - "houseperson") does this every time she moves a piece of furniture or indeed any household item.

Space may also be modified electronically and electromagnetically as with the incursion of electronic energies. This will be explored further. Time likewise can be modified and manipulated since space and time form a matrix, i.e. a continuum or "aether". In fact, time cannot be altered without also altering the fundamental characteristic of space, and vice-versa. This matrix exists in a way similar to the compound of water which exists due to the combining of hydrogen and oxygen. By eliminating either of these two, the principle ceases to exist.

Space and Time (and its rate of flow) can be modified simply by the presence of matter and its accompanying gravitational field. When space is thusly modified, time is likewise modified as a direct consequence. When space-time is modified, one of the first notable anomalies will be a *change in the rate of time flow*. (> article 42-152).

Now as amply discussed elsewhere, all matter is essentially composed of various types of particles which at their most fundamental level, are nothing more than "trapped" quantum Energy (to be expl.). All forms of energy - be they mechanical, chemical, electronic, gravitational, and inertial, etc., are fundamentally only variations and products of quantum Energy. It can be argued that space itself is a quantum Energy field. (> 42-150; *Space-Time is Particulate and Incremental*).

Hence we see that everything that exists is in some way or another essentially an expression of electronic energy. Since this is so, it follows that all things will yield to and be modified by a correct application of such energies. This means inevitably that we may thereby exert greater control over our material environment through electronic technology.

This will include control of such factors as pollution of all types, soil conditions, climate and rainfall, etc.. It is also conceivable that control may be exerted over the winds and even earthquakes by the simple and correct use of quantum-electronic fields. This control will also extend to materials technologies, and here we will find the most amazing and exciting effects including levitation, invisibility, merging, liquification, etc..

123 A ~ RESEARCHERS IN HISTORY, AND ENERGY FREQUENCIES

Wilhelm Reich was a researcher in the 1930's and 40's who spent much of his life researching his discovery of an extraordinary form of energy which he called 'Orgone'. He even exchanged letters with Einstein on this subject. He stated that this Energy would flow from areas of low potential to areas of higher (in a manner similar to gravitational attraction).

He also claimed this Energy could be projected through the atmosphere by means of a device to influence weather patterns and cause clouds to disperse or accumulate. (In fact, he actually built several devices called "cloud-busters" to work with in his researches, (see *Reich's Cloud-buster*).

(It is believed by some researchers that large populations can, by thoughts and emotions, wittingly or unwittingly, influence weather patterns. In fact such an experiment was performed in 2003, evidently with some success it is claimed, by a popular late night American radio personality with the enlisted help of his listening public).

In any event, the dynamic intermediary is E-Energy. *Reich* also claimed that metals and magnets have a strong attraction for Orgone. (Which is probably why many people intuitively find magnets fascinating and are "attracted" (!) to them. As we shall also see, magnetism is sometimes very much associated with quantum Energy. (Evidently, olive oil has a superior ability to absorb and retain q. Energy, which reportedly will remain in the oil for as long as five to ten years).

Baron Von Reichenbach, another energetic researcher of the nineteenth century, found that materials subjected to a magnetic field increased with an intensity of the energy he called 'Od', or 'Odyll', or the 'Odic Force'. Further, he stated that this Energy could be seen (by 'sensitives') to flow along a conductor, e.g. a silk string, from the hand holding it. (Evidently any form of matter is capable of conducting this electronic 'fluid').

Reich also claimed that Orgone functions in a way similar to that of biological organisms in that it pulsates and seeks out recesses and hollow spaces. It is also attracted to bodies of water, moving water, fire, warm places, and other sources of Orgone such as other biological entities, etc..   Again we are reminded that it is not so much the animal, incidentally imbued with q. Energy, who seeks out these natural good places. Rather it is the Energy existing as the active influence, guide, or 'instinct' in the animal and its sub-conscious mind.

One of Reich's major findings was that Energy behaves in a continuing cyclic function with four stages of activity, - Accumulation or Charging, - Tension, - Discharge, - and Relaxation. (See diagram 103). Orgone is emitted from a source but when inside a closed space, tends to accumulate at a point coincident with the geometric and/or gravitational center of the enclosure - and is seen to pulsate in brightness.

When an accumulation of any energy form pulsates, it indicates that the charge or flow is increasing then decreasing in a cyclic manner. Such energy may be a high frequency electric discharge from a capacitor or an electro-static discharge from a Van De Graaff generator, for example. Other examples are a beating heart, alternating current electricity in a wire or transformer, a breathing organism or animal, etc.. In short, *most if not all dynamic activity in the universe is accompanied by inherent pulsation as a vital part of its function.*

So it is with q. Energy. The charge or accumulation first increases to a maximum critical or threshold level where environmental forces cause it to spontaneously discharge and begin the cycle anew. (As will be seen later in the text, this also occurs in atoms and in certain other types of electronic flows). For example, Energy is attracted to magnets and metal, etc., and so an accumulation begins.

This cycle of charge and discharge however does not operate with a 'sine-wave' pulse activity. The activity tends rather to charge or accumulate over a given period of time and then discharge in a fraction of that time. If then this cycle were to be delineated on a graph over time, it would be found that the cycle presents rather a "saw-tooth" profile having a 'forward bias': (See diagram/graph 103). This same activity occurs within a storm cloud during the accumulation and consequent release of electrical energy (i.e. lightning). The electric field of the cloud first accumulates slowly then rapidly collapses. (Diagram 06).

Usually there is in the environment an observer or some other source, or greater accumulation of Energy such as a tree, a body of water, a water-pipe, or a stream, etc.. The Energy is also attracted to these greater sources and when the Energy has reached an accumulation limit or threshold in these biological, metal, or magnetic objects, etc., it then "loses interest" in the object. It discharges and flows or 'radiates' away, thereby producing a cyclical pulsation in the metal or magnet, etc.. The same 'forward bias' occurs in other biological cycles such as the slow accumulation of hunger and the rapid assuagement of it. Another example of this dynamic geometry is the inflation of a balloon. It takes a given period to inflate it but a fraction of the time to deflate.

Reich gave his discovery several names, including, "Primordial Cosmic Energy", 'Life Energy', 'Bio-energy', and 'Formative Energy' (see Appendix). He also stated that Orgone, which he first discovered in 1936, was also present in crystal formation.

(*Nicola Tesla, George Bernard Shaw,* and *H.G. Wells* also believed that crystals such as quartz and metallic crystals, etc., were life forms. It is in fact recognized by biologists and virologists that viruses, although they behave like biological entities - are nothing more than complex crystalline structures, having no brain or nervous system! In other words, they are a completely and truly alien life form consisting of simple animated crystalline structures). (Animation will be covered in depth elsewhere).

While working on Orgone research, Reich found that the Energy emissions gave him conjunctivitis and tanned his skin in a manner similar to short wave ultra-violet rays. Further, he found that the radiations would cause his rubber gloves to become electro-statically charged, (another effect attributed to short ultra-violet rays). (The radiations of Orgone are blocked by glass but allowed by quartz, - again like high frequency ultra-violet light). Hence the Energy emulates frequencies of light. [note 2].

Interestingly, it is known that ordinary light, when reflected from glass (as are certain wavelengths of Orgone), contains a disproportionately large amount of very short wavelength Uv light. Further, the higher the frequency of Uv light (including of course q. Energy), the more it is reflected. A mirror then reflects not only visible light but also virtually all of the q. Energy 'radiation' which strikes it. This is perhaps why mirrors have a peculiar fascination for the occult minded, and fantasy writers such as Lewis Carroll and 'Alice In Wonderland'. (> 130: *The Crystal Ball*).

Because glass blocks the passage of q. Energy, those people remaining indoors behind windows cannot receive the real benefits of sunlight. If it were possible to have these windows made of (fused) quartz panes, these benefits could then be received. (However, the admitted Uv radiation would not be so beneficial to furniture, clothes, paper, pictures, etc.).

> Because q. Energy interacts powerfully with either free electrons or those bound in electron shells, they excite energy levels in the valence and deeper electron shells of the atoms which then emit high-energy Uv rays. The highest frequency of light emitted by an atom is about 4×10^{15} Hz. (i.e. 4,000,000,000,000,000 cycles per second) or a wavelength of about 1000 Angstroms (i.e. "hard" Uv light). So by simple mathematics it can be calculated that q. Energy in the wavelength range of 1 A. would have a frequency of 4×10^{18} cycles per second. (The frequency of electrons is forty times higher than this, so it can be seen that the frequency range of quantum Energy is from 4×10^{18} to 1.6×10^{20} cps).

Orgone also stimulates enzyme activity, aids indigestion, increases libido, causes de-hydration (hence "mummification"), retards oxidation, and neutralizes acidity (in water, etc.). It also has the ability to cause a slight expansion of a volume of mercury, alcohol, or water and other liquids as if they were being affected by heat. (Hence, a thermometer placed in the vicinity of an Energy source will show an apparent increase in temperature).

Water treated in this way can be purified since the expansion of its molecules and Van Der Waals' forces allows any impurities to precipitate out as sediment.

This could of course also be applied to wine preparation, etc.. (In fact the benefits of the ageing of wine so lauded by connoisseurs may in fact be affected instantly with electronic radiations).

This will be a bone of contention for purists, no doubt - but the fact remains that we are here discussing the very fundamentals of the physical universe and the manipulation of <u>time</u> itself, - so blow the ageing! [see 3].

< other energies in history >

Another form of energy in this genre has been called 'N-Rays' (Scientific American, May 1980). These unusual rays were first discovered by *Rene Blondlot*, a physicist at the University of Nancy in 1903. He found that his so-called N-Rays had characteristics much in common with other unusual energies herein discussed. Other discoverers were *Gustave Le Bon, P. Audolet*, - and the existence of these rays was also confirmed by twenty other physicists, including the famous *Becquerel* and *Charpentier*.

Another original discoverer called his energy 'M-Rays', or 'Mito-genetic Radiation'. Some sources of these rays were discovered to be freshly cut garlic, onion, and ginger, (the so-called miracle herbs). These rays were also found to be blocked by glass and permitted by quartz (again like Uv rays) and are able to stimulate biological activity.

According to *Von Reichenbach*, other effects of his 'Od' are as follows: -When bells are struck and made to ring they would glow - as observed by his sensitives. The brightness was proportionate with both the amplitude (loudness) and the pitch or frequency. Since smaller bells ring at a higher pitch, they also emanate a brighter glow.

The evidence presented here and elsewhere strongly suggests that Orgone, N-Rays, M-Rays, Od, and other names given by various researchers to this mysterious and elusive Energy are all equated with one and the same form of dynamic energy. Furthermore, that it is actually an electronic wave form having a specific range of frequencics parallel to the electro-magnetic spectrum. Further evidence of the existence of these radiations/Energies is statements made by prominent physicists.

S. Millikan himself proved that there is a "mysterious ray" which passes from molecule to molecule. Biologist *Kammerer* stated, "the existence of a specific life force seems highly probable to me, ... "it is present certainly in the formative process of crystals"

Finally *Dr. Albert Abrams*, a San Francisco physician at the turn of the century discovered that the mere proximity of a diseased tissue can affect a healthy one. Evidently the disease is broadcast like an <u>electric wave</u>.

Some students and practitioners of 'Radionics' and 'Radiesthesia' (e.g. dowsing) are also familiar with a subtle energy or ray. They claim that these emanations from such substances as medicines, biological substances, and inorganic matter, etc., are sufficient to bring about a desired (biological) effect rather than by ingesting them.

(If a screen is constructed containing a solution of the substance and then placed in front of an 'emitter' of some form of electro-magnetic or particle radiation, and the whole apparatus then being 'aimed' at an infirm body, measurable effects over a period of time are manifested relative to a scientific control).

In Switzerland and Italy, a device designated 'RS-25' is used in public hospitals to affect cures. It is simply a very weak source of radioactive particles "beamed" at a patient through a filter soaked in medicine. The output is very weak and calculated to be 1.2 milli-roentgen per hour or less. Doctors who have used it are very enthusiastic with its success and believe the healing effect stems from secondary radiations of unknown nature. ('The Cycles Of Heaven', Playfair and Hill).

A similar device may easily be constructed in the home. Older television tubes are known to emit high-energy electrons from the screen as well as mild X-rays, and this radiation conforms to a cone-shape covering a wide sweep of the room due to the geometry of the tube and its function. The tube therefore is an emitter of radiation particles and may consequently be substituted for the device mentioned above. Fluorescent tubes also emit ultra-violet light and mild X-rays. [see note below].

> "You can recognize truth by its beauty and simplicity", (Richard Feynman, Nobel Prize winner in quantum electro-dynamics).

NOTE: in all probability, the healing process does not necessitate a source of particulate energy to carry the Energy to its intended target. It is highly likely that any radiation, such as sunlight streaming through the open window will do a perfectly adequate job.

In other words, it is only necessary to hang a cloth soaked in medicine, etc., over an open window on a sunny day to affect a medical change! As you can see, this technique is beginning to sound like remedies of the Middle Ages! Maybe the ancients were not so stupid after all!

(Researchers of alternative methods of treatment such as the above mentioned homeopathy, reflexology, acupuncture, etc., have been obliged to conduct very thorough and intensive research since they are in competition with a very powerful and well entrenched medical establishment).

NOTE 2: here we are presented with three examples of an electronic radiation behaving just like short Uv light - an electronic radiation. All three examples given are phenomena describing surface activity. This is typical of short-wave Uv light. We are not talking about sound waves or radio-waves or a saucer of milk. It is fairly obvious that these effects must be the result of Uv light itself, *or another electronic energy form having the same frequencies*. Nothing else will produce these same effects.

(Incidentally, over seven hundred medicines now in use by the medical community have been selected by the method of <u>trial and error</u> with no understanding of their biological reactions).

NOTE 3: quantum Energy is intimately involved with time, ageing, and so on. The mystique of ageing wine is based partly upon the observation that ageing in humans is accompanied by maturity and wisdom. Q. Energy has a profound effect not only upon wine, etc., but also upon the mental condition and abilities of living creatures. Hence the increase in maturity can be applied in more senses than one. (More on this later).

123 B ~ REICH'S CLOUD-BUSTER

Wilhelm Reich developed several devices designed to control and manipulate 'Orgone', i.e. q. Energy. One of these devices was his 'Orgone Accumulator'. This was a specially constructed box having each side made with alternating layers of conductive and insulating materials, - and has been discussed. Quantum Energy would be absorbed by the outer cover of insulating material from the surrounding atmosphere, conveyed to and accumulated within the cavity. Sensitive fluid type thermometers would indicate that the 'temperature' within was higher by one or two degrees than the surrounding air.

Although a thermometer is designed to measure the degree of heat energy, its principle of action is based on the expansion and contraction of a volume of liquid, usually alcohol or mercury. Since Energy is also a primary source of molecular expansion, it will affect this behavior of matter, - especially water. The thermometer may therefore be used to indicate the presence of q. Energy. In any event, Energy is the Primary Source of all other energy forms - through the via of matter.

Another device created by Reich was his so-called "Cloud-Buster". This was an arrangement of long copper pipes clustered together like a bunch of drinking straws, but much larger. This device was then mounted on a tripod and aimed at the sky. The lower end of this cluster was capped and connected by a metal cable to a source of running water.

Reich had determined that the moving water was a source of Energy and that this Energy would be attracted to the metal cable and then to the copper tubes. In turn, this cluster of tubes aimed at the sky would conduct this active energy and project it into the sky, as if the tubes were shooting light beams. This experiment was repeated thousands of times by Reich and others and has since become an iconic device to those who research his work. Apparently and evidently, the projected Energy results in the formation, dispersal, or 'desiccation' of rain-clouds, depending on how the device was manipulated. The dispersal of clouds in one portion of the sky would result in the formation of rainclouds in another part.

By manipulating the copper tube array, he was able to cause an accumulation of clouds in whatever part of the sky he wished. Accumulation of rain-clouds could be developed to such a degree that actual rainfall would be produced at will*!* Another method would be to substitute a brazier of glowing coals for the running water, since this too is a generator of Energy.

Knowing what we now do about the behavior and characteristics of the Energy, we are able to update this device by substituting electronic technology in place of the running water.

124 A ~ ENERGY THRESHOLD PHENOMENA

An electromagnet is designed to become a magnet when an electric current is passed through its coil. An electromagnet may be suspended above a piece of iron oxide (F^2O^2 - otherwise known as hematite or ferrite) and when the coil is energized, a magnetic field is established between the coil and the oxide. If the current flow is too weak, the magnet will not produce any mechanical effect. If the current flow is slowly increased, the magnetic field will increase in strength.

Eventually a critical threshold point is reached where the oxide is suddenly motivated and drawn to the magnet. (Iron oxide is better for this demonstration since it will respond immediately to the variations in the magnet's field. (This is why the antenna in a transistor radio is usually made of iron oxide, as is the composition of recording tape).

This sudden motion may be likened to a quantum effect. In fact, it _is_ a quantum effect! If the electric current in the coil is too strong, the coil will overheat and be burned out. Likewise when the coil is operating, the iron oxide is attached to the magnet. If then the current is slowly diminished, there will be another threshold point in the procedure where the magnet will suddenly let go the iron oxide. One can readily observe here then that threshold behaviour, the square-wave ("on-off") phenomenon, and quantum mechanics are related. [see note 1].

It can be seen also that there is a "zone of operation" or "window of efficacy" where the current is of a correct and proper value or strength to perform its task adequately. Above or below the flow parameters, the electromagnet will not function properly.

Similarly if there is not enough ambient light, we will not be able to see objects. If the light is too intense, it will be too bright to see anything. Hence, it is understood that there is also a "Window of Efficacy" for visual perception. This "Window Of Efficacy" (W.O.E.) abounds everywhere in every situation and we therefore live our lives bounded by upper and lower barriers and "windows of operation".

The same is true in the quantum world and with quantum Energy itself. Too little will not produce the desired results, and too much will produce results which are not desired.

The Light Emitting Diode (LED lamp) is a solid state quantum electronic device. It operates on e.m.f. input values of three to nine volts. In an ordinary tungsten filament light bulb, - as the voltage drops in value, the light bulb will become dimmer so that its emitted light has a decreased intensity. No matter how low the voltage becomes, the filament will always carry a current, whether the bulb can shine or not.

With an LED bulb, the situation is different. Its light will shine brightly at a constant luminosity while the voltage diminishes until a threshold limit voltage is reached. Below this voltage, the LED suddenly ceases to operate, the electron flow ceases and the light shines not. In effect then it performs like a voltage-sensitive switch: [see note 2].

Energies associated with quantum phenomena are also associated with threshold phenomena. This threshold limit works like a naturally occurring switch, and is of great value in computer design. Quantum Energy when emitted from a point source, will also obey this threshold rule.

If the source generates quantum Energy at a given power density, the Energy is emitted to a threshold limit of distance from the source. Hence the emitted Energy 'extends' a specific distance from the source, the distance depending upon the strength of this source. The Energy thus extends no further and forms a well-defined limit.

This limit is naturally a spheroid one surrounding the source. If the source is located near the ground level, this creates a circular limit on the ground surface. If the source is located in space, this limit would naturally delineate a spheroid like a large bubble, - or indeed a force-field. In fact it would actually be a force-field, - and if powerful enough, this field would become visible, - resembling an actual bubble. (refer to the "Oregon Vortex": 115). (This spheroid is actually an '*Orasphere*' - to be explained).

NOTE 1: a good example of quantum behaviour in materials is displayed by an ordinary piece of iron which has ben exposed to a strong magnetic field. This iron will retain the magnetic field for a short period and may then be attached to a vertical piece of steel. As the magnetization effect wears off, the iron will suddenly 'let go' of the steel. This is due to quantum effects within the iron's atomic structure.

NOTE 2: a similar situation exists in the case of a digital watch. While the battery in a watch is running low, the watch will continue to give accurate time. At some point however, a critical threshold limit is reached where the battery can no longer support the function and the watch suddenly stops - even though the battery still retains a minimal charge.

124 AA ~ ELECTROMAGNETIC and QUANTUM ENERGY

All electromagnetic radiation depends ultimately upon the existence of atomic matter and dynamic (moving) electromagnetic particles. Electrically charged particles are the primary source of Em radiation. However, atomic matter is necessary to support the existence and relative motion of these particles. For example, radio waves are generated by moving electronic particles which are in turn supported by the material conductor, - even if the conductor is a plasma. [See note 1].

It may be said then that matter plus energy is the ultimate source of Em radiation. Now when Em radiation is detected or has an effect, this detection and/or effect is necessarily and essentially manifested by these radiations impinging upon matter and electric particles. Matter and particles then are the essential cause and effect of Em radiation. We may therefore consider that Em radiation travels between the two terminals consisting of particles and atoms: [see note 2].

Without particles and atomic matter, Em energy is not created, nor manifested, nor registered. In other words, without these two - Em energy cannot exist! Between the two terminals - one being the source or cause, the other being the effect, - there is only space (i.e. space-time). This space-time is the bridge which supports or carries the radiation. It may be considered then that Em radiation is the effect of - and is created by matter and dynamic particles in space-time. Likewise then we can say that Em radiation, i.e. energy, cannot exist without space and time also.

We may therefore further consider atomic matter, particles, and space to be cause to Em energy since this energy depends upon these for existence. It is important also to understand the role played by motion and time since time is the effect (illusion) generated by matter and particles in motion. We may therefore consider that Em energy is subordinate to matter, particles, space, and time, and that these are required for the existence of Em energy.

Quantum Energy however is the superior force since it is the source of all. It is the source of matter and hence Em energy, and indirectly even of space since space depends for its existence upon matter (> 9000+). Quantum Energy is generated by matter and dynamic particles but is also subject to generation by the mind/will combination which completely by-passes the requirement for any matter, energy, space, and indeed time!

As we shall see, q. Energy does not depend upon time for its existence.

NOTE 1: this is easy to understand if we use the analogy of a clock. The moving particles are represented by the internal wheels of the clockwork. However these wheels are completely useless without the frame to hold them in juxtaposition so that they can work in harmony as a team. The frame of the clock (the passive element) then represents the conductor which supports the moving particles - the dynamic or active elements.

Hence the static frame and the dynamic wheels represent a dichotomy, and this dichotomy represents the whole dynamic system! This dichotomy brings to mind the old adage that, "the whole is greater than the sum of its parts". (As previously stated - this should say, "the *organized* whole is greater than the sum of its parts". This becomes obvious for when we compare a clock with a boxful of random clock parts, there can be no doubt of the truth of this adage. Another way of expressing this is to say, "organization is greater than randomity"). (Similarly, "the whole is simpler than the sum of its parts"… Willard Gibbs).

NOTE 2: now we see that an interesting relationship appears between matter and energy. Evidently, Em (electromagnetic) energy is secondary to matter, while matter is created from quantum Energy: (more on this in section 2). Hence we have the sequence, - q. Energy > Matter > Em energy. This means that quantum Energy is converted to electromagnetic energy *through the via of matter*. There is therefore no *direct* conversion of q. Energy to Em energy or vice-versa. (Actually as shown later, there are a few cases where electromagnetic energy *can* generate q. Energy).

124 AB ~ TWO FORMS OF ENERGY

Essentially, there are two fundamental forms of electronic energy that we now know of. They are electro-*magnetic* energy and quantum Energy. In our present state of technology, we are very familiar with Em energy. All energy manifests in the form of moving waves which are supported by the appropriate medium. It is the medium which determines the type of wave and its nature.

Em energy is supported by the general space-time continuum and as its name suggests, it is composed of two types of dynamic field - a magnetic field and an electric field. The electric field is the Primary field and constitutes the wave's <u>Function</u>. The magnetic field is the Secondary field and provides the wave's geometric <u>Form</u>. The expanding Em waves assume the form of an *orasphere* which expands uniformly through space. It is the magnetic field which gives the wave this form: (see diag. 25).

The other form of energy is "quantum Energy" which is a <u>purely electric</u> wave having no magnetic component. Furthermore, this electric wave is generated in a different way and has different characteristics. Since it has no magnetic component, it is therefore not obliged to assume any geometric form. This is especially true at the very fine wavelengths between X-rays and those of the electron. The lower the frequency however, the less is this ambivalence manifested.

This quantum-electronic energy form therefore - having no organic structure, does not expand (nor contract) according to any geometric pattern in the same sense that Em energy expands and contracts, i.e. oscillates. (One may use the analogy of an empty flaccid balloon which has no form and does not have any expansion or collapsing force within it). It will not assume any regular structure or organic geometry unless impinged upon by Em radiation or field, i.e. a 'carrier wave'. [see note].

When the q. Energy is completely free of any constraint, it may be subject to control and manipulation by more subtle influences such as the mind/will combination. The finer the frequency, the more easily it is thusly controlled. The fundamental and essential dynamic nature of free quantum Energy is to follow helical and spiral paths and vectors, just as electro-magnetic radiation has the fundamental property to expand as waves. (> 198).

NOTE: imagine a wave in the sea approaching the shoreline. On the water there are insects buzzing their wings thereby generating vibrations which create very small waves on the surface. This provides us with an analogy of larger electromagnetic waves supporting q. Energy waves.

124 B ~ SPIRALLING ENERGY

In many ancient cultures - such as the megalithic period in Europe, - in later cathedral construction, and the Amerindian culture, etc., the spiral (and the helical) form seems to have been given great religious significance and is found in abundance carved upon significantly arranged stones. (These arrangements usually have astronomical significance). In Gothic cathedrals, this geometric form in various styles may be found decorating the stone-tiled floors. This is likely a result of the Pagan influence on the then growing Christian Church. (The helical form is prominent on carved columns in the Vatican Cathedral).

These ancient cultures were very much in tune with the forces of nature and the natural processes, - and the Life-force which evidently permeates all things. It was recognized also that this Energy was generated by living creatures and by other sources - such as flowing water, underground streams, sunlight, the megaliths themselves, and so-called "Ley lines", etc.. ('Earth Magic', Francis Hitching). [see note 2].

Evidently this Energy form, when conducted along flat surfaces or interfaces, does not travel in straight vectors, - especially the higher frequencies - as does electro-magnetic energy. It tends rather to follow a curving path. However this Energy is not related to magnetism although it may be influenced by magnetic fields, - and in matter and metals especially it can facilitate magnetization.

When generated from a point on the ground such as from the base of a megalith or Standing-stone, the Energy will follow curved, i.e. spiral or helical paths: [note 3]. Having reached its natural perimeter of influence, i.e. along the ground (the perimeter being determined by its intensity and other factors), - the Energy then follows another spiral path returning to its source. In this respect it is somewhat like a polarized magnetic field emanating from a bar magnet.

An analogy of this would be the seed pattern of a sunflower. The seeds appear to follow a spiral path from the center to the perimeter of the flower's seed-disc and then continues the same path on another spiral which returns from the perimeter back to the center. [See Note 1]. This is constructed order and to explain the seed pattern is likewise to explain why quantum Energy spirals.

It has also been observed by researchers using *ultra-violet* photography, that on certain occasions this Energy - rising upward from the base of a standing stone - follows a helical path around the stone. Because of these spiral and helical flow patterns and behavior, this Energy (which the ancients apparently and evidently knew about) was called "Serpent Power".

It naturally prefers to follow a curved path as stated, and the geometry of the interface determines what the characteristics of the flow path will be. If the surface is a horizontal one, the path traced will be spiral whereas if it is cylindrical like a standing stone, it will follow a helical path. (A helix is simply a standing spiral). If one cares to investigate further, it will be found that the helical path described will be a right-handed one. In the case of "bad Energy" however, the helix formed will be a left-handed one. [see note 5].

We know that Energy when emitted, accumulates around a source. We also know that it is the nature of this Energy to seek out concentrations of itself - and since the emitting source is also *de facto* a point of Energy concentration, the emitted Energy will seek to return to its source. We also know that Energy will accumulate to a finite distance from the source at which point it establishes a limit as a circular or spherical field. By examining the seeds of a sunflower, we may understand this function.

If we observe the natural universe, we will see that most dynamic flows and movements are approximately circular. Examples of this are the planetary winds and the ocean currents, hurricanes, tornados, and what-not. Beyond this of course we have the standard pattern set for the universe, which is rotation and orbiting of bodies around their primaries.

The rotating-force then establishes the limit to which the Energy will accumulate and sets a boundary for the Energy, inside which the intensity remains uniform. This rotating path of flow is necessarily a spiral path - like the seeds of a sunflower. This results in a circular motion of Energy around its source point as it accumulates and to return to its source.

In effect then this entire chain of events results in the establishing of a "force-field" or Energy vortex (recall the 'Oregon Vortex') which is actually oraspherical but for practical purposes such as at megalithic sites like Stonehenge, forms a circle or vortex of Energy on the ground with a finite boundary. If this Energy field and its source were extremely powerful, this force-field will become quite tangible with its inherent strange and unusual effects. In effect, it will resemble a bubble. [note 4].

An ordinary water fountain, a bonfire, or an energized electric coil are alike sources of quantum Energy, and as such will establish the same kind of Energy vortex and its accompanying Energy limit or barrier. (In these cases, the generation of Energy is very subtle and its effects are not noticed). In fact this is true of any such Energy source such as a power generating station, a transformer station, and so on.

This would also be true of sources such as the planet itself, the Sun, Black Holes, etc.. In the case of a Black Hole, this Energy field is <u>disc</u> shaped in form and extends to the galactic limits. (More on this later; - The Energy field vortex and the material galaxy are intimately related and are a unified phenomenon).

NOTE 1: There is incidentally, a precise mathematical relationship which governs the ratio of the number of seeds in an outward spiral to that of an inward one. This relationship is revealed in the *Fibonacci Series* and always approximates the ratio of 1: 1.61803399, - the reciprocal of which is 1: 0.61803399! This relationship can also be found in the seed cones of the pine and other such evergreen trees.

This was a proportion greatly favored by the ancient Greek and Egyptian builders, and the Renaissance painters who called it the "Golden Mean". The Greeks called this proportion 'Phi' (Φ). Even more interestingly, this proportion is found to have intimate and exclusive control of the construction of the pentagram! It is also hidden in the diagonal of a one by two rectangle – the floor-plan of the King's Chamber in the Great Pyramid).

> In the case of the Energy's flow path, the fact that this spiralling flow pattern brings the Energy back to its origin displays the way in which it forms a circular boundary at the rim of the seed pod. In the generation of an Energy field, this typifies the circular boundary created by an Energy source. Briefly said, this behaviour of q. Energy constitutes a so-called *'force-field'*. (Of this, later we more shall speak).

NOTE 2: we refer to the ancients as being familiar with this arcane knowledge. In fact, the majority of the population was quite ignorant of its existence, and this situation persists to this day. This knowledge was a secret knowledge passed down to successive generations of Magi, priesthoods, and royalty. They were able to use this knowledge to impress and control the general populace. Control *is* the name of the game!

NOTE 3: one may take note that the higher frequencies of q. Energy deviate from the normal patterns of energy flow in at least two respects. First, it tends to follow curving, helical, or spiral paths and secondly, it is interactive with the mind, both as cause and as effect. The paths of flow are apparently along interfaces and surfaces. Evidently then there is a link between the mind and the tendency for spiral/helical and rotating vectors. Thirdly, the Energy - when not interacting with matter, has no geometric structure, but does tend to form spirals and vortices - like cigarette smoke.

Lastly as we shall investigate, this energy is emitted having a *square-wave* form. There does not appear to be a preceding cause for this apparent behaviour, and it must therefore be assumed that it is a fundamental intrinsic property of q. Energy. Just as an electric flow in a conductor will generate a magnetic field, so we find that freely flowing quantum Energy will tend to follow curving paths. One may refer to other articles which deal with vortices and helical flow patterns.

It is apparent that this tendency to form spirals and vortices is due to no other reason than the Energy's interaction with the space-time continuum, in the same way that rising cigarette smoke will do so when interacting with air. As we shall also see, the STC is evidently subject to influence by various forms of electronic energy: (to be explained).

Furthermore, quantum Energy is extremely sensitive to other, essentially electronic fields - such as interfaces between differing substances, electric or magnetic fields, gravity fields, and even the rotation of objects and of the Earth and other celestial bodies.

Beyond these arguments however, quantum Energy in a free state will tend to follow curved paths simply because that is its nature; it is the formative/creative Energy. You may as well ask why a rainbow has seven colours.

NOTE 4: if one draws a circle, he will first establish a center point. This is done almost intuitively, - and in fact a circle cannot be drawn without one. In geometry, this point is known as a construction point. Likewise, if one finds a circle either on paper or on the land, etc., he will intuitively seek out its center point. Why should one do this? Because this is the nature of the link between the will and the physical, the spirit and the universe, the mind and matter. The link with dynamic spirals and helices is a similar one.

NOTE 5: *'Bad Energy'*: - we are not concerned much with 'bad Energy' in this book, however it does evidently exist, just as does apathy, grief, fear, anger, hostility, and so on. It will have sources which are considered to be anathema to life. These would include positive electric charges or water which contains harmful chemicals, offal, feces, decaying matter, - and other undesirable sources, perhaps 'Heavy Water' (Deuterium Oxide), and so on. Hence, we can see that the more such chemicals etc., are put into the ground, the more 'bad Energy' we shall receive in return!

As elsewhere described, quantum Energy is interchangeable with electromagnetic energy through the exchange medium of matter. If the matter is anathema to life, which is to say if it is a poison, a toxic metal such as lead, aluminium, or arsenic, etc., a solution of 'left-handed' enzymes and sugars, feces, decaying matter, and so on, the Energy produced will be "bad Energy". 'Bad Energy' is detrimental to health just as 'good Energy' is beneficial.

If you go to the seediest parts of town, you will find the general population there is in bad health and their bodies and buildings in a state of bad repair. This is due solely to the bad emotions and Energy which are rampant in these quarters. It is truly a sad fate for anyone who is committed to this lifestyle. (Unfortunately, this too is where you will find drug dealing, stolen property fences, pawn-shops, and so on).

Governments seem to be determined to giving them handouts of food, clothing, and so on. This is not bad of course, but what they *should* be doing is improving their lifestyles by giving them meaningful employment, organized group activities, responsibility, and good participatory entertainment *with coercion if necessary*. In this way, their psychological well being is assured, and from this will follow good and worthy citizens.

124 D ~ THE VORTEX EFFECT

As we have seen, quantum Energy has a natural tendency to accumulate in certain conducive areas and to pulsate in intensity. Furthermore, it has a propensity to form dynamic vortices within the accumulation or field, especially when associated with energies produced by electronic flows.

We can imagine a large energized electro-magnetic coil having exposed windings (diagram 31). The coil will produce a magnetic field (indeed, an *oraspherical* field) which then surrounds it and will also generate a dynamic field of quantum Energy. This field will form a space-time vortex which rotates or revolves around the coil. If this generated Energy is intense enough, it will actually exert a physical effect upon its environment. For example, if a handful of confetti is then thrown into this area, the confetti will respond by moving as if being carried along by a rotating whirlwind. (see 'Oregon Vortex'). There's no doubt that dynamic vortices such as tornados and whirlpools are fascinating objects.

The very ubiquity of the vortex and its static counterparts, the spiral and the helix, must raise some speculation as to their primeval origins. We can also find these geometric forms in clock-springs, screw fasteners, various types of mechanical components, in plant growth, and in galaxies, etc..

Circular and spiralling dances were a common feature of ancient cultures and whether intended to or not are clearly an intuitive representation of these dynamic forms and forces. It is the very ubiquity of this geometry which seems to suggest a connection beyond the purely physical into the non-physical, - as if it were an important element in some great abstract plan.

As the Reader by now must be aware, the boundaries between what is - and what might be are tenuous at best. Indeed as should also be clear, we do not yet know very much about what reality truly represents.

> In the previous articles, we have seen that quantum Energy of very high frequencies is peculiar in at least two respects. Firstly, it is both created by the Being and is interactive with the mind/will. Secondly, this Energy has the property of following a curved and ultimately spiral or helical path. If we observe the universe at large, we will note that the dynamic geometries involved almost always include circles, spirals, and helices. The Being/will creates this spiralling and helicating Energy, and here we find it as a ubiquitous and intimate component of the universe.

>Briefly put, we have entities (Beings) such as you and I which can create Energy. Matter is composed of Energy. The Energy tends to follow spiralling paths. The universe is constructed upon this spiral, rotating pattern. The Reader may care to give some thought to this.

Here is an experiment which will demonstrate how q. Energy will influence even the most mundane objects in your environment. Imagine that you are in a moving car. This car will represent an entire vibrating molecule. Now stretch a string, thread, or rubber band tautly from side to side (or top to bottom, etc.) of the car. Midway along this thread fix a small weight such as a button to represent an atom within this molecule.

At first you will observe the bead vibrating from side to side, but *inevitably* it will have the natural tendency to translate this sideways oscillatory motion into a

<u>revolving</u> or rotary one. This natural tendency to revolve or rotate is inherent in all such vibrating systems, and is a result of the influence of quantum Energy. It is therefore reasonable to expect this phenomenon in actual molecules.

125 A ~ STONEHENGE & E-ENERGY EMISSIONS

Q. Energy exhibits other peculiarities which further distinguish it from ordinary electro-magnetic energy forms. It will extend its sphere of influence to a certain distance - which is sharply defined geometrically, and does not diminish in intensity at any point within this radius. In this sense it is like a solid sphere of matter which does not change in density with distance from its center but is bound to a finite radius.

As explained elsewhere also, Stonehenge is a powerful 'generating station' of Energy due to its vast network of underground water. ('Earth Magic', by Francis Hitching). (also > *The Great Pyramid*).

This Energy extends inward and outward from the stone ring to a well-defined limit. The purpose of the circular ditch and mound surrounding the stone structure was to define this limit and further to contain the Energy within the circle, - (and perhaps at the same time to serve as an effective defensive mechanism, i.e. a *double entendre*): [see note 4].

At the time that Stonehenge was actively in use as a temple, community center, healing station, observatory, communications crossroads, etc., this ditch may even have been filled from an underground water source to create an effective energy accumulator. It resembled very much the moats of later castle structures. In fact the builders of Stonehenge may very well have been the originators of this type of construction (at least in Britain and Europe), then perceived by castle builders to have meritorious benefits.

(In retrospect it appears that wherever possible, the ancients did things with *double entendres*, i.e. 'double intentions', - in other words - with multiple purposes. (Diagrams, phrases, constructions, etc., having double meanings or intentions would automatically be seen to have a magical significance). [see note 02]. One possible reason for this technique was to convey esoteric information among the ruling elite while appealing to the proletariat on a lower level. The Tarot Cards are likely one such device.

A twelfth century British historian - *Geoffrey of Monmouth*, recorded the general belief of that time that megalithic sites and the stones associated with them had healing powers. (It is now believed by many researchers that all megalithic sites served mainly three important functions: - 1. as holistic healing centers where the mind, body, and the spirit were treated as a whole integral unit); 2. As astronomical observatories; and 3. As centers of religious and spiritual awakening, - i.e. temples and community centers.

It has recently become apparent also to researchers [Note 1] that these centers were intimately associated with the science and art of sound manipulation, as in song and chanting. Further that many megalithic structures were very precisely designed in order to create *standing waves* of such sounds. [see note 6].

This technology of sound and stone was passed on and the Cathedrals of medieval times were also engineered and constructed with the specific purpose of utilizing sound vibrations. In this context it was simply a more modern application of an ancient technology. (> section 1.6).

Hans Holtzer is or was until recently a famous researcher into strange energy manifestations and paranormal phenomena such as Poltergeist activities, etc.. He has taken many photographs of locations where such activities are known to exist.

Apparently, unusual energies will manifest in environments like churches and cathedrals - especially when there is energy and physical vibration present such as hymn singing congregations accompanied by organ music, - (or afterward, when the structure has become properly "energized"). (The entire concept of hymn singing by a large group in an enclosed space which vibrates is in fact an ancient technology designed for the benefit of all and especially for the priesthood). [see note 5].

It has been observed in photos of church interiors that condensed Energy forms such as "ectoplasm" become visible during services. Such manifestations are further intensified by the sunlight streaming through the colored stained glass windows. (> article 141 A). It has also been observed in such photos that dull surfaces, as of stone and wood - become highly reflective of ambient light.

In 1979 researchers and scientists investigating the 'Rollright Stones' (a stone circle of 77 stones in Oxfordshire, England) detected ultrasonic vibrations emanating from the stones. This phenomenon has been found to be present at virtually <u>all</u> megalithic sites and is often connected with the seasons, the phases of the Moon, and the rising Sun. The mechanism involved is not yet properly understood.

It may in fact be understood by realizing that the Sun and Moon have gravitational effects on the surface structure of our planet. As the surface bedrock is stressed and distorted by this flux in gravitation, it will generate rock movements and friction - resulting in such ultra-sonic vibrations. These are then radiated to the surface through the clay, soil, and bedrock, and thus transmitted to the standing stones.

It is also known - as described elsewhere, that the planets Saturn and Jupiter will influence the contraction and expansion of crystalline rocks deep beneath the Earth's surface. The created piezo-electric waves in turn interact with the piezo-electric qualities of the megalithic stones and are then translated by the stones back into mechanical ultra-sonic waves. With this in mind, it seems entirely feasible then that these vibrations are being transmitted to locales wherever there are natural rock formations.

For example, lunar eclipses are known to interrupt dowsing currents, ('Earth

Magic': Francis Hitching). These currents are also modified by the proximity of buildings, trees, standing stones (megaliths), etc., - as is also understood by practitioners of Chinese '*Feng Shui'*. ('Cycles of Heaven'; Playfair & Hill).

It is now known that ultrasonic radiations have some unusual and curious effects. For example they can speed up healing and growth in animals and plants, - and humans.' They stimulate chemical and electrical activities in the nervous system and brain, as well as stimulating glandular (e.g. the P-P glands) and psychic functions, - and Alpha and Theta wave production by the brain.

Alpha and Theta wave activity (3~10 cps) is associated with clarity of mind, sensitivity, increased awareness and insight, and paranormal perceptivity and abilities - (again children and animals). [see note 3]. The Pituitary gland is regarded as the master gland of the endocrine system, - it resonates like a tuning fork to ultraviolet light (received through the eyes), - and also to the emitted ultrasonic frequencies of quartz: (> *The Crystal Ball*).

It is known that ancient Egyptian and Babylonian priests would wear magnetized medallions and trinkets or jewels at the center of their foreheads. (Jewels such as rubies, emeralds, etc., are predominantly quartz or aluminium oxide, with inclusions of metal ions to give them color). (Incidentally these ion inclusions will resonate to microwave radiations of 10^{13} Hertz).

The properties of quartz, as noted - would serve to stimulate the Pineal gland. *Edgar Cayce* - a famous seer and psychic of the 1930's, proved this when it was demonstrated that his psychic faculties improved greatly when a semi-precious stone was taped to his forehead.

NOTE 1: such as *Paul Deveraux,* and *David Keating* of Reading University in England exploring at Maes Howe, - also *Bob Jahn* of Princeton U.),

NOTE 2: this is partially why the temples of the new Christian religion were built upon the older pagan sites. In fact there are many reasons why this was done. Firstly, it was believed that the magical energy generated by the older sites would permeate the new church structure, thereby satisfying the new congregation, - *ergo*, it was an enticement to the old worshippers to join the new religion.

Secondly, by building over the old sites the pagan worshippers would have no choice but to go to church in order to receive the benefits of the old site and to covertly follow the old gods. Their children and grandchildren would of course become the new Christian congregation, - not truly cognizant of why they were there, but paying their tithe anyway.

In short, by using this technique of building upon the old sites, the new priests were exercising a form of mental magic (i.e. psychology) to capture the minds and hearts (and purse strings) of the population.

NOTE 3: Interestingly, ultrasonic frequencies of sound will also refresh and "rejuvenate" metal which has undergone work stress and mechanical strain. (> 305

B (2)).

NOTE 4: in fact the ring and moat were constructed hundreds of years before the stone megaliths were erected, indicating that it was already a holy site long before "Stonehenge". What made this site holy is now known to be a vast underground network of streams and waterways unique to all of Britain. By constructing the surrounding circular berm and inner moat, the ancients plan was to capture and intensify the natural Energy thus generated.

NOTE 5: have you ever wondered why churches and cathedrals always use an organ as the principle musical instrument? The reason is simply because this device is capable of generating tremendously powerful vibrations in the surrounding masonry and the ground, and in turn creating strong emissions of q. Energy in the enclosed space. The clergy of today have little concept of why organs are used, - it has simply become a tradition without meaning. The interesting thing about it is that almost no-one seems to question it.

NOTE 6: the ancient worshippers were very fond of song and dance, and these activities were somewhat formalized by the ruling priesthoods who had access to a technology which has long since been forgotten. Usually these formalized dances would take the form of a large number of people dancing in a circle with hands joined. The purpose of this is to create an altered space-time within the circle concentrated with quantum Energy. Various activates such as vibrating the ground, vibrating the dancing bodies through movement and song, creating a dynamic circle of motion, all these factors work together to create this circle of power. The priest involved would of course occupy the privileged central point.

Moreover, as will be explained, any revolving circular mass will generate q. Energy which tends to accumulate at its geometric center.

125 B ~ REAL MAGIC

Many Readers have no doubt heard it said by an acquaintance or a friend that "nothing is impossible", or that, "nothing surprises me anymore", - and other such statements. Why would someone say such a thing? Is there then - as is implied - a place in this vast universe for the incredible to occur?

It is somewhat of a paradox - yet understandable, that as modern technology improves and advances - and produces yet more electronic wonders, - that the concept of real magic as known to the ancients recedes ever more into the mists of obscurity and is regarded as the impossible.

Just two hundred years ago - let alone a thousand - the concept of a colored picture which moves and speaks would have been ridiculed. (I refer of course to the television or DVD player). And yet in that same period not so long ago, people still believed in witches, wizards, giants, dragons, demons, goblins, fairies, water nymphs, sprites, - and what have you.

The interesting question is, - did these things really exist?

I simply wish to impress upon the Reader the notion that real magic is possible and is not such a bizarre idea. People who have witnessed the strange and incredible occurrences during Poltergeist activity have indeed witnessed real magic. The only difference between Poltergeist activity and Magic proper is that one is created by subconscious, random energy projection while the other results from conscious control and intention.

The concept of 'magic' then is really only one of viewpoint. We may accept something as incredulous and "mind blowing", - or we may accept it as "common-place", - or at least something which has a rightful place and belonging to the phenomenal universe.

If you have ever watched the movies, 'The Shining' - with Jack Nicholson or 'The Exorcist' (based on a true ocurrence), you may have wondered if such bizarre phenomena are possible. Let us set aside for the moment that Hollywood movies today exploring such subjects are generally well researched. We may also ask why should anyone conceive of such ideas unless they had some - however remote - basis in actuality.

The stories presented present a case for the existence - essentially - of an unknown form of energy. Furthermore, that this energy is interactive with the mind so that its ability to perceive realities and create its own realities is enhanced. (As we read elsewhere, perception and creation are very closely related).

126 ~ ANCIENT MAGIC

This article is included here since it is interesting as an academic study. It is somewhat parallel to the theme of Energy and will serve to involve the Reader more in a comprehensive understanding of how Energy relates fundamentally to both the mind and to physical energy and matter. Furthermore, it is even interesting.

> "Magic is the traditional science of the secrets of nature which have been transmitted to us from the Magi. By means of this science, the Adept becomes invested with a species of relative omnipotence and can operate superhumanly – that is, after a manner which transcends the normal possibilities of man", (Eliphas Levi, 1810 ~ 1875).

He knew what he was talking about.

The word Magic derives from the Latin '*imagis*' and from the Greek '*magos*', meaning Magician - from the old Persian word '*mag*' meaning 'great' (from whence we derive our words 'magnificent' and 'majesty'). This in turn is related to the Sanskrit '*maya*' meaning "the illusion of reality received by the physical senses". This suggests that what we call objective reality is in truth a subjective interpretation of what we have *agreed* to term 'reality'. Hence, Energy works as much on a subjective level as on the observed or perceived objective level.

Ancient magic was not the popular *legere-de-main* that we are presented with today. Magic in its original and true sense was just that, - Magic, in a very real sense

- and it was serious business - not for children. This ancient magic was none other than the production and manipulation - by a Master of Energy - which was controlled by the will of the magician, Adept, and Master. It also included a good understanding of psychology and of the mind - both conscious and subconscious - and its hidden potentials.

There is a great deal of such lore and legend passed down to our modern Western world from its source points in the Middle East, - much of it transmitted to us by the Crusaders and Templars in the Gothic period. Such places as Babylon, Assyria, Chaldea, Egypt, Israel, and Phoenicia, and so on were - thousands of years ago - "hot-beds" of magic, sorcery, wizardry, witch-craft, the black arts, necromancy, and so on. Stories passed on to us from these areas such as "Sinbad The Sailor", "Aladdin's Lamp", "Ali Baba", etc., abound with references to magical adventures.

> (The Jewish legend of the *'Golem'* is another example of this type of technology handed down to the present time via the Cabbala, - having its source in Babylon (a birthplace of occultism, mystery, and magic). Incidentally, the Greek word for sorcery is "Pharmacopeia" from which we get our word pharmacy. A large part of the arts of a sorcerer included the use of drugs and mind-altering agents, - i.e. pharmaceuticals.

This would set the sorcerer apart from a magician or Adept, - one who is a master of the generation and projection of Energy. Another interesting word is "Abracadabra" (originally "Abrada-kadabra" which actually means "perish the word") and is a word of Chaldean origin). [See note 1].

Altogether then, this lore has been passed down to us on a quasi-religious and cultural level by early Christianity, Islam, and Judaism. (Incidentally all three of these religions were replete with magical practices in their formative years and this is probably due to these practices being extant long before these monotheistic cults. After the Dark Ages however, Christians were castigated for such involvements).

(It is also interesting to note that it is these three religions (all originating in the Middle East) have a tradition of a single omnipotent God overlooking us all. By the creation of a single God, the hearts and minds of the population could be captured and administered to by a single controlling elite. The presence of many gods would make it difficult for the priesthood to control everyone, whereas a single god for everyone makes social control much easier. It was a form of the 'globalization' that we see at work today. The 'God' of today however is finance and trade, - money).

Our once very popular 'Ouija board' is another example of this, - as are also the crystal ball, Tarot cards, the pendulum (used in dowsing), the 'magic wand', etc.. (In fact the traditional conical and brimmed hat that we today associate with witches and wizards is a holdover from the middle ages when Jews in Europe were required to wear this style of headgear - along with a yellow circle, - to distinguish them. [see note 7].

The origin of such conical headgear goes back to the ancient Middle East, - countries such as Persia, Babylon, Assyria and so on, - and were worn by the magicians. Moreover, it is even likely that this headgear was actually used in ancient Babylon and/or Assyria by the ruling priesthood.

This means that the traditional wizard's conical hat and the brimmed version associated with witches had the same origins. The Assyrians and the Babylonians had similar cultures so that there would be a sharing of many of these ideas and practices.

> "The literature of Chaldea—especially its religious literature—teams with references to magic, and in its spells and incantations we see the prototypes of those employed by the magicians of medieval Europe. Indeed, so closely do some of the Assyrian incantations and magical practices resemble those of the European sorcerers of the Middle Ages and of primitive peoples of the present day that it is difficult to convince oneself that they are of independent origin ...'

'That Chaldean magic was the precursor of European medieval magic as apart from popular sorcery and witchcraft is instanced not only by the similarity between the systems but by the introduction into medieval magic of the names of Babylonian and Assyrian gods and magicians". ('Myths And Legends Of Babylon And Assyria': - *Spence*).

The image of the old, invariably white bearded wizard with the black cloak and conical hat, - shooting bolts of lightning from his fingers (kept alive today by children's comic books and fairy tales) is a remnant from the primeval and medieval world of *Merlin* (the magician of King Arthur's court). This period of magical lore was generated by, - and had existed for several hundred years from the twelfth century (the Crusaders) until the Renaissance period. A European (Celtic) form of magic also existed long before that - as it had in all countries of the world.

Actually, the lore of magic has existed as far back as Man has a history. The stories or fables connected with Camelot and King Arthur go back to the sixth century, i.e. the "Dark Ages". [see note 2]. During the Renaissance period, i.e. the "Age of Enlightenment", this culture seems to have been relegated to the periphery (or the 'underground') of our lore.

The cloak and conical hat were also magical accoutrements. Cave paintings found in Spain depict women dancing around a large phallus and wearing conical "witch" hats. Pre-Christian Aztec witch queens are known to have worn conical, brimmed hats and even to "ride" broomsticks.

Apparently, there is a magical lore or body of knowledge more ancient than is presently thought and seems to have been parallel to and contemporary with the global megalithic era and culture.

The most interesting thing about this is the apparent ability of these people to communicate across vast distances. To the ancient peoples, magic was a very real force and one that was generally held in awe and feared. Some 'magic' as practiced then - and even today by some, was simply the generation and projection of quantum Energy by Adepts.

The black cloak worn by wizards or magicians was simply a device to insulate and help in absorbing and retaining the accumulated Energy by the Adept, - (black leather or wool is best).

o

The cloak is often depicted as black with white zig-zag designs on it along with pentagrams and crescent Moons. Where on Earth was such an image conceived? (Interestingly, the symbols of the pentagram and the crescent Moon are a collective symbol for Islam - from whence much of this knowledge and lore was taken and transmitted to Europe. They are to be found on many Islamic national flags: - see note 3]

The Zig-Zag diagrams decorating the cloak were usually geometric devices called 'Sigils'. The pentagram also is a type of sigil, - as are also the seven and eight pointed stars since they are symmetrical geometric devices, and can be drawn in one continuous motion of the pen. (They were thus deemed as "magical"). ("Zig-Zag" is actually a Babylonian phrase meaning approximately, "coming and going", "left and right", "here and there", "this and that", and so on, - and may actually refer to the universal dichotomy).

In ancient times, certain diagrams were considered to have magical power and value, especially in relation to spirit entities, - either in the practice of calling them up, controlling them, or warding them off.

Such diagrams were sometimes 'Magic Squares'. A magic square is a square drawn on paper, etc., divided symmetrically into smaller squares or 'cells'. Numbers placed within these cells will always give the same total when added up in a straight line horizontally, vertically, or diagonally. They were thus deemed to have magical properties.

If one now has a magic square - then by joining each number sequentially with a straight line until all numbers have thusly been connected, a symmetrical 'zig - zag' pattern will be produced. Such "mirrored" patterns or 'sigils' were by themselves also regarded as having magical significance and power by association with their origins.

The conical hat and the cone shape in general have the specific property of generating Energy and also of focussing the Energy downwards towards the base, - somewhat like a dish antenna. Hence the wearer of such a hat receives the benefit of Energy focussed onto the areas of the pineal and pituitary glands, and was hence of interest to the magician.

L. Turenne, an engineer, several decades ago wrote a popular book in French called 'Les Ondes Des Formes', (meaning "Waves From Forms"). He, among other scientists acknowledged in this present book, suggested that all objects radiate a type of energy wave not found in the electro-magnetic spectrum. He also indicated that these emitted waves are <u>compressional waves</u> corresponding to the shape of the object itself. This would correspond to the objects 'aura' or self-generated dynamic force field constituted of radiated q. Energy: [please see note 4].

This technology of shape was also utilized and understood by the priesthoods and nobility of later times. In the periods of castle building from ancient times up to the late Renaissance and thereafter, the castle - whether used as a defensive structure or simply as a stately home - would invariably incorporate a high tower having a conical roof or spire, - (especially those castles of Europe).

This tower with its rooms would be a place where the local or resident magician, sorcerer, initiated castle owner, priest, or rabbi, etc., would perform his rites and ceremonies. In fact, it is likely that these mystical practitioners, magicians, or sorcerers had influence in the construction and geophysical location of such large buildings. This tower with a pointed spire became a traditional and recognizable structural element in all later churches and cathedrals, and is still used to this day with stately homes. Today its meaning has been lost (or suppressed) and its use is merely as ostentation, now mindlessly identified with "class".

Many castles incorporate, and are surrounded by a moat or water filled ditch. The water level of the moat was maintained either by a stream or spring of underground water. In any event, running water was an essential ingredient in magical technology. In this sense then, a castle constructed of stone emulated somewhat a megalithic site. Hence, we see a connection in Europe between the megalithic sites, castles, and cathedrals, - all of these had a magical function in some sense. (> *Stonehenge*; - and *The Great Pyramid*).

Castles built upon high ground or hill-tops or even mountain-tops were often cunningly constructed around sources of water such as springs. Again, this is a type of "*double-entendre*" since water is also necessary to any domicile. [see note 5].

It seems evident that from the dimmest reaches of history (from Babylon and probably from even before that) - up until and including the present, the holders of secret and occult knowledge have been allied with the wealthy and super-wealthy classes, each finding a mutual benefit, - the desired coinage being ultimately power, wealth, and control of the 'lower classes'. The same is true today and is of course a very secret and well guarded knowledge.

The so-called 'Dunce Cap' used in schools as late as the prewar years is another such holdover. The child who was a poor achiever in a classroom was made to sit in a corner of the room and wear a conical hat, - with the idea of being able perhaps to make the child smarter. The hat in due time consequently became the associated mark of a moron, i.e. a "dunce", something to be feared and ridiculed by children. Hence has come the ungraceful demise of a superb piece of technical equipment.

The magician's 'wand' or staff was usually made of freshly cut wood (i.e. hazel, oak, or sycamore) and when pointed by the Adept would serve to focus the projected Energy, - directed by the will of the operator. This is the reason why many people intuitively and instinctively object to being pointed at, especially if the pointer is a rod of some kind. The Tarot cards, the pendulum, and so on, are other devices based on the manipulation of quantum Energy via the subconscious mind and physical body: [note 6].

NOTE: Sympathetic Magic; - this includes the idea that similar objects "communicate" and that there is 'sympathy' between them. Modern day 'Radionics' is a revivification of this once arcane science. Another belief is the idea that cause and effect in the natural order are not irreversible and in fact can be interchanged so that the natural sequence of occurrence is actually reversed. *Effect* becomes the *cause* and *cause* becomes the *effect*. As we shall see later this is a common occurrence in quantum mechanics.

This is why "sympathetic magic" works when we accept its functionality. For example, if we place a living plant in a dark room together with an image of the Sun, i.e. a photograph, a painting, or a drawing, etc., it will thrive better than it would without the likeness. This is of course a bizarre idea, - nevertheless, it is nothing else but magic! (You may also call it quantum physics).

If the experimenter creates such a situation, it means that there already exists in the operator's mind the predisposition to succeed in the mission. In fact the picture, etc., is simply a via to reinforce the operator's intention. The question inevitably arises then as to whether it is the Sun's image or its reproduction alone which serves to stimulate the plant, or if it is the involvement of the experimenter or observer which brings about this stimulation. In effect, the image is a parallel to the biological placebo.

In actual fact, it is not the image of the sun which promotes the plant, it is rather the intention and "borrowed Energy" of the experimenter which is doing the job. This quandary is resolved immediately when we realize that the real answer lies with both factors.

In other words, the solution to this problem is that the plant, the image, and the operator are a "holistic dynamism". The Sun and the image are in communication - as are the operator and the image. Hence, both subjective and objective realities are in fact the same reality. Here we have several causes resulting in one effect: (a typical feature of quantum mechanics)!

This is also to say that if skeptics are involved in the experiment, the "positive holistic dynamism" is reduced in effectiveness and failure is possible. In short, successful experiments where the outcome is unknown is entirely dependent upon the operator's "positive - creative attitude".

Another form of "sympathetic magic" explains much of the architecture of the ancient magical societies which was built according to their cultural belief systems. The Egyptians and the other civilizations lived their lives in the absolute conviction that they co-existed with a spiritual counterpart world or universe and that when they died, they would inevitably join with this alternate existence. Often included in this architecture were 'false doors' carved into stone or rock or built into tombs and temples, etc..

NOTE 1a: "Abrada Ke Dabra" means "perish like the word" and was originally written as a diminishing word spell. This spell was found written on an ancient Sumerian tablet. [see note 1b). Magic originally was used for good, but became corrupted into evil use and Black Magic. In Roman times, this incantation devolved into "Abracadabra". Magic words were invented by the ruling priesthoods to gain power over the general population. In Egypt, a belief survives to this day that special words of power arranged correctly and uttered properly have actual magical force.

A form of magic known as "sympathetic magic" was also practiced. An example of this is where the ink washed off a written charm and then drank had the same power as the original charm. Such intuitive ideas seem ludicrous to we industrialized moderns, and yet in the world of quantum Energy such things are possible*!*

Another such example is the construction of false doors. Many tombs and places of initiation had - as part of their architecture, structures resembling doors, but which were actually solid walls. The prevailing belief was that when the deceased or initiate entered the "spirit world", these false doors would then become real doors. The disembodied entity may then pass through this door into another realm or state of being, a created reality, - a hidden chamber: (or perhaps, a parallel space-time*!*).

Statues resembling gods would come to life, thereby permitting the spirit wanderer to communicate directly with the god in question. Similarly, statues of fierce animals would serve to guard and protect the tomb or sanctuary. Similar practices were affected in the ancient Orient.

Note 1b: the idea was based on the belief that describing something with the use of a name, a phrase, or word was to give it life and existence. Hence, to use a magic spell which diminished such a term was to take away the life or existence of the object or life-form in question.

NOTE 2: the 'Dark Ages' were so named because they really were literally dark. In the year 535 AD, the notorious volcano Santorini erupted in the biggest such explosion known to Man. (Can you imagine a volcano with a caldera thirty miles across?*!*). Ash and dust clouds were spread over the entire face of the globe and caused the equivalent of a "nuclear winter" and a mini ice age, - which lasted for about two years.

Death, pestilence, and disease was widespread and civilization took a terrible downturn. This event was recorded by most cultures including China, Indonesia, and Japan, and it signified the final collapse of the Roman era and civilization, - which never really recovered from this blow; (and indeed, the same for many other civilizations around the globe). There is little doubt also that at this time, much knowledge was lost - as well as the arts of civility and good social order. This period was also the time of the famous English King Arthur and his magician Merlin.

NOTE 3: It is true that the pentagram is a very ancient magical symbol, - as is the six pointed star (hexagram) composed of interlocking triangles.

(Evidently this hexagram form of star was also used as a symbol for Venus, the 'Morning Star'). Furthermore, the Hexagram (from where we get our word 'Hex') was originally a *Christian* symbol, and was often used as a device in large windows in Gothic cathedrals - still seen to this day.

This symbol proliferated in cathedral design and perhaps this is the message whose conveyance was attempted in Gothic stonemasonry. The planet Venus was regarded as a symbol for Christ's return to Earth and the hexagram was used as a symbol for Venus, - hence the connection between the star and Christ).

(Later on, Jews wishing to build synagogues were obliged to hire Christian architects who naturally applied their structural and architectural arcane knowledge to the construction. Since the important heretofore Christian hexagram device was therefore often incorporated into the synagogues' style (with Christian fervor), it became one of the reasons why it was adopted as the symbol for Judaism.

There is also the fact that the "Seal of Solomon" was a magical device consisting of two solid triangles - white and black - one overlaid upon the other. This became a contributing factor towards the choice of a hexagram. (Incidentally Solomon was schooled and raised in the Chaldean and Babylonian magical arts and mysteries. Judaism found its beginnings in the Babylonian mysteries roughly five hundred B.C.. The word Judaism derives from the name of Judah, the Northern division of geographical Israel which was divided after Solomon. The tribes of Judah were captured by the Babylonians, and it is in Babylon that the religion of Judaism was formulated and eventually returned to Judah. In fact, the writings upon which Judaism is largely founded is called the Babylonian Talmud).

NOTE 4: Columbia University scientists *I.I. Rabi, P. Kusch, and S. Millikan* have proved that a form of "strange energy" passes from molecule to molecule, each molecule being a transmitter and a receiver of this energy.

(A solid cone made of transparent material - if placed vertically under a light source, will focus the light to a point immediately below the base, - in this case it is in fact a lens. Any non-conductive, transparent, or opaque material made into a solid or indeed even a hollow cone shape will thus act in a similar way as a focussing lens for visible light, heat, radio-waves, or microwave radiation, or indeed quantum Energy.

NOTE 5: (*Double entendres* it is found, are a reference to - on the one hand the physical world, and on the other, the magical, - and by extension the spiritual. In this way, a construction employing the two could be considered by the magician and others to be a physical 'gateway' to the world "beyond". Indeed this is very possibly how the custom of burial began, - the dead seeking to somehow enter the spirit world by means of this 'gateway'). ('Gateway Of The Gods', by *Z. Sitchin*).

NOTE 6: The pendulum is an extremely versatile piece of equipment. This is to be expected in view of its utter simplicity. Not only is it useful in technical applications such as with clocks, surveying, and experimental gravitational/inertial work as described, - it can also be used as a via between the mind and physical aspects. It is popularly used by dowsers and other diviners for discovering items to their satisfaction.

In ancient Egypt, the pendulum was known and used both as a 'practical' technical device and for divining. In fact, the pendulum was improved upon here by attaching it to a piece of wood about a foot long. This wood formed an extension of the hand and arm, thereby increasing its sensitivity. It was called the 'Mer-khet' or 'instrument of knowing'.

The pendulum seems to work better for some people than for others, and this is somewhat predictable. If the individual is of an unreasonably skeptical or cynical nature, he will have a predilection also to be sullen, morose, or generally negative. Such individuals are inclined to ridicule and usually do not generate much Chi energy of their own, - and this is why they have this temperament. People of a more cheerful disposition are more likely to be successful with the pendulum and other such devices simply because they generate more Energy. It is this Energy which is the dynamic medium enabling the successful operation of the various devices.

Typically, the pendulum is held in the hand and its activities are predetermined. For example if it revolves in a clockwise fashion, this can be taken to mean a positive or an affirmative reply to the query posed. If it swings simply in a straight vector, this can mean something else depending upon the decision of the operator.

The conventional wisdom is that the hand which holds the pendulum is the unconscious tool which then moves the pendulum for the operator. This may well be true, but we must also consider that the hand is a conduit and source of Energy which is then transmitted to the device itself. To a small degree then it becomes alive and is able to form the communication between the operator and the sought after item.

According to what we have now determined, it is then primarily the Energy which is the principle dynamic force moving the pendulum, which is then guided consciously or unconsciously by the operator. In other words, if a sufficient supply of q. Energy is available to the pendulum, it will behave according to the wishes of the operator even if he does not have bodily contact with it. This implies that the pendulum may in fact be suspended from an overhead point such as a tree or rafter, etc., and the question asked of it. The pendulum will then - like magic - oblige the operator by indicating the reply. In Poltergeist effects, we may refer to such items as swinging chandeliers, etc.). This is quantum Energy at work!

NOTE 7: after the First World War, many Jews migrated to the USA (at New York) seeking refugee status. Many of them were illiterate and could not write, and hence were required to put an 'X' beside their names. Many of them refused to mark an X since they regarded the 'X' as a Christian symbol which they objected to, and instead drew a circle. The circle in Greek is called 'Kyklos' and consequently Jews were called 'Kikes'. (It is possible however, that they were also called 'Kikes' in the Middle Ages due to the identifying circle).

Similarly, there was a large influx of Italians, likewise seeking immigration (also at New York). They had no passports or documents and were 'Without Papers' or 'W.O.P'. Thereafter Italians became known as 'WOPs', which became a derogatory term through generated perception.

127 AA ~ BIOLOGICAL BODIES AND E-ENERGY

Every biological entity is an engine which by definition inflows and outflows matter and energy, i.e. $(M_1 + E_1 = M_2 + E_2)$. Any engine may be defined by this equation wherein M_1 and E_1 are matter and energy inflowed, M_2 and E_2 are matter and energy outflowed.

The manifested (output) energy is usually a combination of muscular and physical movement and the emissions of electromagnetic radiations such as radio-waves, micro-waves, heat energy, even visible light photons, - and sometimes even electrical energy.

The mind/body combination also generates and emits emotional energy, i.e. quantum Energy. On occasion, this Energy is emitted to a degree where it can be experienced by others in the environment. Children and animals are keenly aware of this Energy projection.

This Energy and its various frequencies can manifest as fear, anger, joy, etc., - and in these cases, extreme emotion can actually be transmitted, felt, and experienced by others. Examples of such are joy, fear, anger, hysterical mob behavior at rock concerts, riots, political rallies, religious activity, and so on, etc.. In the case of emotions as anger, hostility, boredom, cheerfulness, etc., this Energy is projected at the subject.

Emotional energy (i.e. quantum Energy) exists in a wide range of frequencies and it is these frequencies which manifest as emotions by individuals. It follows then that by diligent self-discipline and meditation, one may easily master his production of emotional energy and switch from one frequency to another.

Obviously this does take some mental power and will, - and practice to achieve this kind of ability. (Great actors and actresses have been known to be able to do this). The frequencies of this Energy can extend well beyond the range of emotions with which we are familiar, into areas which may not be recognized as emotions at all.

Much evidence has been collected which demonstrates that this Energy is emitted and radiated from the hands and eyes especially. (See *Geller's Frequencies*, 121 B). The magi and Adepts of ancient times were able to generate and project prodigious quantities of Energy. This Energy is the dynamic force manifested in Poltergeist activity.

It may be observed that children and animals generally have higher body temperatures than the average adult human. This indicates that they are by some degree more "alive". Such energized bodies have a faster rate of healing and are less susceptible to illness and malfunction. They are also more physically active, alert, and aware of their environment.

Other manifestations of the Energy presence are complexion of the skin and its tone, quickness of movement, and general vitality. In fact people with copious amounts of Energy find it difficult to sit still. They feel that they must be active and doing something! (> *Ageing and Health*, 150).

127 AB ~ AETHER TRAILS

All matter and objects - due to their molecular, atomic, and quantum-electronic activity - generate and emit q. Energy. In fact this Energy is both in-flowed and out-flowed by the object/mass in order to sustain its physical existence; this inflow/out-flow is a universal principal.

Anything physical, i.e. matter or energy, which is inflowed must also be out-flowed, and vice-versa. In an engine for example matter is - in a practical sense, translated more or less into energy. Hence all objects and matter inflow and outflow, (and emit) q. Energy.

Quantum Energy as elsewhere explained, interacts with the space-time continuum. This means that when an object moves, it leaves behind it an Energy trail in a similar manner to the way in which a ship leaves a wake in the water. Hence when you walk along a street or across a field, etc., you are leaving an Energy trail behind you in the Space-Time Continuum and the physical environment.

Other living organisms perceive this trail in various ways, such as smell, infra-red, or ultra-violet perception, etc.. This would explain the existence of animal and cattle tracks across meadows in rural areas of Europe and other countries. Humans likewise, to a lesser degree are also able to 'sense' or perceive these trails on a sub-conscious and intuitive level.

For example when you walk to work or to the bus-stop etc., you leave an Energy trail or 'wake' in the "aether". If you follow this route regularly, the Energy imprint upon the space-time and environment accumulates in strength. Over time and with repetition, this route becomes more strongly established. If you follow a short-cut or cross the road at a favorite point, this will also be recorded on the 'aether'.

If animals (or children) are then allowed to travel from your home to your rendezvous point (or in the reverse direction), it will be observed that they tend to follow the route established by you, unless other routes of equal or greater strength have also been established by others. A very sensitive individual would be able to follow this trail by the use of '<u>dowsing</u>' technology. He may even be able to follow it without dowsing equipment and moreover - do it even if he is blindfolded. Animals of course are very sensitive to this form of Energy deposit.

Now we are discussing biological organisms which leave Energy trails. Let us suppose that an artificial device is able to generate q. Energy in very large quantities. If such a device were to travel along a specified route we may assume that it will leave a strong trail or imprint on the space-time condition and in the ground over which it travels. Further than this, the device, while generating its own quantum Energy field emission - will easily detect again such a trail previously left. Hence a device like this having its own motive power would be able to physically follow this trail as if it were a living creature having awareness of its environment: [note].

In effect this means that such a device could be taken to a location from a base point and then under its own power it would find its way 'home' again by following this trail! This may seem far-fetched, however this marvelous example of quantum physics will be explained.

NOTE: as I have repeatedly tried to explain and will explain further, this is not really as fantastic as it may seem. Such a device is only behaving according to the same universal principles with which we as biological organisms follow. In effect it is we as bio-organisms who may be considered to be Energy generating 'devices' following these same principles.

127 AC ~ EFFECTS OF Q. ENERGY ON MATTER AND BIO-ORGANISMS

The effects of quantum Energy upon matter can now be well understood. This Energy works at the quantum level, i.e. the sub-atomic level, so that this activity becomes largely electronic. Today's electronic technology is based upon the conductivity of certain materials such as metals, semi-metals such as silicon, arsenic, gallium, and semi-conductors such as carbon, graphite, and other materials which have been rendered artificially conducting.

In all this complexity however, it must be noted that almost the entire field of electronics is based on the simple principle that the electrons used are those from the outermost shell of the atoms involved. In quantum-electronic technology, electronic behavior will extend to all the electrons in all of the shells of every atom. This means that in many cases, virtually all substances, be they plastic, oil, wax, wood, ceramics, and substances which are usually thought of as insulators - are potential electrical conductors. This is the basis for true superconductivity: (to be explained).

If one is able to manipulate most or all the electrons in an atom, many interesting possibilities become available. To begin with, we have the potential for room temperature superconductivity; since objects are rendered visible by their surface atom's electronic behavior, we may render them invisible to Radar. It will be possible to give them a chameleon-like ability to evade visual detection or even real invisibility.

Further, it will be possible to cause materials to "melt" or liquify. This will not be by the application of heat energy but with that of quantum-electronic fields. This means that it will be possible to liquify substances not previously considered subject to melting, such as wood, stone, diamond, cloth, etc.. This may be referred to as "quantum-atomic melting": [please see note].

There will be the potential for real transmutation of elements (e.g. 'lead into gold', etc.). Beyond this there will be other marvels such as the merging of objects so that they occupy the same space in the same time.

Since quantum-electronics is the dynamic bridge between the physical environment and the *mind*, we will observe and experience even greater wonders. The body is of course composed of atomic matter. This means that such biological matter may also be subject to quantum-electronic manipulation. The benefits of exposing a body to 'quantum fields' are many. We must first realize however that a living biological body is not just another random lump of matter.

The body is a dynamic entity and is maintained and controlled by a resident mind, will, or Being. Hence any attempted modification by an outside source may also be in conflict with this entity. It *may not* be in conflict however, and this depends entirely upon the agreement and determinism of the controlling entity. The determinism of the Being is usually in favor of the improvement of his body condition.

Virtually all biological bodies are conductors of electricity to some degree. The human body has a measured electrical resistance of roughly 12,000 Ω (Ohms). This resistance may fluctuate according to the emotional and thought state of the owner's mind. These fluctuations can actually be measured on a very sensitive, simple electrometer.

Now as amply explained, the organism is an interactive combination of biological matter including the Pineal and Pituitary glands, chemical energy, electrical energy, and of course q. Energy. When a modification of one of these factors occurs, the other factors are included in this modification. This electrical resistance has been shown to be associated with negative or positive mental energies and influences which may retard (or enhance) the abilities and natural state of the individual. If a conductor (including biological bodies) should be saturated with accumulated quantum Energy, this electrical resistance will diminish. In fact, it can be diminished to a degree where the body suffers no ill effects.

With this diminishment of resistance, we will find other factors being likewise modified. The physical tone, condition, and the abilities of the body improve, electrical energy increases, the natural abilities of the P & P glands are released and enhanced. The ability of the mind to generate, direct, and project this Energy also increases. Most organisms have some small ability to do this and it is done mostly on a sub-conscious level.

This natural ability is enhanced by the presence of q. Energy, and may be objectively and consciously controlled. If a (human) body is suffused greatly with q. Energy, noticeable effects will be objectively observed both by the one directly affected and by an independent observer.

One of the first things noticed will be a subjective realization of a joyous and "devil–may–care" feeling and attitude. Increased activity in the glandular system (including the Pituitary and Pineal glands) will be experienced. This will result in greater perception and awareness, quickened mental processes, improved visual abilities, and overall physical improvement. The physical senses will be improved, sharpened, and heightened, with increased libido. In other words, a literal rejuvenation of the senses. (This state of mind is usually associated with Theta brain waves). Muscular tone and skin rejuvenation will occur and missing body parts such as hair, teeth, etc., will regenerate.

In fact it will seem to all observers that the individual is regaining his youth and vitality in a miraculous and even "magical" way. The aged and elderly will regain physical youth, and lifespans may be measured in the several hundreds of years*!*

Beyond this, we will find even more astounding and incredible abilities. For example, broken bones will be healed instantly. There will be other abilities manifesting such as the ability to see in the dark and communication with wild animals. These are only a few of what we may expect. Yes, I know this sounds fantastic - but alternatively, consider the fantastic living drudgery that we are obliged to put up with in today's world. Surely there has to be a better way, n'est-ce pas?

> "I can believe anything, provided it is incredible", (Oscar Wilde).

NOTE: generally speaking, we may expect that more complex substances such as biological tissue will be prone to "melting", while simpler materials such as metals will experience the opposite effect of becoming stronger and harder. How is electronic melting achieved?

Melting as a result of the application of heat energy is due to the molecules resonating to the frequencies of infra-red and microwave energies. The molecules are vibrated so violently that they dissociate from other similar molecules. This causes the substance to become fluid or to oxidize, etc.. the fluid then is a molecular fluid. With electronic melting, it is the atomic electron shells which are so strongly energized that the atoms dissociate from other atoms so that an *atomic* fluid is thus formed from the substance.

127 C ~ WHAT IS EN/ENERGIZATION?

When a substance is endowed with a large amount of q. Energy, the atomic, and consequently the molecular ~ crystalline structures undergo profound changes: [see note 1]. The atoms are affected in that all electron shells are given q. Energy or electromagnetic energy sufficient to raise their 'energy levels'. This means that electrons in the electron shells are given extra En/energy. This in turn results in the shell either expanding physically in diameter, or the electrons gaining velocity within the shell.

When an atom absorbs electromagnetic energy, it means that this energy is in the form of a quantum, i.e. a photon of definite wavelength and energy content. When it absorbs Quantum Energy however, this Energy does not have to be of any particular wavelength, and that is the fundamental difference. Let us use a simple analogy; if we wish to fill a glass with water, we can do it primarily in one of two ways. We can take another similar glass full of water and pour it into the first glass. This measure (a glassful) can be called a quantum of water. The second method is to fill the glass under a tap. This latter method would be akin to "filling up" an atom with a steady inflow of quantum Energy. In other words, a quantum of Em energy is a precise measure of quantum Energy.

In this event, the glass may be partially filled to any level depending on how long the tap is turned on. When an atom's electron shell absorbs En/energy, the shell tries to expand. It will do this if the atom belongs to a liquid, gas, or a plasma. If the atom is locked into a solid material, it cannot expand and consequently the absorbed energy is expressed as an increased velocity of the electron having a higher frequency.

In either event, the electrons may be raised to a higher energy level. Conventionally, the raising of the energy level is achieved by the electron receiving a 'quantum', i.e. "packet" of electromagnetic energy, such as a light photon, and this gives us the recognizable 'energy level' in the energy band. In the case of quantum Energy however, the Energy received is an arbitrary, non-measurable amount of quantum Energy. Therefore, the energy level of the electron shell is raised arbitrarily by this quantity of Energy. (see diagram 16).

The positions of the 'energy band levels' are decided by the photons which the electron shells can resonate to and absorb, and since science only recognizes electro-magnetic quanta/particles - it therefore only recognizes the so-called energy band levels. The electron shell however responds to and resonates to <u>all</u> wavelengths of quantum Energy, and can therefore absorb any arbitrary amount of q. Energy.

By 'Energizing' an atom this way, we can increase its (e.g. valence shell) energy level to an intermediate 'non-quantum' level, i.e. a level not yet recognized by atomic physicists. In this case, Em energy is not re-emitted directly as a "quantum" (photon) by this shell since the 'stored' Energy would have to constitute a complete electromagnetic quantum (photon) having precise measurable and recognizable wavelengths.

However, if the atom has stored sufficient quantum Energy to constitute a full quantum (photon) of energy, that Energy can be released as a photon of electromagnetic energy. In this way, quantum Energy is converted to electromagnetic energy through the agency of the atom (i.e. matter). It is assumed that the electron will only respond to known types of quanta, i.e. photons of one type or another. These photons/quanta exist and therefore the electron responds to them. The electron will also respond to the finer wavelengths of quantum Energy.

This intermediate state of En/energization is thus retained indefinitely and is not stimulated to be released by any electromagnetic force (e.g. photon) since the energy level of the electron shell is now an *intermediate* ('non-quantum') one and does not correspond to any full quantum level. We may say that the atom then is En/energized. In order for this retained Energy to be released, the shell would need to be stimulated by q. Energy of a wavelength corresponding to the one stored.

> Summary of the above:

Energization - as understood by physicists, is the raising of the energy level of an atom by the absorption of a photon of electromagnetic energy or energy sufficient to cause the atom to emit a photon. Energization by absorption of *quantum Energy* may be referred to as Energization with a capital 'E'. Hence in this book we shall use the term "En/energization".

Sometimes the atom will receive sufficient quantum Energy to raise the energy level of the electron shell to its next higher 'energy band' causing it to emit a full quantum of energy, i.e. a photon. In any event, when an atom is energized to some degree and quantum Energy is involved, the atom is "En/energized".

A single atom may have many levels of En/energization at various times. Some of these levels are recognized 'energy bands' which represent the atom's temporary retention of a quantum of energy such as a light photon. There are many intermediate levels however, which are not yet recognized by science and which are levels of excitation. These levels are almost infinite in range between the ground state and the energy band level required to spontaneously emit a photon. Furthermore, all of the atom's shells may be involved in this En/energization. *An inner shell can be Energised* and will retain this higher energy level indefinitely until this shell is stimulated by impinging Energy of similar frequency.

An atom may retain a level of excitation indefinitely if it is lower than the energy band level of a 'stored' photon. In this event, *the atom is a different form of matter*. Such atoms are normally scattered randomly throughout a piece of matter and constitute a small percentage of the overall population. If the mass is completely or sufficiently composed of such atoms, it becomes *a new form of matter.*

A greater population of these En/energized atoms will be found in the biological mass of children, young animals and vegetation. The atoms which compose the mass may all be slightly Energized and so the mass itself may be said to be slightly Energized. If the atoms composing the mass are all *highly* energized, the mass then may be said likewise to be highly energized.

It may be seen then that a mass or object may have a myriad of levels of En/energization. The percentage of atoms in the mass which are En/energized may be large or small. Furthermore, the degrees to which the atoms are En/energized may be high or low. Hence depending upon these factors, a wide variety of effects may be achieved upon the structure and behaviour of the mass in question.

For example, if a large part of a mass (i.e. a Momentary Population of atoms) is composed of such Energized atoms, its coefficient of friction is lowered. In one instance it becomes more malleable; in another it becomes much stronger. Its specific gravity is lowered, and its reflectivity of Em radiation is increased. If the mass is transparent, its index of light refraction is also increased. Likewise if the mass is opaque, it may be rendered transparent and vice-versa. If this Energization occurs in a fluid, the fluid will likewise have increased fluidity and reduced flow friction and viscosity, - its light refraction increases, and so on. These are all variable qualities depending upon the state of Energization of the mass involved…

The main points to remember are that;

1: An electron shell can absorb any amount of q. Energy to bring it to an intermediate energy band level (i.e. a non-quantum active level).

2: The atom (electron shell) can retain this intermediate energy level indefinitely since no incident or ambient electromagnetic energy in the environment will stimulate a release (due to lack of resonant wavelength).

3: q. Energy can be absorbed by the shell sufficiently to bring it up to a full 'energy band level' so that a full quantum of energy (i.e. a light photon) can be released. This implies of course that quantum Energy is responsible for the generation and release of (light) photons.

An electron shell then can absorb q. Energy at any sub-atomic wavelength and the shell may then be En/energized at any level between its ground state up to and including the emission of a photon and beyond to the point of the atom's ionization (emission of an electron).

(An interesting parallel of this last can be drawn with a Japanese garden water device which collects flowing water. The water accumulates in a container until eventually the container tips over from the weight of water, thus spilling its contents. It thereby releases a "quantum" or specific measure of water. Until this threshold point is reached however, the container can hold any lesser amount of water without spilling it).

This is all possible because of the essential difference between energy quanta and E-Energy. Quanta of electromagnetic energy exist as measurable 'energy packets', whereas quantum Energy is not bound up in such discreet wave-packets. An analogy would be very much like comparing bottles of water to flowing water.

NOTE 1a: if the electron valence shell of an atom expands, this expansion will naturally affect the bonds connecting the atoms constituting the molecule. The most obvious change will be a slight change of the molecule's geometry. Such changes are more noticeable in simple molecules such as the water molecule. The molecule changes slightly and itself expands, taking up more space. If the Energization of the atom is 'permanent', the molecule too is permanently altered.

NOTE 1b: This concept is really very simple, and we may use a balloon as an analogy for the atom. If we inflate a balloon, the inflowing gas forces the balloon to increase in diameter. This means the balloon now contains an increased energy content (the input gas contains heat energy). On the other hand, we may restrict the balloon's diameter to a smaller dimension while inflowing the same amount of gas/energy. In this case, we have a similar amount of energy now in a smaller space.

128 A ~ QUANTUM TECHNOLOGY AND THE MIND – Part 1

We often see animals - either domesticated, on the farm, or in the wild, and observe how carefree and full of the enjoyment of life they are. Apparently they do not worry generally about how they will eat, sleep, or survive. They most certainly do not have any cares regarding taxes, medical bills, paycheques, rentals, lawyer's fees, transportation, unemployment, old-age security, social anxieties, car maintenance, and so on. They are generally free to come and go as they please.

On the whole then it must surely be said that animals by and large are better off, although they own nothing, - not even a pair of socks! They are <u>free</u> of the call of the material world. This is the state of Buddha and Tao that we all long for secretly: (see *The Fool*).

In such a state we cannot even fear death, for the only difference then between life and death is the having and the not-having of a piece of meat (the body). The only possible fear would be for the *loss* of a body which in such a Buddha-like state of being does not constitute much import. This is the state of being that was and is sought after by followers of Zen. Such a state of being is available and attainable today, more so with quantum technology.

As has been repeated many times, this technology is as much a science of the mind as it is a science of the physical and the material. The physical and quasi-physical links between q. Energy and the mind is as follows; - quantum Energy > the endocrine system (including the Pituitary and the Pineal glands) > the brain > the mind > the Being. (This sequence also works in reverse).

There are in fact more direct (subjective) routes however and these will be discussed. Suffice it to say for now that with quantum technology, the Being will find happiness and freedom, the state sought after by so many through history.

128 AB ~ EINSTEIN AND "SPOOKINESS"

What is the nature of a "spooky" feeling? Primarily this experience is the result of an increased awareness: [see note]. This awareness is usually associated with the unusual, the unexpected, and the unfamiliar. We are at all times constantly surrounded by and interacting with energy flows and patterns which are free or governed by some intelligence. It is the awareness of these energy flows guided or seemingly guided by intelligence which we call "spookiness". The cause of the "spooky feeling" can be easily explained. (*Spook* is the Dutch word for 'ghost'). [note 2].

When humans or animals form large groups, there is generated significantly large amounts of quantum Energy. This is so especially if these organisms are physically active or in song, etc.. However if large quantities of Energy are present in a situation where all is stillness and quiet, and there are very few individuals, - this stillness coupled with an excess of Energy is interpreted by the individual on a subconscious and emotional level as being "unnatural". This feeling is called "spookiness", - especially when it is dark or foggy, etc.. Others may describe the feeling as "the air is electric", "energy in the air", or "a tense atmosphere", etc.. With 'spookiness' however there is a definite element of fear.

Another name for this emotion is 'fear' or anxiety. In other words, an individual feels the presence of life and life activity when all is still and there is apparently, and assumed - none such in the environment. Animals are especially sensitive to such primitive and natural emotions and will thus demonstrate behavior characteristic of fear.

Einstein himself expressed reluctance to venture deeply into the field of quantum physics saying that he felt it was "too spooky". But of course - it is "spooky" since we are dealing with Energies which interact not only with matter, energy, space, and time, but with the mind and awareness.

Any force originating with an outside source and interacting directly with the mind is by definition an intrusive force (which must ultimately be electronic in nature, - including chemicals and drugs) and the mind reacting to this subjective/objective intrusion refers to it as "spooky".

We may call this sensation by any other name such as fear, but tradition has given us this special word. Indeed some mind-altering drugs do create a sensation of 'spookiness' in the victim, and even outright fear - sometimes bordering on hysterical fright.

Some people say they "get tingles up their spine", or their hair "stands on end". These are all psychological reactions to this unusual type of sensation. We may also say that any essentially objective (i.e. physical) force which impinges upon and interacts with the mind results in what must be termed subjective sensations.

NOTE: this increased awareness is the result of the mind/brain combination being stimulated by unfamiliar energies such as q. Energy. This Energy interacts with the centers of consciousness and perception such as the Pituitary and Pineal glands and the endocrine system.

Personnel involved in the generation of and manipulation of quantum Energies will be constantly exposed to situations where this feeling or awareness manifests itself, and must therefore be psychologically prepared and trained. The emotion which may sometimes strongly insinuate itself will be fear. Fear itself is useless to Man - for fear weakens, and the strong survive by conquering fear. Fear is actually *anxiety* and this in turn is physical tension electronically stimulated.

NOTE 2: modern children are very familiar with battery operated toys such as dogs which bark, dolls which talk, and so on. They do not find these 'spooky' or 'creepy' because they readily accept anything which is given to them or is accepted by adults. If such an item however were presented to an adult of say, the twelfth century, he would possibly have a heart attack, - moaning in a fetal position, "By the Blessed Virgin, here is a metal contraption with life! – Heavenly Father, save us all!"

129 B ~ "STRANGE PARTICLES"

The question remains, if X-rays and electrons are particles, then why do we not find particles completing the Em spectrum between these two? Well actually we <u>do</u> and they are given names such as "quarks", "muons", "pions", "charm", "strange", "mu-mesons", and so on. However, they are not stable particles and exist for only a very short time.

Why then do they exist at all? This is where philosophy comes to the fore and where a simplistic explanation cannot be scoffed at, - but on the contrary, must be accepted. In fact, these particles exist only because they are <u>postulated</u> into existence. In other words, physicists *wanted* to find them, *needed* to find them, *tried* to find them, *hoped* to find them, *expected* to find them, *agreed* upon finding them, and therefore <u>did</u> find them! It is this expectation and hope, intention and purpose by many minds which created them! This is how it works! Naturally, physicists don't like this idea since it undermines everything they stand for.

One should expect to find them in any event since one would suppose that there must be particles to complete the gap in the Em spectrum. This however does not invalidate the existence of a parallel spectrum of electric wave energy, and as we shall see this parallel energy also really exists.

The gap in the electromagnetic system exists only because these strange particles are so short-lived and unstable. They are unstable simply because they have have no place in the established order, and this too is a mystery. They are created and then they convert quickly back into the q. Energy from which they were formed. In a sense, they are like the eddies and vortices which form in a flowing river, impermanent, unstable, and of little consequence and no value.

129 C ~ A HIGHER SET OF PHYSICAL LAWS

It has been seriously suggested by physicists involved with quantum physics and mechanics that there are alternate realities or "planes of existence". It has been often suggested in the esoteric media that a different reality exists wherein for example, Germany won the Second World War in Europe! Now if such alternate realities are possible and do exist somewhere - somehow, then we should make a probing analysis of our own so-called "reality". (In fact, some theoretical physicists actually subscribe to such ideas and can even back them up with mathematics!).

Indeed our own 'reality' can readily be shown to be subjective, elusive, impermanent, shifting, and mirage-like. Further it can be shown with quantum physics technology that there are alternate rules and laws of the physical universe which we have thus far chosen to ignore.

Initially we would consider such an unstable reality to be a "dream-like" condition of the mind. In fact, the everyday experience which we now call reality is actually more akin to a dream-like state wherein reality is created by the dreamer himself. Dreaming is the actual creation of a personal and private reality, and it will be remarked by many dreamers that their experiences in this state are more "real" than in waking life (and certainly more adventurous!).

As is amply demonstrated elsewhere, alternate realities or 'planes of existence' can be created by quantum technology. This is so since this technology obeys laws and rules which belong to a more liberally structured (i.e. "user friendly") physical reality.

Hence it will be seen that experiences and conditions of the mind are closely linked to quantum technology. If this is difficult for you to grasp immediately, it will become progressively clearer to you in other parts of the text, since this one of the themes. The bottom line here is that quantum physics and physical technology interacts with the creative potentials of the mind at a very fundamental level.

There are two sets of rules or laws governing the behavior of the physical universe, i.e. the "lower" and the "higher" laws. At present on Earth, we are involved with and immersed in a physical condition which operates with the lower set of rules. With the introduction of our new technology, Man will be exposed to the second and higher set of laws. Briefly stated one set concerns masses, i.e. bonded atoms, - the other set concerns independent particles, whatever their nature may be.

In our present state of affairs, we have agreed to rules which state, for example - that all matter obeys the law of gravity, that every object has an unchanging specific gravity or measure of inertia, or that only one object may occupy a given space or location at a given moment.

Likewise, that time has only one rate of flow; that two sequential events may not have a single simultaneous effect; that a single event should not produce similar effects simultaneously in two or more separate locations; that a single event should not produce similar identical effects at separate moments in time; that nothing may travel faster than the velocity of light; that 'time tunnels' and 'wormholes' in space don't exist; or that the velocity of light is constant. That an object cannot "teleport" instantaneously from one location in space (or time) to another; or that atomic elements may not transmute into other elements; that a given space cannot change its intrinsic volume to become larger or smaller; etc..

All of the above listed phenomena do not and cannot happen in the mundane world of today's physicist. These phenomena follow the second or higher set of rules to be established by quantum-electro-dynamics and quantum physics. Hence it will be seen that the so-called immutable physical laws we are familiar with are shown to be in fact *mutable*.

129 D ~ QUANTUM AND COSMIC UNIVERSES

The 'Cosmic Universe' refers to the universe of matter, energy, space, and time that we are familiar with. The Quantum Universe refers to the 'matter', 'energy', 'space' and 'time' of the atomic and sub-atomic - i.e. quantum, condition.

The atom is the smallest unit of cosmic matter, and as such may be regarded as the bridge between the two universes. Sub-atomic (i.e. quantum) particles should not be called "matter", but may be referred to as "quantum matter" or "quasi-matter". Cosmic matter, as most of us know - operates according to cosmic principles, which for the present we may call *Einsteinian* principles. These include the principles of Relativity, gravity, inertia, time, electromagnetic radiations, fields, and so on.

Quantum Matter, however - such as electrons, protons, and so on, are composed - not of particles, but of pure energy, i.e. quantum Energy. Thus, electrons are composed of Energy having quantum wavelengths.

The Quantum Universe is composed of 'quantum matter', 'quantum Energy', 'quantum space', and 'quantum time'. Time as we know it (i.e. cosmic time) does not exist at the quantum level, - as will be explained.

If we refer to quantum "time", we must refer to it as "<u>Now</u>"; it is always Now, and has no flow rate. This explains why particles and Energy at the quantum level have such peculiar modes of behaviour. (*Elvis Presley* sang a song called, "It's Now or Never". I wonder if he ever realized the awful and profound truth of this statement - for indeed, if something is not done *Now* then of course it is *Never* done. The only time which exists is Now).

Quantum space is not the space we are familiar with. The 'space' which we find within and between atoms and particles is actually a matrix or 'sea' of quantum Energy in which particles exist: [see note]. Hence within and between atoms, particles are in constant and intimate communication.

Atoms in matter are held together through the agencies of magnetic and electric fields. These in turn are manifestations of quantum Energy (elsewhere expl.). Hence, the atoms within a mass are likewise in communication, just as are all the nodes (knots) in a net.

If we were to stretch a net taut and cause one node to vibrate, then all the nodes will likewise vibrate in response. Since electrons are composed of dynamic quantum

Energy (explained in section 2), we will find that they communicate in a similar manner. Not only this, but in fact they can "dissolve" or decompose into their original quantum Energy, and even reform again instantly at another location. (> *Thomson Experiment*).

This type of behaviour is very common in the quantum universe and this peculiarity of particles (including photons, etc.) occurs simply because there is no time flow at the q. level to impose a temporal restraint. The electron (and other particles) exists always in a moment of <u>Now</u> and hence its position is not restricted to one locale either in space or in time.

By contrast, an object or atom in the cosmic universe is normally required to change its location by moving through space and time in sequential moments of 'now'. It cannot relocate instantly and without the passage of time. Quantum particles can do this because they don't exist in time; they exist in the 'Now'.

We may use a crude analogy to illustrate this. If you wish to communicate with a friend, you may send him a written letter by mail. The letter will move through time and space and your friend will eventually receive the letter - after a passage of time. However, you may also communicate by means of e-mail using electronic technology. The moment you press the 'Send' button, your friend receives the message virtually instantly without the passage of many hours or days.

Not only this, but the message can be sent to hundreds of friends in the same instant, thereby placing the same message in many locations at the same time! The electronic message device is actually paralleling the behaviour of quantum mechanics.

NOTE: one may liken this matrix to a kind of violent sea or river. Within this sea are vortices or whirlpools and these vortices represent the patterns of quantum Energy which become the particles of quantum matter. (> 207 AB (1.5)).

129 E ~ AFTER-IMAGES

If you have a white sink or bathtub, obtain an overhead incandescent light-bulb having a clear glass envelope. If you now fill the tub with water, you will be able to observe some interesting phenomena. Place a model boat or piece of wood, etc., on the water's surface. You will observe the shadow of the object displayed on the surface of the white porcelain. (A white plastic bucket will also serve the purpose).

Now you may create a disturbance of the water's surface. The resulting dynamic fluid patterns will also be displayed on the white tub surface. (These reflected patterns are due to the *Schlieren Effect*, after the inventor of *Schlieren Photography*; see note]. The waves and vortices, etc., will eventually disappear apparently from the water's surface. Interestingly however, these disturbances will still be visible as shadows reflected from the container's white surface and may be referred to as "after-images". (They are not actually after images *per se*. They are the images of still remaining slight disturbances on the water's surface, made visible through the refraction of light. They are referred to as "ghost images". [see note 2].

Hence the dynamic motion of the water exists although it is no longer normally visible on the water's surface. This demonstrates that subtle dynamic effects remain in a fluid medium after the more obvious effects have disappeared. Not only this but we also know that the structure of the medium itself has been permanently changed. We know this since if a drop of ink is placed in the water it will disperse and eventually migrate throughout the entire body of the fluid. (This effect will accrue in any event, due to the incessant molecular '*Brownian*' motion of the fluid.

Water may be used as an analogy of the space-time continuum. (Space itself may be thought of as a fluid in that it permeates everywhere where there is no matter. It also supports dynamic energy patterns within its structure). On a mundane level, these disturbances are unimportant and indeed are not detectable, but in events wherein large masses or energies are involved, these distortions become significant so that space and time are likewise affected.

In the case of large masses, the greater the mass with its gravitational field and the higher the velocities involved, the greater will be the distortions of space-time: (refer to Black Holes, and section 900). So-called "ghosts" and Doppelgangers, etc., may therefore be considered to be forms of energy "after-images" and dynamic recordings in the space-time continuum made visible by the application of quantum technology and techniques: (more on this will be found elsewhere).

NOTE: this method of photography is used to photograph or film the shock waves or heat waves emanating from bodies which are not normally visible to the eye.

NOTE 2: this "ghost-like" activity in the fluid is a good analogy of disturbances in the space-time condition which are detected by 'sensitives' having the ability to perceive subtle influences which the majority do not.

130 ~ CRYSTAL BALLS

"the only things we know about the quantum world are the results of experiments", (from Paradoxes And Possibilities).

< please refer also to 303 A > Substances may be divided into essentially two categories, - opaque and transparent (and of course semi-transparent). There are several types of transparent substances. Included are glass, water, alcohol, gasoline, ice, diamond, quartz, plastics, Aluminium oxide, oils, gases, cellophane, organic solutions, and biological substances such as those constituting the eye's lens, etc.. (The human eye lens is a substance transparent to all portions of the light spectrum!).

These substances have one fundamental difference from opaque substances.

Their constituent molecular atoms are spaced further apart, thus allowing photons of light to pass between them, - according to theory. Opacity in a crystal results when the atomic valence shells are too close or overlap. The exception to this rule is diamond, which is pure carbon (C). This implies that diamond has a low density and hence a low specific gravity.

All metals are opaque since they are not composed of molecules, thereby eliminating any large gaps between atoms. Most transparent substances are molecular in structure - such as water (H_2O) - composed of a molecular combination of oxygen and hydrogen, - or quartz (SiO_2).

Light photons are hundreds of times larger than the distances between the atoms of opaque substances. (The inter-atomic spacing is normally in the range of 10^{-7} ~ 10^{-8} cm. This of course implies that opaque substances are generally denser than transparent ones.

Quartz or 'rock crystal' - especially the clear quartz, is quite an interesting substance. As is commonly known, the almost exclusive use of clear, high quality quartz in its natural state is with the manufacture of 'crystal balls'. This technology and its use has existed for thousands of years and these balls are in use today by certain people, - 'psychics' and so on. The crystal ball is an excellent example of quantum physics in action.

There is also a large industry in the Far East, - notably China, Taiwan, and Tibet, where priests and monks use them for "scrying" or "gazing". Some very expensive ones can reach enormous sizes, - up to eight or nine inches in diameter, or more - and are of perfect and flawless material.

It is said by those who are familiar with their use that quartz has a "vibration" (i.e. atomic vibration) which is compatible with that of the "third eye", meaning of course the Pineal gland, - and strengthens one's seer-ship ability. (See Glossary: - see also note 1 below].

Further, the pineal gland itself also responds to these frequencies, so that Uv emanations and Energy passing through the quartz impinge upon the pineal gland indirectly through the eyes and the endocrine system.

Some semi-precious stones such as amethyst are of quartz, which include a small amount of a coloring agent, i.e. ions of metal. Such crystal balls, or indeed any such crystal formed into a smooth somewhat spherical shape, - also have a peculiar property. If such a piece of quartz is dropped or hit sharply, a star shaped mark or 'wound' will appear on its surface. After a few days of "resting", this mark will completely disappear - as if the crystal were alive and has healed itself! (> 121 A: *Materials...Memory*).

The reasons for this include the fact that all objects - regardless of their constituent substances, possess a local space-time continuum which surrounds and penetrates them, - and this in turn constitutes an electronic memory field. This of course also includes biological bodies composed as they are of atoms, molecules, and electronic bonding forces in their structure. This is especially true for quartz since its constituent atoms are generally in a higher Energy state than most other substances. This is because quartz is transparent to Uv light and q. Energy of all frequencies.

F.F. Evison, PhD., and member of the American Geophysical Union, announced in 1963 the discovery that during certain seasons related to the revolution of the planets around the Sun - crystals such as quartz, garnet, agate, etc., deep within the earth's crust, would alter their crystal structure, - resulting in a gain or loss of up to three percent of volume.

The great volumes of these crystals within the rocks far below the surface cause tremendous displacements in these rocks when their volume changes. (In a kilometer of crystal for example this displacement would therefore be as much as thirty meters). These expansions and contractions seem to be related to the planets Jupiter and Saturn respectively. (Locations of deposits of such crystals would naturally be prone to earthquakes).

These two planets have long been known by the initiated for their strange properties related to expansion and contraction, ('Astrology, The Space Age Science', by Joseph Goodavage). These stresses and strains are also responsible for the vibrations in rocks and the consequent generation of ultra-sonic emanations (and earthquakes). [See note 2].

We are now familiar with certain facts about the body's production of Energy and its projection through the eyes and hands, and also about the properties of quartz crystal. Hence it is now possible to discuss the mechanism by which seers or psychics are able to interact with the crystal sphere in order to "see" or perceive events within it.

The ever-popular scene from occult lore of a seer gazing into a crystal ball is based on a factual one whose existence has been with humanity for many thousands of years. As far as can be determined, this historical scene goes back to events in ancient Egypt, Babylon, Sumer, Chaldea, Tibet, and China, - and perhaps beyond these into unrecorded history.

In these ancient times, polished metal or stone mirrors, braziers, and bowls of water, etc., were also used for gazing or 'scrying' in order to see events not normally revealed to the profane. These mirrors were made by polishing the smooth surface of a flat, usually circular piece of stone or metal, such as onyx, jade, obsidian, gold, silver, copper, bronze, or brass, etc.. They may have been anywhere from a few centimeters up to a meter or more in diameter. They were used also by lamas in Tibetan monasteries as divining aids: [see note 6].

By means of staring at the highly polished surface of such a mirror in a darkened room, they were able to perceive those things they wished to see. Such a highly polished metal surface is very attractive to quantum Energy, - especially if it is of gold or silver as these surfaces are reflective, of pure metal, and always electrically active. (To be explained).

By concentrating on the mirror, the seer imbues the metal with his own Energy which is emitted by his eyes: (see 121 B). The mirror is a solid state electron~photon device - just as is the crystal ball, and therefore lends itself to the same interaction with the observer or scryer.

Modern glass covered mirrors do not lend themselves so readily to scrying since glass is an effective barrier to both Energy and ultra-violet rays. Some wavelengths of these energies however are able to penetrate - as is evidenced in *Geller's Frequencies*. Furthermore their surfaces are not photo-electrically active as are those of some metals. The very best reflective metal to use is gold, the next, silver - and this will be explained.

The crystal ball is a very ancient perceiving device and like the mirrors, is used for the manipulation of Energy. The Energy source is supplied by the seer and the respondents seated around gazing at the ball - often with a source of light and fire such as a candle or brazier, - which are also strong sources of q. Energy.

Evidently in ancient Mexico and other Central American cultures, the crystal ball took the form of crystal skulls. However, they were essentially polished spherical blocks of clear quartz and doubtlessly served the same special purpose. The sphere functions best for a 'psychic' or 'sensitive' who has superior abilities of receptivity, perception, and awareness, - someone whose brain-waves are being generated strongly in the Alpha~Theta ranges, (i.e. 3 ~ 10 cps). [see note 3 below]. In fact children would respond and interact easily with this device and if a child were raised with this in mind, he would develop a superior ability to scry.

The spherical shape is the best for this object because of the sphere's greater volume in ratio to its surface area - making it the most efficient accumulator and retainer of quantum Energy.

During the sitting, the ball is receiving Energy from those seated around the device gazing at it, (and from the flame or glowing coals), - and its Energy accumulation increases. The atoms of the crystal quartz become increasingly Energized - becoming more susceptible to subtle, unconscious impulses from those seated. These atoms are influenced in two ways. They are able to shift slightly in their position within the crystal lattice - and molecules and single atoms may expand or contract. (The movements of Saturn and Jupiter may also have an effect, as outlined).

Also, Energy may be unconsciously directed to or from the atoms so that their electronic energy level will be thusly influenced, and atoms themselves (being Energized) will contain increased electrical activity and will also try to expand or contract: (explained in section 2). This in turn influences the inter-atomic bonding and electronic forces, etc., and thus affects the way in which light photons are deflected or refracted within the crystal. This means in effect that areas within the quartz sphere can become more or less Energized, which in turn affects the density of areas of quartz within, and the refractive qualities of the transparent crystal.

If such affected atoms are Energized to a sufficient degree, they will even emit light photons which may then be detected by the keen eye of a good 'sensitive'. These light emitting atoms however are normally very small in number but may be enough to form subtle yet discernable images. ('Psychic Discoveries Behind The Iron Curtain': Schroeder and Ostrander). This density and refractivity change is very slight - perhaps no more than one percent of the average, but may be as high as three percent (as explained) and may be perceived by the sensitive eye.

Q. Energy, being of extremely short wavelength, is refracted by quartz more strongly than is ultra-violet light. This means that if a quartz prism (or diffraction grating) is used, this Energy will be located on a spectrographic screen well beyond the violet end of the light spectrum (as with a rainbow). The ultra-violet light and radiant Energy entering the crystal sphere will be focused toward the center of its geometry (since the transparent sphere is after all a lens). Then this internal light is partially "trapped", reflected and re-reflected by the inner surface back into the central area of the sphere. Hence accumulation of Energy is "encouraged" to proceed.

These subtle changes in density, photon emissions, and light refractive properties may then be interpreted by the seer or psychic as actual images or messages influenced and created by the minds of the seer and of those seated around the device interacting with the crystal structure. A way to describe this process is to liken it to a holographic image or a three-dimensional television screen.

It is said by some that the ball at first becomes misty or "cloudy" when being used, then shortly becomes clear again - revealing the looked for scene or manifestation. This misting effect merely serves to substantiate the energizing process and is the result of the initial energizing of atoms within the crystal before any dynamic organization and imagery takes place. (Older black & white television sets using vacuum tube technology used to behave in a similar fashion - giving a jumbled and fuzzy picture before the circuits were fully warmed up and operating properly).

Water is sometimes used as a substitute for a quartz ball. Water has its own interesting properties and if placed in a transparent container, preferably spherical and of quartz (i.e. fused quartz) - will behave in a similar manner to a crystal ball. (Fused quartz spherical containers are obtainable from any good scientific supply shop. A further advantage of such containers is that they are inexpensive and come in a variety of sizes). However although water is not a static crystalline substance, its molecules and atomic bonds are similarly very sensitive to subtle energies.

If the water is chilled to reduce random molecular motion, better 'reception' will be obtained! It is a good bet that if the ancients had had the technology of today, they would have more often used globular fused quartz spheres containing clear water as a substitute for solid crystal balls. [see note 5].

(Nostradamus - whose fame as a seer and psychic is well known, used a bowl of water as a scrying device. He also used a brazier of glowing coals, since such a mass of glowing coals or wood is a good source of E – energy. Further, the composite atoms and molecules, being in a state of agitation are easily influenced by the seeking mind).

We may correctly say - in the case of a crystal ball or any other such device, - that bigger is definitely better. In the case of a sphere, the greater its diameter the greater is its volume in ratio to its surface area (the surface is where energy is radiated away) and hence the more efficient it becomes as an accumulator and storage device. In this respect at least, it is useful to liken the behavior of quantum Energy to that of heat energy. Following the d^3 - d^2 rule - whenever we increase the diameter of a sphere, the volume increases more rapidly than does the surface area.

This becomes more evident when we think of a solid sphere of matter as a finite number of atoms. There is also the fact that the larger a sphere becomes, the less curvature its surface has. This of itself tends to retard the loss of radiated electronic energy. This is quite true for any regular geometric volume, such as a cube, etc., but a sphere is the most efficient.

Furthermore, the smoother the surface and the more reflective it is, the more efficient it becomes since internal energy within, upon contacting the interface (i.e. the internal 'sub-surface') is partially reflected back into the interior - whether the energy be heat energy, light photons, or q. Energy. Hence the q. Energy will accumulate more rapidly with proportionately less radiant loss at its surface and furthermore, the Energy is held within the sphere's mass for a longer period of time after the Energy input has ceased. This effect increases the larger the sphere.

> There is another method by which the efficacy of the scrying device may be improved. If the ball is impinged upon by a vibrating or oscillating electric field, this oscillation will affect the quartz structure and obtain the required results as described. Water of course will respond similarly, and so will all transparent substances and reflective surfaces. (refer to *Kirlian Photography*).

NOTE 1: quartz occurs naturally in many forms, most of which are opaque with inclusions of other minerals, thus imparting a variety of colors to the stone. If a clear piece of quartz is formed into a lens, it will permit and focus visible light, ultra-violet, and higher frequencies of quantum Energy. Since these types of quartz are plentiful and easy to purchase, large pieces may be obtained which may then be made into lenses, - albeit being aware that the focal length for Energy is shorter than that for visible light and even ultra-violet light.

Because a crystal ball is transparent, the incident Uv light is able to interact with every atom within the quartz mass, - thereby raising them to higher energy level states and rendering them more susceptible to influence by thought and emotion.

NOTE 2: it should therefore be possible to devise a method of earthquake prediction based on this knowledge of ultrasonics and the effects of these two planets on crystal expansion.

NOTE 3: It is known that virtually five percent of the population is in this category and that ten percent can see in a range beyond visible light, i.e. ultra-violet. In fact, 4.99 % of all people have a higher psychic ability and are natural social leaders. This fact has been exploited by the priesthoods and ruling classes throughout history. Great men such as *Christ, Buddha, Lao Tzu, Thomas Jefferson, Ghandi, Mohammad, and Moses*, etc., would also fall into this category.

NOTE 4a: the physical properties of water are quite unusual and contradict the theoretical calculations that apply to a perfect liquid. Chemist *Duval* has stated that "water is a liquid that still remembers the crystalline form of the ice from which it originates". *Bernal, Fowler*, (1933), and *Frank*, in 1939 forwarded the idea that the molecules of water were organized to a degree not found in a perfect liquid and that it had a "pseudo-crystalline" structure similar to that of solid bodies.

Water is so structurally unstable that it will respond to even the slightest electric or magnetic field variations. It is so sensitive in fact that it will respond to even the movements of celestial bodies. *Professor Frank* called this characteristic the "Trigger Effect". *Even the approach of a biological organism to a glass of water will initiate changes in the molecular structure of the liquid.*

Mrs. Olga Warrel, a renowned American psychic, is able to mentally affect the crystalline structure and color of the formation of metallic salts with the evaporation of the water from solution. It has been found that water "treated" by her has *a changed molecular structure*. It has a lower surface tension - diminished by about ten to twenty percent. This is an effect which increases evaporation and dessication as noted: (113 B).

Interestingly, magnets have the same effect on water and on sublimating crystals. Evidently also they have an effect on the germination of seeds. Russian scientists using spectroscopy have found that magnetic fields affect the hydrogen bonding in water. They found that three percent of the hydrogen atoms (i.e. a momentary population) were unbonded, - higher than normal: [see note 4b]. This relationship of quantum Energy to hydrogen is of paramount importance and will be referred to in future articles: - please take note*!*

NOTE 5: Water molecules are actually slightly electrically polarized - one end being positive, the other negative. Further, a volume of still water in a container below thirty degrees Celsius behaves, - due to the way in which its molecules are connected, as if it were just one huge water molecule. (*Bernal* and *Fowler*, 1933; and *Prof. H. Frank*, University of Pittsburgh in 1939). In fact such water in a container could also be considered to be a single large liquid crystal). [See note 4].

NOTE 6: if you wish to try this yourself, the best way is to view the reflective surface while it is reflecting a dark or shadowy part of the room, preferably with a plain undecorated wall, etc.. Do not look at your own reflection for this will give you no advantage. Further, the room should be dimly lit, perhaps with a small window.

131 A ~ PERIPHERAL VISION

The human eye is more sensitive than we normally give it credit for. At the back of the eye behind the lens opening or 'pupil' is the retina. Directly behind the lens in the focus area of the retina is where we will find the light receptors called 'rods' and 'cones'. The rods enable us to see the darker areas of feeble light in shades of gray, indigo, violet, also ultra-violet (*Reichenbach*). Cones enable us to discern brightly lit areas and colors. Surrounding this area of the retina is a field of rods which are more closely spaced than in the color section.

This area of the eye is more sensitive to light changes and motion in the blue, violet, and ultra-violet ranges. It is also impinged upon by such light coming from the sides or periphery of the eye. So-called 'sensitives' and psychics have a larger percentage of these rods in the color area of direct vision and are more sensitive to movements in the Uv ranges.

As wonderful as the biological eye is however, only about ten percent of light entering the eye through the lens actually gets to the receptors at the retina. This is due to absorption and light scattering within the eyeball itself, and is also why looking at objects through a short tube will give a clearer view - since the tube will reduce this scattered light. [See note 2].

One common example will illustrate how peripheral vision is demonstrated. On occasion, energized fluorescent lighting tubes are faulty but appear quite normal when viewed directly. When looked at using peripheral vision however, such tubes can be observed to flicker noticeably. This is because the visual rods in the peripheral area of the retina are more sensitive to light and relative darkness, changes from light to dark - and motion, especially in the Uv ranges (which are predominant frequencies emitted in fluorescent lighting).

Cats for example have rods almost exclusively covering their eye retinas, and their lenses and pupils are of course very large, - with anywhere from five to seven times the area of a human pupil. This gives them a tremendous ability to see well into a range of vision which is all but completely denied to human eyes. (Cat owners will affirm this).

This, coupled with the fact that their brainwaves are almost exclusively Alpha, makes them very "psychic". It also makes them amazingly perceptive animals. What wonderful stories we might hear if only we could communicate intellectually with cats (and other animals, too*!*). That which some humans are only able to glimpse rarely and fleetingly in their peripheral vision, cats are able to see directly in clear and bright detail. (Indeed, if it were possible to see the world as a cat sees it, one might suffer a severe strain to his grip on reality*!*). This is why cats sometimes seem to have that smug, "Heh, heh, I know something you don't", look.

Cats were usually chosen to be the "familiar spirits" of witches and warlocks.

Up until quite recently in history, cats in East Anglia, England, were shut up in attics to ward off "evil spirits" which might be lurking there. (Evidently evil spirits shun the light, preferring to hide and "lurk"!).

As explained elsewhere, q. Energy is manifested in the higher Uv frequency range (about 1000 Å) and beyond, into the finer frequencies and wavelengths, - and so any source of this Energy or its movement will be detected by these rods.

Surrounding the pineal gland in the brain is an area which produces melatonin, a chemical normally associated with the production of skin color. However this substance is also associated with visual perception and an increase in its supply or an enlargement of this area can stimulate the visual receptors to perceive ranges of the light spectrum beyond the norm. A small percentage of individuals are endowed thusly and can perceive energies which the majority does not. It is known that psychics with a well developed ability to perceive ultra wavelengths of energy can also see the electronic field representing the missing limbs of an amputee , - thus giving us a clue as to the wavelengths of this Energy.

Furthermore, such fields are sometimes <u>recorded on photographic emulsion</u> when simply in the mere presence of such a psychic. This would indicate an Energy link between the psychic, the photo-emulsion, and the bio-electric field. (Again - 'holistic dynamism').

(*Kirlian Photography* - made famous in the 1970's by Schroeder and Ostrander in their book: 'Psychic Discoveries Behind The Iron Curtain', is a special type of photographic (and filming) technique using high frequency electric fields, and was developed in Russia for capturing energy forms on film which may not be seen by the naked eye).

There is further evidence for the existence of this 'bio-electric field'. The fact that amputees often report that they are experiencing an itch or pain in the missing limb indicates that that something unusual is going on. It would seem to indicate that there is a memory energy field still existing where the solid flesh and bone has been removed. (> '*Synchronicity...*', Section 03). (Refer to the glossary. See Bibliography: 28 & 38).

Psychics are known to be able to generate greater than normal amounts of Energy. This in turn when focussed, affects the film and/or bio-electric field. (The bio-electric field is the dynamically active electronic Energy influence which surrounds the body, and is known by sensitives as the 'aura'. This energy field pattern is the electronic 'blueprint' which provides a 'mold' for the living protoplasm).

This increased Energy interplay improves the sensitivity of the film chemistry while causing stronger emanations or radiations from the bio-field. In other words, the complete effect is an holistic one where the machine, the bio-organism, q. Energy, and the mind are integrated and interdependent, - like the wheels of a clockwork - or the leaves of a tree.

In the case of such a multiple level interaction, it is a difficult - perhaps impossible task to say which is the primary or true causative phenomenon, and this is typical of quantum mechanics. In referring to such a circumstance, one may use the term "holistic dynamism". We see again as in quantum physics that there can be multiple causes and multiple effects, but if one of these is eliminated - nothing at all may happen!

When a doctor heals his patient, there are many factors which interplay. There is the doctor's will, his attention, care, the medication, the patient's will and belief system, - and even the physical environment. Indeed even the doctor's <u>own</u> belief system must be included. Whatever happens to the patient is a result of these factors combined - and their interplay. One may even say that the physical 'medicine' is merely incidental, - as many doctors themselves believe! (This of course explains why homeopathy and placebos work. Indeed as elsewhere explained, the medicine *itself* may be regarded as actually just a 'placebo'. In this event, a placebo emulates a placebo!): (> 300 A-12).

In this respect, it is likely that ancient Oriental medicinal philosophies are more efficacious since their origins are more intimately related to the natural world and creative forces, including likewise 'Chi'. These techniques originated at a time in the past when Man was more "in tune" with his natural environment, - and more interested in it.

The bio-electric field or "L-field" is described in *Edward Russel's* 'Design For Destiny'. In his book, he explains that the <u>L-field may be controlled by thought</u>. (See also, 'Psychic Discoveries… Iron Curtain'). [See note 5].

While on the subject of Holistic Dynamism, it is interesting to note that this principle may be applied to any dynamic energy system. For example, a tree depends for its survival on its root system - yet these roots in turn depend upon the tree also for their survival. It is an interdependent system and the same is true for the leaves and all other parts of the tree.

It can be seen also that each root depends upon all the other roots for its survival, so does every leaf depend upon the other leaves, - and indeed upon every other part of the entire tree. This relationship extends between every leaf, every root, every branch, and the tree itself. In fact one may say that a tree is nothing more than a structure combining leaves and roots! By extended analogy it may also be seen that every human is likewise interdependent and intercommunicating with each other member of his society. With this in mind then it becomes clear that indeed the whole of any integrated dynamic energy system really is more than the sum of its parts! It will also be noted that the root arrangement is the input system, - the leaves are the output system.

< perception and attention >

You can perform a simple experiment which demonstrates that visual perception (and other forms of perception) is more than a simple physical interaction of photons impinging on optic nerve endings.

Try looking at an object in front of you. Focus your attention on it and then without moving your head or eyes, switch your attention to another object towards the side of your line of vision.

To a degree you will be able to visually perceive the target object. What has changed in this experiment? There has been no physical movement of your body or visual apparatus, and yet now you are able to perceive a different object in a different location.

The answer of course lies in *your ability to change your direction of attention without using any physical via.*

There have been cases on record where for example people have been able to smell with their feet or other portions of their body, and even the blind "seeing" with their hands and detecting colors with their fingertips. Such cases have been investigated by prominent physicians and psychologists such as the Italian *Cesare Lombroso, Dr. Angona, Petetin*, and others. Investigations are more vigorously pursued by Russian researchers and much can be learned from them.

Cats again have the wonderful ability to perceive from the 'sides' of their eyes - and since their eyes are apparently so much better constructed for visual detection and perception, they are better able to actually see peripherally since the focal area of the lens is wider. This, coupled with the fact that they have uncluttered, simple minds and are hence more able to focus their attention on targets - gives them a superior ability to see peripherally. Since they are naturally hunters, this of course gives them tremendous advantage. [see note 4].

In fact if you are observant and are blessed with the ownership of a cat, you may sometimes detect them watching you from the periphery of their vision. Dogs also can often be seen to use this trick. Children also to some degree have this innate ability to see and perceive things which are not normally noticed or seen by adults. Their worlds and perceptions are more akin to those of animals: (this will be further discussed).

The ability to shift direction of attention and thereby to perceive things using this attention (sometimes from other parts of the body) indicates that the animal or human is something more than just an "evolved" chemical machine. One is also obliged to question just what is this "thing" that is focussing its attention from behind the eyes. [see notes 3a & 3b].

Here is another test you can perform which will give further evidence that a non-physical agency (i.e. you) exists. Find a friend who is willing to help you perform this experiment. Both of you sit down in two chairs facing each other. Then do nothing except look each other in the eyes without fidgeting or using a via such as drinking tea, etc.. Don't fold your arms since this can be a via, - just put your hands on your knees.

You will at first find this an uncomfortable experience and difficult to do. Your breathing may become laboured. You may find yourself and your friend compelled to look away from each other.

You would not find this difficulty with his nose, his chin, or his ears, etc.. In the first place, you do not look "into" his chin, hands, fingers, feet, or what-have-you. Why then have you this difficulty looking "into" his eyes? It is because of your conscious recognition that there is something there which is difficult to confront, - something which exists but which is not physical. The difficulty you have is proof of this.

It is said that "the eyes are the windows of the soul". This experiment should demonstrate to you the truth of this. You may accept this as evidence of a non-physical agency, you may even accept it as 'proof'. In any event, it is something you cannot ignore.

When you socialize with someone you may like him, be indifferent to him, or dislike him; how does this happen? If you dislike him, it is not because you don't like his feet or his hands or his nose, etc.. You may not like his manner, - but primarily - you make this decision by what you see in his eyes, - and I'm not talking about the colour!

NOTE 1a: as a final note, it should be emphasized that the physical eye is an extremely specialized organ. It has one function and that is to record light photons emitted by active electrons. Therefore anything seen which may appear as unusual or even "ghostly" is recorded on a level transcending the physical eyes.

Quantum Energy is also emitted at visible light, ultra-violet, and other frequencies as explained, and can therefore *emulate* photons: [see note 1b]. Sensitives having perception capable of registering such light will, as a result be able to perceive any emitted Energy in this frequency range. In fact - knowing the true nature of q. Energy, we can say that light photons of any frequency can be emulated - including other Em energies, radio, radar, television, heat, X-rays, electrons, and so on.

NOTE 1b: this question of emulation is somewhat of a paradox since - as we shall learn - photons - and indeed all quantum particles - are composed of quantum Energy frequencies dynamically arranged in a specific type of geometry. Hence to say that a photon emulates quantum Energy or vice-versa is like saying that a cat emulates an animal!

NOTE 2: The same principle is utilized by camera buffs when they place a glare protector on the front of their camera lens. This prevents peripheral light from entering the lens and being dispersed, thereby spoiling the sharpness of the image on the film.

NOTE 3a: one extremely important factor in the abilities of humans and animals is the *ability to focus attention*. Those who can do this are superior individuals and will succeed in many endeavours. This focussing is not a physical thing, it is exclusively an ability of the will. The focussing of attention may include the visual perception or the perception and understanding of mathematics, a document, a job or task, driving a car, relationships, dealing with people, and in fact anything you wish to do or be. The focussing of attention is that which gives one the ability to project his Energy, as in the case of 'Chi' practitioners. This focussing of attention is that which gives the eagle its ability to see small animals in the grass from a great height, - and so on.

NOTE 3b: are we nothing more than pieces of meat? Whence comes the abilities of mathematics; to compose, play, and appreciate music; to design and build jet airliners; etc.? Meat cannot do these things.

NOTE 4: since q. Energy is not normally visible, how is it perceived by living creatures? The perception is not wholly visual as with the physical eyes. The eyes are used, yes - but the perception is more to do with the perception through the endocrine channels connecting the eyes with the pituitary gland. If you have the experience you will understand its unusual quantum-electric nature. This is a material explanation, however. In fact the physical body is not needed for perception.

NOTE 5: try this simple experiment. Find a large place having fluorescent lighting where there are plenty of people moving around - such as in a bus station, a mall, etc.. Fluorescent lighting contains Uv light which is essential to the experiment. Find a blank wall which is white or some pale colour. When someone walks past the wall, watch him carefully. You will see an energy trail or energy pattern resembling his own body following closely behind him. This may take a little practice since you are using vision of a higher frequency. With a little practice however, you will soon see it. In fact - if you do this often, you will develop an ability to see with this higher frequency vision. It will become second nature to you and you will wonder why you never saw this before; - it's a natural ability.

132 A ~ PARTICLES & QUANTUM E-ENERGY– FURTHER NOTES

All physical phenomena - indeed the entire physical universe - are composed of electric, magnetic, electromagnetic fields, and their oscillations, - (and quantum Energy). Em energy is essentially combined electric and magnetic fields in motion and combined in various relative proportions. For example, radio waves are composed predominantly of electric fields in motion combined with relatively weak magnetic fields.

As the frequencies increase and the wavelengths shorten, the magnetic field component increases in strength relative to the electric according to the m^3 - n^2 rule: (see Glossary). As the frequency increases, the wave energy graduates to a more particulate nature. With X-rays (Roentgen rays) for example, the particulate nature is dominant and extreme.

The magnetic field binds the electric wave component into a tightly bound Em particle. Electrons are much higher in frequency than X-rays, and even more particulate (x 40). Midway through the spectrum, we find that visible and infrared photons are completely ambivalent, having both particulate and wave type radiation natures.

Pure electric waves also exist as a spectrum parallel to the Em spectrum but since they have no magnetic component, they have a different behavior pattern resulting in different phenomena. We may call this form of Energy "quantum Energy, "E-Energy, "E-wave energy", etc.. However Energy having wavelengths shorter than 1000 Å. are classed herein as "quantum Energy".

This is of course especially true in the gap between one Ångstrom wavelength and that of one fortieth ($^{1}/_{40}$) Å. (the wavelength of a ground state electron). In actual fact, one fortieth Ångstrom is not the smallest wavelength which can be generated since electrons in excited energy bands of atoms have higher frequencies than this, according to theory.

Electric fields are expansive in nature thus permitting electric and Em field waves such as radio waves to expand through the space-time continuum. However expanding *pure magnetic* waves cannot exist due to the contractive nature of magnetic fields in general. (Magnetic fields are of secondary value to electric fields since electric fields are generated by the mere presence of electrons. Magnetic fields however rely on both the presence of electrons together with their dynamic and organized motions).

Because pure electric waves have no magnetic field component, they logically do not have any tendency to form electro-magnetic particles, i.e. with "magnetic binding": (more on this later). As a result of this, the limitations of electric wave frequencies are diminished. In fact, the limit on electric wave frequencies reaches well beyond X-rays to the electron particle, and probably beyond this also into the realm of thought and aesthetics, etc.. [See note 1].

Particulate energy would be the logical and reasonably expected energy form to be found in this range of the electromagnetic spectrum, - hence electromagnetic particles would be the manifestation looked for by physicists. (Apparently the proton, existing at the nucleus of the atom, has not been assigned any frequency, - although all criteria in the fields of physics demand that it logically should have one). [See note 2].

At these extremely high frequencies, this Energy behaves like a fluid, flowing in much the same manner that heat energy flows within all substances but with a higher velocity. Like heat also, Energy is able to carry and transport electrons through a conductor - (like a sea-wave carries a surfboard). It is also capable of "loosening" electrons from atoms in conductive and non-conductive substances thereby rendering them with increased conductivity (> 199 B & 221 A: *Superconductivity*).

> All gross physical phenomena that we are familiar with in our own sciences and everyday experiences are the results of the combined effect of electric and magnetic fields, - and quantum Energy. Expanding Em radiation such as radio-wave

energy has a magnetic field component. This component has a tendency to contract and condense as a field - while the electric field has a similar parallel tendency to expand. *Hence we find that Em wave radiation represents a combination of two forces which tend to counteract each other.*

Without the electric field component, we would find the magnetic field collapsing and "condensing", since this is its intrinsic nature. In other words, the magnetic field component exerts a "braking effect" upon the expanding Em wave. It would follow logically then that a purely electric wave, unhampered by a magnetic field component should have a higher velocity. However this is only speculative since other factors come into play. One such factor placing a limit on the velocity of radiation is the "density" of space, i.e. the space-time continuum (STC). (It is apparent that electric waves are capable of carrying and sustaining a magnetic field. As we shall see they are also able to generate magnetic field components. This is expected since a moving electric field will create a magnetic field).

This so-called 'density' places a natural limit on such radiation, (explained later). Hence we shall find in this book that all physical phenomena from the subatomic to the galactic are generated by four factors. The primary factor is the dynamic q. Energy which is found everywhere permeating all matter, energy, and space. Secondary factors are the dynamic electric, magnetic, and Em fields, - as will be explained.

Gravity and inertia are the results of dynamic quantum-electric fields. Quantum Energy of extremely high frequencies, i.e. wavelengths of one to one fortieth Angstroms behaves in a conductor as described - like a fluid. A metal wire in fact behaves like a waveguide for this Energy. Exterior to matter in free space, and if *highly condensed*, it becomes visible and simulates actual structured atomic matter.

However this simulated matter does not consist of atoms or particles of any kind and has therefore no "binding force". Consequently this "quantum-matter" will quickly dissipate like a gas and fade from visibility. It can however can take on recognizable forms if influenced by the mind, consciously or sub-consciously.

Such visible forms of quasi-matter have been observed and termed "ectoplasm", etc.. If a volume of such Energy were condensed and influenced so that it assumes the image of a human (or animal, etc.) and animated by a non-physical agency, one could conceivably refer to it as a 'ghost'. Indeed one could not call it anything else*!* (> "Tulpa", 306 A).

Such a volume of organized Energy - being of a dynamic quantum nature, would actually pass through a physical barrier or body unimpeded with no modification to its own coherency; (this is of course reminiscent of Poltergeist phenomena).

NOTE 1: it is possible that even higher frequencies than that of electrons do exist. Such frequencies would be capable of negating, i.e. not being subject to, even the 'density' of space-time. In this case they would not be subject to any limitation on their velocities. In fact such frequencies would be from our viewpoint, "instantaneous".

They would not require a passage of time for their delivery and indeed could hardly be said to have any connection to the physical universe at all! Such frequencies would conceivably constitute the Energy form of thought itself! We find such occurrences in quantum mechanics (> *Thomson Experiment*). Further articles will explain these latter concepts.

NOTE 2: there is only one conceivable explanation for the existence of protons. The electron will be explained elsewhere as the interaction of quantum Energy with the space-time continuum and its 'density'. Protons, being larger - would be expected to have a longer wavelength and a lower frequency than electrons. This frequency would lie within the expected range between X-rays and the electron. This means they consist of an oscillating electric field. Since electrons have a negative electric charge we may then ask why the proton has an opposite and positive charge. We can indeed also ask why it should *not* have a positive field.

The only reasonable answer is that - just as a negative charge is a characteristic of the electron and its frequency, then this *positive* proton field is simply a characteristic of this particular *frequency* of q. Energy! No one is obligated here to explain why the proton is positive or why the electron is negative. This would be like explaining why light of a certain frequency appears blue while that of another frequency appears red! The fact that they *do* is sufficient for our purposes.

132 B ~ ELECTRONIC MATTER

Just as there are two fundamental kinds of emitted energy, so too there are two kinds of matter: (See 132 AA). The matter we are all familiar with is atomic matter and since the atom is an electro-magnetic particle, we may therefore call this substance electro-magnetic matter. However, there exists naturally in the universe vast amounts of "quantum matter". This is often referred to as "Dark Matter". There is also "Dark Energy".

The existence of these two phenomena is easily explained. They are both one and the same 'substance'. The reason why it has two distinctions is simply that Dark Energy when condensed in large enough quantities and density, is referred to as Dark Matter. The distinction is one of density, like clouds and water vapor.

It is possible to create a form of "matter" which is *not atomic* in structure and is less dense. This matter consists of pure unorganized electronic Energy. This so-called "etheric matter" can emulate atomic Matter just as q. Energy can emulate electrons in a conductor. It consists of quantum Energy and is highly susceptible to control by the will. (As we shall see quantum Energy of the correct frequency becomes electrons, and the two are in fact interchangeable. This vacillation between particle and wave occurs in all active conductors. Indeed - as will be seen, even atoms can vacillate between wave and particle).

In certain esoteric fields of research, this Energy manifests as "quantum matter" and may take on the appearance of humans or animals, etc., by coercion of the will. In these fields of research, it may be called "Ectoplasm" or "Etheric Matter", etc.. When taking form, it may appear transparent or even solid, - thereby resembling a so-called "ghost". Such forms of quantum matter can be animated by a capable will. ('Tulpa': 306 A). Such organized and animated stuff is usually seen only by 'sensitives', but under suitable conditions can actually be photographed.

133 B ~ GRAVITY AND SPACE-TIME

If the gravitational force of the planet were somehow increased, i.e. if the specific gravities of all its composite materials were increased, it would contract in volume. (The planet *can* contract and expand, and this will be explained later). (Remember, the planet is almost all liquid iron!).

We may use our useful balloon analogy once again. If we immerse an inflated balloon in a pool of water, we will observe that the balloon will contract in size and volume. Further, if we push the balloon deeper into the water, the increasing water pressure will cause it to continue to contract and diminish in size. (This process could of course be taken to its logical conclusion in the deepest seas so that the balloon would theoretically be reduced to a tiny fraction of its former volume, - perhaps to the size of a ping-pong ball or pea.

In this event, the contained gas would become very dense and liquefied and perhaps incapable of raising the balloon to the surface). We may see from this that the air transits through a 'threshold' point where it is buoyant one moment and non-buoyant the next.

This teaches us an important lesson in that if enough coercion and force is applied to something, eventually that thing will become 'self-programmed' to obey the coercion, even though the exterior coercion no longer exists. The coercion has become "interiorized".

< what is gravity? >

It is commonly agreed upon that gravity is that condition of space-time created by the presence of mass. Gravity is a quantum-energy condition generated by atoms of matter. This condition may be modified using the appropriate quantum-electronic technology. In other words, gravity is a condition of space-time created by its interaction with certain forms of emitted energy. Gravitation can be modified with quantum-electronics.

Scientists have concluded that gravity does not belong in the known electromagnetic spectrum. If it did, it would of course be known by now! It is therefore a force which exists either completely independent of this spectrum, or it exists parallel to it in some way, - or it belongs to an unknown portion of it. Indeed we have seen that there *is* an unknown portion, and further that not all electric energies are electro-magnetic. Therefore we may conclude that the gravitational force lies parallel to the spectrum and/or within the small unknown portion of it.

As has been amply discussed throughout (being a major theme), - the underlying nature of all matter is composed of sub-atomic energies, i.e. quantum Energies. It is likely then that gravity - being an essential property of virtually all solid, liquid, and gaseous matter, i.e. atoms themselves - is itself a quantum Energy form.

Quantum Energies - as elsewhere explained, are essentially electronic and include the electric fields, electron shells, the bonding forces between atoms and sub-atomic particles (such as protons and electrons, etc.), - and molecules. Since gravity is a field effect similar to heat radiation, and further since this field interacts with electromagnetic radiation and electronic particles, e.g. light photons, electrons, etc. (Einstein), - and yet further, since this dynamic field must be sustained and maintained by the universal principle of inflow/outflow, one is therefore obliged to conclude that gravity is a dynamic quantum-electronic force-field.

In other words, Energy is in-flowed to sustain the gravity field. This is true for all emitted radiations and fields, regardless of their type and nature. One fascinating aspect of gravity is the fact that it affects all objects and yet all objects are transparent to this force. This would seem to indicate that not only is gravity an electronic force but is also a condition of space-time, as pronounced by Einstein. As we have learned so far from the nature of quantum physics, it is likely that the truth includes both eventualities, - i.e. an effect having two causes. This further suggests that dynamic electronic fields interact with the space-time continuum.

134 B ~ QUANTUM ENERGY AND THE SPECTRUM

From a technical point of view, there is nothing bizarre or impossible about the existence of quantum Energy. It is simply an electric wave. The only fundamental difference between q. Energy and electromagnetic (Em) energy is that quantum Energy has no magnetic field component. (The purpose of the following section - (1.6) - is to demonstrate that not only is this a very practical energy form but also that its longer wavelengths were known and used by the ancients in electronic communication).

Electromagnetic *particulate* energies existing between X-Rays (i.e. 1 Å. wavelength) and ground state electrons ($1/40$ Å) do not to form naturally - as we have observed. Only electric waves of these frequencies and wavelengths can exist stably. Particles do exist however: [see note 2].

The proton is one such, having a wavelength somewhere between the two wavelengths mentioned. It is more than likely that all the strange, bizarre, and apparently useless particles (such as "quarks, pions, bosons, mu-mesons, muons, strange, charms", and the like) resulting from experimental nuclear particle bombardment (as in cyclotrons, etc.) are also just such random Em particles: [see note 2].

X-ray particles - like other photons, are formed by the combined electric and magnetic fields resulting from the movements of electrons and electronic particles. The electron itself is an electromagnetic particle like a photon, but its formation (also like the photon) is the result of the dynamics and geometry of Energy (explained later). X-rays are particles produced by the actions of electrons impinging upon atomic shells in the target material. In structure and internal function however <u>all</u> these particles are *virtually identical*, - apart from size - as will be seen.

The electric Energy wave of $1/40$ Angstrom is short enough and energetic enough to self-generate a special kind of magnetic field having suitable strength and geometry: [see note 3]. In fact electric waves moving in circular or vortexial motion will generate magnetic fields as well as interact with them. This occurs in the atom where circulating electrons - themselves electric waves, generate the binding magnetic fields for the atom: [see note 1].

The electromagnetic spectrum consists of a wide range of frequencies and wavelengths. Each segment of the spectrum has its own special attributes, idiosyncrasies, and peculiarities. However there are no sharply defined divisions between the characteristics of one segment and those of another – there is always some overlap, - as in the colors of the rainbow.

For example, there is overlap between violet and ultra-violet; ultra-violet and q. Energy, and so on. As an electron in a valence shell generates a photon of ultra-violet light, the valence shell is set into a vibratory mode generating a purely electric field pulse of quantum Energy.

Visible light and heat are the only two forms of radiation which may be immediately detected and consciously sensed by the human body - (under normal conditions). Generally the higher the frequency (and the more particulate) and/or the amplitude of the Em radiation, the more dangerous and disruptive they become to the biological organism. Q. Energy has its own particular attributes. Firstly, it does not usually manifest as particles.

When E-waves are compared together with hard X-rays, the two seem to present us with a dichotomy of characteristics. For example X-rays are biologically destructive whereas q. Energy can promote regeneration and correction. Further, X-rays are particles which are not electrically "polarized" and hence are apparently not affected by electric, magnetic, or electromagnetic fields. E-waves on the other hand are electrically polarized pure waves and hence are influenced by these fields.

In fact q. Energy is capable of so many beneficial wonders that it seems as if their existence is just too good to be true! [Note 1]. Because of their special range of wavelengths, they are able to interact perfectly with atoms' electron shells and electrons (and even nuclei and the very particles therein). Since atoms and electrons are themselves the fundamental particles and building blocks of matter (not forgetting protons and the questionable and arbitrary neutrons), it follows that if one wishes to tinker with matter on the most fundamental of levels, one need only to generate and manipulate quantum Energy of these frequencies.

We have seen that q. Energy can interact with atoms in a manner similar to Uv rays and X-rays. They can also behave in a manner similar to moving electrons by emulation and hence can simulate electricity as their frequencies approach those of electrons, (> *Poltergeist Phenomena*).

In certain situations, ambient and atmospheric E-emanations can, like X-rays, affect the emulsion on a photographic film so that areas in the photograph will appear as haziness. (Q. Energy can emulate any form of electromagnetic energy, such as heat, etc.).

This phenomenon has been observed in connection with Megalithic structures and Standing Stones, etc.. (See *Wilhelm Reich* and Orgone).

< particles >

As one progresses along the electro-magnetic spectrum from radio waves to X-rays and electrons, the wave nature of radiations become more and more particulate. This is why X-rays for example behave more like particles than waves. The wavelengths become increasingly shorter meaning that the magnetic component of these waves gains increasing influence over the electric component and eventually causes these waves to infold upon themselves to become particles of dynamic energy.

As previously discussed, the magnetic field strength increases inversely with the cube of the distance, i.e. wavelength of the radiated energy, while the electric field increases inversely only with the square of it. Therefore these radiations having relatively stronger magnetic fields would be expected to be of particulate form. ($M^3 \sim N^2$ Rule; Glossary).

However since E-waves do not have any magnetic component and are purely electrical field waves, (although they can generate magnetic fields) they are therefore not "obliged" to become electro-magnetic <u>particles,</u> - i.e. dynamic infolding particles of self enclosing energy. (They become actual electrons when their wavelengths reach the critical, threshold dimensions of these subatomic particles, i.e. $^1/_{40}$ Ångstrom. In this event when Energy becomes particulate, it has much to with the space-time condition: [see note 2].

NOTE 1: actually this is an erroneous viewpoint. It is a viewpoint "normally" adopted by Mankind about such promises because Man lives under such abominable conditions of betrayal, corruption, deceit, and general suppression, - and is rarely permitted to realize his dreams. Rather than perceiving quantum Energy as a magical cure all, - an "impossible dream", - we should view it as our wonderful <u>birthright,</u> - and should instead be wondering <u>why</u> the state of Man is so pathetic, and why he is so comncerned with trivialities!').

NOTE 2: The question remains - if X-rays and electrons are particles, then why do we not find particles completing the Em spectrum between these two? Well actually we <u>do</u>, however they are *not stable* particles, and exist for only a very short time. Why then do they exist at all?

Incidentally we may say that the proton is one of these "strange particles", with the exception that it has found its own important role in the atom due to its special dimension and its positive electric charge. Again, the neutron is another such particle.

NOTE 3: in fact, it is the quantum Energy at this frequency impinging on the space-time continuum which causes the Energy to infold and form a particle. More on this later.

NOTE 4: the idea of an electric wave generating a magnetic field may seem strange but in fact, this is just how magnetic fields are created. In an electrical conductor, the moving electrons generate a surrounding magnetic field. It is not the electrons *per se* which create the field, it is the moving *electric field* which does the trick. An electric wave is simply a variation in an electric field so that even an (moving) electric wave will generate a magnetic field. This will be explored further in section 2.

134 BA ~ THE WAVELENGTHS OF E – ENERGY

The wavelengths and frequencies of q. Energy (and associated energies) have been suggested - by researchers into para-physics (including *W. Reich* and *Von Reichenbach*), to be in the range of ultra-violet light. Ultra-violet light is conventionally regarded as being in the wavelength range of 1000 ~ 4000 Ångstroms. X-Ray radiation is generally accepted to be in the range of 1000 Å. to 1 Å. Hence due to this spectrum, we have terminology such as "hard" and "soft" X-rays and Ultra-V light. (see 'Wavelengths': Glossary).

In physics, customary usage allows for a considerable overlap between these two types of radiation - the nomenclature being one of simple convenience for the occasion. Also the diameter of atoms ranges from about 1 Angstrom (Hydrogen), up to 30 A. for the heavy elements. So then, one can see that X-rays of appropriate frequencies would be able to interact strongly with atoms. Such X-rays are able to penetrate deeply into the layers of the various electron shells of atoms, disturbing the constituent electrons thereof and even releasing them from their shells.

> It has already been suggested by two physicists, (the Nobel Prize winning *Dr. Millikan* who discovered the electron, and *Dr. I.I. Rabi*), *that atoms and molecules constantly radiate electric waves, - the frequencies depending on the atom's size or energy level or substance, i.e. the number of electrons and electron shells contained.*

X-rays shorter than 1 Å. do not exist due to the manner in which they are produced. Electro-magnetic waves which are many times larger than the dimensions of atoms such as light photon-waves, microwaves, etc., - when impinging upon an atom, will behave as if the atom is opaque and will not penetrate it.

However the shorter EM waves (e.g. ultra-violet and X-rays) whose wave-lengths and frequencies correspond with the outer and internal structures and particles of the atom, will "view" the atom as somewhat transparent - and an open structure allowing entry. They will resonate with those internal structures and particles causing them to oscillate. They may even impart enough energy to electrons, thereby ejecting them altogether from the atom. Actually, ultra-violet light regards the outer shell as transparent, but not the inner shells having higher energy content.

Quantum Energy of even higher frequencies will carry enough energy so as to penetrate to the very nuclei of atoms and interact easily with their internal structure, including protons and neutrons. This is a major reason why quantum Energy is important to nuclear fusion. (Amply explained later to the diligent, the eager, and the patient).

E-energy however consists of fundamentally mass-less electric waves and hence does not introduce into the atom the more destructive aspects of particulate energy forms such as X-rays. The differences between the effects of X-rays and q. Energy on matter are somewhat like the differences involved in dismantling a sugar cube by the use of a hammer - or by using a drop of water. The one is primitive and mindless brute force, while the other is the intelligent use of gentle, simple, natural forces).

Q. Energy however, because of its intrinsic high-energy content and frequencies, has the capability of influencing the structure of internal electron shells, - and adding or subtracting electrons thereto. The Energy is also able to separate and take apart the nuclei and its constituent particles. This ability would depend upon the E-frequency and amplitude, both of which may be controlled in application: [see note 1].

As explained elsewhere, q. Energy comes in a variety of frequencies and wave-lengths. Q. Energy of Uv light frequencies can behave like, and cause effects similar to, Uv light itself, (*W. Reich*). Frequencies which are parallel to X-ray frequencies will create effects similar to those radiations, and hence will have the same powerful penetrating abilities.

Using such radiations, the same X-ray effects may be obtained without however necessitating the use of a fluorescing screen. Effects will be directly visible. The reported effects by *Reich* and others (such as sunburn, conjunctivitis, electro-static charging of rubber, etc.), are possibly caused by actual 'hard' Uv radiations, which are in turn excited by the actions of Energy on the electron shells of atoms, (e.g. atoms of air molecules, etc.). This would mean of course that the emission of Uv rays would be a secondary effect of q. Energy. It is also possible that these emissions are q. Energy of these particular frequencies. Once again we see that one effect can have multiple causes.

Frequencies beyond those of X-rays, approaching the wavelengths of electrons, (i.e. q. Energy proper), will naturally exhibit properties which are as yet unknown to physics and which may seem quite unusual, even bizarre and whimsical. (To be discussed). (Diagram 01).

Q. Energy consists of electrical waves, and as such are <u>in part</u> subject to some already established laws in physics. For example, those radiations travelling in space may be absorbed, blocked, transmitted, reflected, refracted, and focussed, etc.. It must be noted that the longer wavelengths of Energy can be reflected and refracted similar to the behaviour of Uv light. The shorter ones however have no electromagnetic parallel and will follow their own laws of behaviour.

They may influence other electromagnetic radiations such as TV signals, (i.e. microwaves), radio waves, and light waves, etc.. They are able to do this by modifying existing Em waves in a way similar to that in which radio waves can be modulated, as in FM (Frequency Modulation) radio transmissions. The electric component of these waves is thus modified. Hence, q. Energy having X-ray frequencies can be carried by light photons or other forms of electromagnetic radiations.

The main difference between FM modulation and E-modulation is that the frequencies modulating the FM carrier wave are generally larger in wavelength than the carrier wave, and are themselves electro-magnetic in nature, whereas the Energy modulating the 'carrier wave' is actually of a shorter wavelength than its carrier. This carrier wave may of course be of any wavelength as short as 1 Ångstrom. [see note 2]. That is, quantum Energy interacts with both particulate and with radiant wave-type energy.

This is entirely possible and feasible as can be observed in nature. If we watch a water-logged bee buzzing frantically on the surface of a pond we can observe that it generates tiny waves. If we drop a stone in the water we can see that the waves generated are much larger in dimension. Nevertheless, we will note that the smaller waves generated by the bee are visible within the larger waves and are carried along by them. The same is true of a ship vibrating with its engines. The small vibration waves will be visible upon the larger sea waves as they pass the ship.

NOTE 1: this ability to interact with subatomic particles means that among other things, the Energy is capable of separating individual neutrons into their two components, i.e. an electron and a proton: (see *The Neutron*). A variety of quantum frequencies may be obtained due to the electronic tension of the atomic inner and outer shells and their En/energization levels.

NOTE 2: it is very easy to imagine how Em energy may be modified by using a simple analogy. Let us suppose we have two musical flutes. Each one is made out of a different type of wood, for example bamboo and walnut. In all other respects these flutes are identical.

To many, the sounds emanating from these flutes are likewise identical and cannot be distinguished apart, - the differences are subtle. However a highly trained ear will be able to tell the difference. Hence we find that the vibrations created by one wood type are different from those of another type. Such differences will also be present in stringed instruments, percussions, and so on. The same would be true for most musical instruments, including brass winds.

In this analogy we can say that a particular vibration is carried by one instrument and not by another, and yet the difference is indistinguishable.

134 C ~ RANDOM EFFECTS OF QUANTUM ENERGY

Quantum Energy may be likened in some ways to "hard" Ultra-violet light. This form of Em radiation (~1000 Å wavelength) is invisible and has a powerful effect upon the surface atoms and electrons of materials and objects. Further, this light form only affects the surface electrons of atoms, i.e. the valence electrons. Furthermore, hard Uv radiation has a strong tendency to behave as a particle – not as a wave. In this respect it is more like X-rays, so that the interactions with electrons are more akin to the interactions of billiard balls – not as those of sound or water waves, etc., upon each other. As for its effect upon atomic structure, quantum Energy has also the combined characteristics of both Uv light and X-rays.

These radiations however are particulate, and q. Energy being a wave-type energy is able to interact with *all* particles at a very fundamental level. Particles such as electrons and protons, etc., - as with atoms also - may be constructed or deconstructed, or otherwise modified. [see note 1].

Hence we will find that quantum effects will be manifested in the vicinity of a quantum Energy source. Such effects will be for example the "fogging" of photographic film or 'scrambling' of digital chips, an increased rate of chemical reactions, the rapid de-charging of batteries, capacitors, and other such electrical storage devices, an increased growth rate of crystals and biological organisms, the increased healing rate of wounds, sores, broken bones, etc., shortened pregnancy gestation periods, etc.. (Batteries are 'de-charged' through internal short-circuiting).

Further, we may expect the regeneration of faulty bio-tissues, the mending of cracks and flaws in various objects and materials, the sharpening of knife and razor edges, crystalline re-structuring of materials such as metals, quartz, etc., an increase of the maturing rate of alcoholic beverages, the re-growth of tissue potentials such as hair and teeth, etc..

Further to be expected are the restructuring of aged skin and other organs, improved general health, the alleviation of unwanted conditions such as cancer, tuberculosis, etc., the increased reflectivity and smoothness of surfaces, increased purity of substances, the improved taste of food, the improved lubrication properties of oils, etc., the increased fluidity of liquids and gases, etc.. Indeed such a list is of course virtually endless and is limited almost only by one's imagination. [note 3].

Some unwanted effects will also manifest themselves, however. For example, insulating materials will become electrically conductive so that batteries, switches, and wiring of conventional use will become more or less useless. (It may be regarded as a general rule that all materials are virtually "transparent" to and subject to pervasion by quantum Energy, thereby rendering them electrically conductive). Hence anyone familiar with electronics, particle physics, materials technology, etc., should be able to intelligently predict the effects of such an energy form. [note 2].

NOTE 1: as is discussed elsewhere, all particles are composed of wave-type energy which has become "trapped", i.e. 'infolded'. Hence, q. Energy of a suitably fine wavelength will interact by resonance with the structure of these particles.

NOTE 2a: this futility of trying to "capture" such a pervasive energy form has a parallel with the futility of trying to measure its quantity and strength (504 B). This again illustrates the personal and subjective nature of quantum Energy situations. The physicist will therefore be exposed to a variety of phenomena which are as much subjective as they are objective, and which will not lend themselves to rigorous analysis or measurement.

NOTE 2b: the discharging of batteries and capacitors, etc., is affected by the insulating material becoming electrically conductive. This is due to the well known quantum "tunneling effect". In fact, the electrons are converted into waves by interacting with q. Energy. As waves, they simply "shine" through the insulator. (> 430 A: the *Bose-Einstein Condensate*).

NOTE 3: again, one would have to suppose that the ability to bring about improvement in every aspect and in every corner of life is surely too much to hope for in a single form of energy. This is the mindset of those who have been defeated too often. These people *know* that things can never improve and that anyone who makes this promise must therefore be a charlatan. However, there is another way of looking at this situation.

Life on Earth for mankind is difficult in every respect, but this is because he has been *denied* the benefits of adequate 'Life Energy, 'Chi, 'quantum Energy, 'Orgone', or whatever you wish to call it. Not only is quantum Energy a force to manipulate the physical universe with, it is also a supremely civilizing force - having a profound effect upon the mind and spirit. This is achieved partially through a general field effect and through the endocrine glandular system.

146 C ~ MECHANICS - vs - QUANTUM-ELECTRONICS

Life on this planet is mainly biological and consists primarily of the two "kingdoms", - animal and vegetable. The animal world - although consisting of electrochemical organisms, is a mechanical opera. Virtually all human tasks are accomplished with the use of tools and machinery. The body itself is a mechanical entity "designed" and used to perform mechanical tasks. Evidently, there are three levels of dynamic organization in the universe. These are essentially: the *quantum* or sub-atomic level; the mundane or *mechanical* level; and lastly the *cosmic* level.

The formative forces at the quantum level are primarily the magnetic, electronic, and quantum Energetic. At the mundane level, the forces are primarily mechanical, and at the cosmic level they are surprisingly enough, magnetic, electronic, and quantum Energetic again. It would appear then that for some reason, quantum energies have been denied to the mundane level. The essential difference of this level is that it is the level at which life-forms interact with the environment.

Hence, it is the interpretation of the environment which life-forms place upon it which we perceive as being of a mechanical aspect. In other words, for some reason we are incapable of interacting with the environment at the quantum level and are therefore obliged to deal with it mechanically. [see note 2]. A simile of this would be as if we were denied the ability of speech and were obliged to use hammer, chisel, and stone to convey messages. Another such simile would be as if we were denied fire and were obliged to jump up and down to keep warm.

If we are sitting down in our favorite chair and wish to fetch a book from the book-shelf, we are obliged to raise the body, walk to the shelf, raise our arm to the shelf, grasp the book, take it off the shelf, turn around, walk back to the chair, turn around again, and then lower the body, being careful all the while not to knock over the potted plants and the cup of coffee, or step on the cat's tail.

All this activity is mechanical and all our machines are extensions of our bodies based on our understanding of familiar mechanical principles. The vegetable kingdom is predicated exclusively on electro-chemistry and Coulomb Forces, etc.. All activities are therefore electro-chemical. These include the absorption of food and liquids from the soil, the absorption of gases and liquids from the air and their expulsion. Likewise, the electrical and chemical dispositions of these absorbed nutrients. [note].

The next stage beyond electrochemical life activities would be quantum-electronic. Tasks which are now done by mechanical and electrochemical means would then be done using quantum-electronic fields and flows. In this situation, tasks would be performed using such fields and wave-forms.

Hence, it would be possible for someone to "fetch" a book off the shelf simply by generating and emitting quantum-electronic fields, i.e. without using any mechanical means. (Of course, the family pooch may not easily accept this state of affairs and would likely run howling to the nearest closet!). This raises the interesting point of why life on this planet is generally incapable of using such electronic fields. In fact we actually all do - but on a very minimal level.

NOTE: it would seem that a creature of the vegetable kingdom is more or less completely focussed on feeding and breathing since it apparently does nothing else. It evidently does not engage in other activities such as play (as we recognize it) or nest building, etc.. This would seem to belie a state of mind-numbing boredom.

Perhaps a tree does have some notion of play but it is likely to be of a different order and perception than that of an animal or human. The animal's notion of play is one which necessarily includes self-motivating mechanical activity. Since the tree, etc., cannot or does not self-motivate, it possibly relies upon the winds, birds, and rains, etc., to provide the motion. This gives us pause to ponder. (When we see a tall tree dancing as in a strong breeze, can we be sure it is the breeze - or the *tree* causing the motion? This may seem like a childish question, but give it some thought).

If a tree for example is capable of experiencing emotion, these emotions are likely to be stronger and to include 'serenity of beingness'. This is a far superior emotion than is as yet experienced by the vast majority of Mankind. There is another possibility however, and that is that a tree consciousness is not necessarily stuck in its body*!* (I wonder too, if industrious insects like bees, wasps, or ants - or even beetles have a sense of play. In order to have a sense of play, one would also have to have an autonomous consciousness and awareness of self, - and of others).

NOTE 2: it amuses me to imagine some scenario like the Garden of Eden (*E-Din*, in Babylonian mythology) where Adam and Eve were cast out of 'Paradise' (another word derived from the Persian meaning simply a formal and enclosed garden): ('Adama' means Man). From that point on, things weren't easy for them, since they could no longer simply *create* things; they obliged to use muscle power. They were no longer in a state of grace.

147 ~ MATTER IN TENSION and VIBRATION

Matter in tension is sensitive to vibration. This vibration may be sound vibration transmitted through the air, the soil, rocks, water, and so on. It can be an oscillating light source, and if the frequency of the oscillations is high enough, the matter in tension will respond to it. The commonest forms of vibration that modern man is familiar with is sound and electrical.

A common example of this is the humble umbrella. When extended to its functioning position, the umbrella consists of an area of cloth stretched tautly over a metal frame. This stretched cloth is then sensitive to sound vibrations in the air. If a loud noise is heard nearby, one can actually feel the vibrations detected by the cloth travelling through the handle.

Another method of placing material in tension was used by the builders in stone of many past ages. This is a technology which has been handed down from the early megalithic builders of Stonehenge and other such sites. The cathedral builders also used it in their great structures. This technology however was also known to the very ancient peoples including the Egyptians, most of the middle eastern civilizations, the Greeks, and possibly the Romans. The technology found its last expressions in the works of the *Gothic* builders and later dwindling works. After that, the knowledge was either lost, forgotten, or hidden, etc..

If you have a cardboard or wooden box, it will be subject to vibrations just like the 'hanging stones' or the umbrella, etc.. The interesting thing here however is that the material has not been engineered to be under stress. The very fact that it is a rigid structure of thin material places it in a condition where it will respond to sound vibrations. Furthermore this rigidity appears to be a factor in itself rendering objects capable of responding to vibration.

For example, stone has a rigidity resulting from its own atomic structure and for this reason sound waves can be caused to travel within its mass. A mass doesn't even have to be solid, it can be a liquid or a gas. in these cases we may say that water for example, has a "liquid rigidity" and a gas has a "gaseous rigidity". Water will convey sound waves better than will air or any gas.

Let us imagine that you have a giant pyramid of ping-pong balls all glued together. This would be a rigid mass, and as a result it would capable of relaying sound waves throughout its mass. Such a structure of balls would closely represent a crystalline structure composed of atoms. All substances known to Man are capable of transmitting vibrations, hence we may deduce that vibration is a fundamental activity of the universe, regardless of the frequencies or wavelengths involved.

SECTION 01.6
(ELECTRIC WAVE ENERGY)

A ~ THE GREAT PYRAMID AND QUANTUM ENERGY

PLEASE NOTE: the purpose of section 1.6 is to establish and clarify the existence and role of electric field waves in nature and quantum ~ atomic behavior. This technology is once again indicative of the ancient's knowledge of this peculiar Energy form. < please also refer to 125 A >. Further than this, it should offer fascination to the Reader.

(One may note the inclusion of the Pyramid and the Ark as two of the world's greatest mysteries of antiquity. This is not the reason why they are included however. They are both intimately related to the twin subjects of quantum-electrodynamics and quantum Energy, and this is why they have been included. They are mysteries because this technical information has been suppressed. This fact strongly suggests in turn that this is valuable information retained only for a select few).

Δ

< water > In Viet Nam during the American involvement, the American military used dowsers to locate explosive mines and underground water. Dowsing is a very ancient art and para-science, predating even Pharaonic Egypt, where it was also used. This again demonstrates the practicality and expediency of the efficient military mind. (*"If it works, use it"*). This is certainly a superior attitude than that taken by physicists who are blindly obedient to the *status quo* and peer pressure.

It is said by dowsers and researchers into the field of para-physical phenomena that underground water and metals, etc., emit a kind of energy or radiation which the dowser is sensitive to. The rod or pendulum, etc., then indicates the presence of such substances. The famous *Michael Faraday*, electrical engineer and pioneer in electro-magnetism, concluded in his researches that electrical currents and <u>unknown radiations</u> were generated by flowing water. This was confirmed by later experimenters *F.B. Young, H. Gerrard*, and in 1918 by *W. Jevons* in Dartmouth, England.

In water, the hydrogen bond has only ten percent the strength of most ordinary chemical bonds and hence water molecules are very fragile, constantly forming, breaking and reforming, - even in still water. Water is also very sensitive to electromagnetic fields and is particularly unstable between the temperatures of $35 \sim 40$ degrees Celsius, - which is the body temperature of most active animals (*Bernal* and *Fowler* in 1933, and *Prof. H. S. Frank*, University of Pittsburgh in 1939; - see also 'Introduction to Physical Oceanography' …).

Flowing water especially is a very strong generator of q. Energy, which is why people generally feel so good when they visit rapidly moving streams, rivers, rapids, or places like Niagara Falls, - or when taking a shower! The internal friction of atoms and molecules in flowing water causes atomic bonds to constantly break and re-form continuously in great quantities, thereby generating and emitting electronic quantum Energy.

At Stonehenge in England it has been found that there is an unusually large and intricate underground network of flowing water. This is often also true of other ancient megalithic sites such as stone circles, barrows, and dolmens, etc.. The stones above ground level act as accumulators and 'at-tractors' of this Energy from below, which in turn is emanated into the enclosure of the circle of stones.

Many medieval cathedrals also were built over more ancient, sacred, pagan, and megalithic sites, which in turn were situated above underground streams of water. In fact almost all of the cathedrals of the earliest period were built above wells, ('Earth Magic': Francis Hitching).

This is one of the little publicized secrets of Stonehenge (and of megalithic engineering in general). Places such as these were consequently used as locales of worship, healing, communication, insight, wisdom, and stimulation of the glandular systems, - particularly the endocrine system, including the Pineal and Pituitary glands.

Δ

< the temple > This activity allied with chanting, song, and dancing (which causes the body and its molecules, electronic bonds, and atoms to vibrate and reach higher energy levels) must have been a very powerful influence indeed upon the initiates participating in ceremonies there. The dancing would also cause vibrations to be transmitted in the ground which would in turn affect the stones causing them to quiver also.

The chanting and song within the enclosure of the stone vaulting of cathedrals is a technology descended from ancient times (from Egypt and Babylon, etc.), and there is now evidence that in fact all megalithic sites were used in conjunction with sound vibration generated by choirs or groups of priests and initiates, etc.. (See footnote 1).

The great pyramids of Egypt are perhaps among the first of such great stone (megalithic) structures, 'temples', or 'cathedrals' in pre-recorded history. There appears to be no earlier attempt generally to construct such imposing structures using such dedication, technology, permanent materials, and engineering precision: [see note 2].

Among other purposes, they were used as temples of initiation, meditation, and "awakening of the inner self", (as explained in the ancient Egyptian 'Book Of The Dead': by Wallace Budge. The 'Book of the Dead' itself was actually a guidebook for those upon reaching the underworld in death to find their way out. In effect it was similar to our own Bible with a similar intent and purpose. (As the Bible is to the church and cathedral, so was the Book of the Dead to the Pyramids).

The pyramids probably embodied all the known technology of the ancient system of physics, the more recent megalithic structures being merely faltering attempts to emulate and retain the knowledge of long ago embodied in these great geometric generators.

'Pyr-amid' in Greek means "Fire in the Middle". The word may also derive from the ancient Hebrew/Chaldean, "Urrim-Midden" which means "Lights and Measures". *Edgar Cayce* in his readings has given that the Egyptians called the pyramid 'Khuti' or 'Khufu' which was said to mean "Glorious Light". The pyramid was also called "The Pillar of Enoch", and in the Bible is referred to as "The Pillar of the Lord". (Could it be that the great pharaoh *Khufu* was named after the pyramid?).

Another name given was 'Mer' which literally means "Place of Ascension", (i.e. ascension to the abode of the gods). Another name, 'Merkhet', means "Instrument of Knowing". The Coptic name for the Great Pyramid was 'Piramit', meaning "the Tenth Measure in Numbers". (Indicating perhaps that the mathematics and geometry used to construct the pyramid and other structures of that era were based on the decimal system. In fact, my own five year analysis of the Pyramid and its geometric proportions seems to support this.

[The decimal system is not necessarily a modern one. A numbering system based on ten is very reasonable since we intuitively learn as children to count on our ten 'fingers'. Our ten toes would seem to reinforce this logical choice. Ten then seems to be an obvious selection for a numbering system base (although a base of twelve is actually better)): [see note]. It may be immediately perceived that the various names given to the pyramid constitute another classic "*double entendre*", - a favorite 'magical' word play of the ancients.

It also seems evident that the name and its variations were known to all cultures of the ancient world. The Mayans of Central America called their pyramids, 'Pirhua Manco', meaning "Revealer of Light".

It is also interesting to note that modern scholars now believe the name 'Babel' (reminding us of the Tower of Babel), i.e. Babylon - derives from 'Bab-ili' meaning "Gate of the Gods". (It is also believed by Bible scholars that the name Babel derives from the Hebrew 'Babal' meaning "mixed, confused, or confounded", - from where we get our own word 'babble'), The word Babel then is another good example of a *double entendre*. (The magic intended was to link two ideas using words so that the use of the phrase would invoke the power of both, and at the same time join them in magical association. [See note 4].

According to old records, the Tower of Babel was a Ziggurat (Zig-Gur-At). (Among other meanings, 'Zig' also means 'step'. Ziggurat means 'Stairway to the Gods'). This was a type of stepped pyramid, an example of which today measures 288 feet both in height and along the base. This tower in the Sumerian language was called "Etemenenanki", which translated means, "House of the foundation of Heaven and Earth".

> In reviewing the names and titles given to great pyramidal structures around the ancient world, it becomes apparent that certain words, phrases, and concepts

are attempts to describe the purpose and function of these buildings. We have seen that key words are; "Fire", "Light", "Ascension", "Knowing", "Gate", "Revealer", "Heaven", and "Gods".

With these few clues - taken together with what we now know from the many researchers and writers on the subject, and a cultural awareness that the Pyramid generally is a thing of secrets and mystery, - we can now conclude intelligently that the pyramids were places (i.e. temples) of initiation and spiritual enlightenment, and a pathway to a more elevated existence, - as well as perhaps communication with the 'gods'. (We have always been told thus in any event by dedicated mystics and researchers, such as *Mme Blavatsky* and others). [See note 2].

Furthermore, they were a place where some kind of 'fire' or Energy would be found or generated. If one reads the 'Egyptian Book Of The Dead' this conclusion becomes a certainty. (this manuscript may also be translated to mean "The Book of the Great Awakening").

Those being initiated were introduced to some strange forces and further, were made to undergo a process of ascended mental and/or spiritual enlightenment where some kind of forbidden and exclusive knowledge was revealed to them. This initiation incidentally, was reserved for the elite classes such as the royalty and the priesthood, etc., (and no doubt the very wealthy got their slice of cake too!).

One such Adept or Master was *Moses*. He studied at the great temple of Heliopolis. In Egypt he was known as "*Osarsiph*' or '*Manetho*', (A Search in Secret Egypt: *Paul Brunton*. According to the Bible, Moses was a Man of the temple and was versed in all magical wisdom, (Acts VII, 22)). In Egypt, Heliopolis was the spiritual equivalent of what Lhasa in Tibet, or what Jerusalem would be in the world today, - i.e. a center of legendary religious mystery and occult learning.

(As a prince of Egypt, Moses would have been trained at an early age in the arts of Magic and psychology. Childhood is the best age to start with such training since children's minds are open and they will believe anything. Belief is the most powerful force; it is the essence of Magic).

NOTE: in Britain the numbering system opted for was one based on twelve. This is why a foot has twelve inches, twelve pennies in a shilling, and why much trade was also based on this number. It is possible that this system was derived from the system used by the ancient Phoenician traders who spent a lot of time travelling to England where tin was to be found in abundance. Tin was an essential ingredient in making bronze.

In the ancient Middle East, the numbering system used in Babylon and its surrounding areas was the sexigisimal one based on sixty. The two systems, - based on twelve and sixty - were actually very convenient since twelve for example, may be divided by two, three, four, six, and twelve itself. (incidentally, Brazil, - the country's name, is a word derived from the Phoenician name for iron, i.e. "Brzl").

NOTE 2: Seldom mentioned is the fact that all around the base of the Great Pyramid are to be found sea-shells, as if at one time this land had been immersed and had formed the bed of a sea. It is interesting to speculate whether the pyramid itself has been immersed. This is further corroborated by the fact that when first opened - the walls, floor, and ceiling of the so-called "Queen's Chamber" were coated with a half-inch layer of sea-salt. This reminds us hauntingly of the Biblical story of the Great Flood; 'The Secret Teachings of All Ages': Manly P. Hall).

Δ

< structure > The pyramids are now silent schools of construction technology which required tremendous planning, great intelligence, resources, time, manpower, engineering, purpose, and the will to carry out the project to its conclusion.

It is therefore inconceivable, in light of the ancients' advanced knowledge of megalithic technology, that all this planning and effort would be focussed into such a vast and important structure, while omitting one of the most important and key elements associated with most, if not all, megalithic structures.

This element was non other than _underground running water_ – the central key and generator of that for which all the planning must have been for, namely - q. Energy, - as will be explained.

The Great Pyramid is no less than a solid state quantum-electronic device, - and the underground streams were the generators of both quantum Energy and electricity itself! (after _Michael Faraday_).

If investigative archaeology is done with this in mind, there is little doubt that evidence for underground streams existing or having existed under this and other pyramids - will be found.

In fact, infra-red photographs of the Sahara desert from overhead satellite cameras show that underground rivers have existed there in the past and still exist today. Furthermore, this underground reservoir of water is estimated to be about 150,000 cubic miles in volume! It is also known that the Sahara used to have large rivers flowing across its surface, and had large forests covering many parts of it long ago.

It is now generally agreed that the Sahara was once a lush green and fertile plateau, about five thousand years ago. It is quite reasonable to suggest that the land supporting this lush landscape degenerated and turned to sand which then, blown by the winds, covered the rivers. These rivers still flow beneath the covering sands!

It is also quite possible that the vertical shaft sunk into the floor of the so-called 'Pit' or "Chamber of Chaos" within the Great Pyramid is a finger pointing down in the bedrock as a clue to what lies beneath. (Perhaps it was an aborted attempt to tap into an underground stream! Or indeed, it may still be a functioning channel for the Energy flow from below!

As we have already seen, Energy is attracted to cavities, tunnels, and shafts, etc., in a parallel context to the way in which animals, being imbued with "life-energy", - are also attracted to these same features of topography.

< purpose & technology > the Great Pyramid as a technological device has four primary functions. Firstly, it was designed geometrically to work as an "inverted funnel" to accumulate and channel upward the generated electricity and Energy from the "living rock", i.e. bedrock, below. This accumulated electronic energy infuses the entire mass of the granite and limestone structure.

However, as we have seen in the work of *W. Reich*, such a mass attracts Orgone (i.e. quantum Energy) towards its center of mass/gravity/geometry, - especially when the mass itself is suffused with the Energy. This center roughly coincides with the two great interior chambers, i.e. the so-called "King's Chamber", and the "Queen's Chamber" below it. These chambers, being cavities, are further attractive to the Energy and consequently become veritable "immersion rooms" or 'baths' of accumulated q. Energy. [note 3].

Secondly, the planet generates electricity within its dynamic molten bowels partly as a result of its electro-dynamic field in space, which captures free electronic particles and conducts them toward the planet at the polar regions. A generated electron flow or current percolating up from the molten magma through the bedrock, the cover of soil, and water, etc., eventually is emitted by the surface into the atmosphere. (The wind and ocean spray are effective mechanisms to this end).

These electrons and negative ions are then picked up by the moving air currents and transported aloft to be eventually deposited and collected by water vapor and droplets forming clouds.

The pyramid's base therefore acts as a collector (as does any high mountain) for these percolating electrons, - and q. Energy.

Granite and limestone are usually regarded as electrical insulators, but in fact they are conductors of electricity, albeit poor ones. The electrons percolating from the base-rock of the pyramid, from the two sources mentioned - to its apex, constitutes an electronic flow. (It is very possible that the selection of special stone for the construction of megalithic sites everywhere was based also upon the stone's electrical conductivity and/or its slight radioactive content).

(Furthermore, not only is granite a conductor of electricity, but so are the (vertical) *interfaces* between the stone blocks of the construction. In fact these interfaces are better conductors than the stone itself).

The moving electrons attracted to the apex and coerced upward by the difference in voltage potential, interact electronically with the atoms of the pyramid's stone, thus producing yet more q. Energy. Underground streams of water also are very efficient conductors and collectors (and generators) of electricity from the surrounding area.

The voltage potential at the Earth's surface extends upward from one hundred to three thousand volts per meter so that the gradually accumulated charge at a mountain peak, for example, would be significantly higher than at base level. The result of this activity would give the mountain peaks a potential difference of millions of volts.

When winds and clouds move and form at these peaks, they are also picking up tremendous amounts of electrical energy (i.e. electrons and negative ions) *and* q. Energy, - thus contributing to weather patterns. (mountain climbers cannot experience this difference in potential since they too are accumulating charge as they climb).

The positive charge thus induced at the peak by winds and sunlight, etc., serves to intensify the negative electric current flow upwards from below. In overall effect, the mountain itself is like a point on the surface of an electro-statically charged sphere, attracting electrons which then gather at the tip of the point, from where they are then released into the atmosphere. Thirdly, the Earth's rocky mantle is constantly in a mode of mechanical vibration due to incessant seismic activity, which normally we are not aware of. (Every day there are over a million earthquakes distributed all over the globe, most of them too subtle to be perceived by the average individual). Hence the stone pyramid, being built directly upon a bedrock platform and thereby forming an extension of it, also vibrates, and this includes the composite granite and limestone internal structures. This vibration of the pyramid structure and its internal components is another mechanism for producing Energy. This vibration is ultra-sonic in nature. (See *Stonehenge*). All buildings are subject to the same modes of vibration.

When a semi-conductor of electricity such as granite, limestone, graphite, etc., is undergoing ultra-sonic vibration, its 'work function' increases, which is to say its electrical conductivity increases. In other words, the material becomes a more efficient conductor of electricity.   (This is partly due to the constituent atoms in the crystal lattices being "jiggled" and "bounced" around, thereby "stretching" and "compressing" the inter-atomic electronic bonds. Consequently, the atomic shells are thereby distorted and plucked like piano wires and caused to vibrate. This not only "loosens up" the valence electrons but also generates q. Energy).

Furthermore, its conductivity will again increase due to a phenomenon known as photo-conductivity. This means that a semiconductor (like stone) will conduct electrons better if exposed to electro-magnetic Hertzian radiations such as radio waves and microwaves, (which are always in the atmosphere, being generated by the thousands of lightning storms in existence around the Earth at any given moment).

> Above the King's Chamber are five so-called 'relieving chambers'. They were so named because, for lack of a better theory, it was thought their purpose was to relieve stresses in the pyramid's construction. Upon analysis, this seems like nonsense since there is so much "unrelieved" stone above this construction. Furthermore, there is no evidence of such 'relieving chambers' in other pyramids. Likewise if there are no other examples, why would the engineers incorporate this type of

construction on a first-time, one-and-only basis; especially in a pyramid which is apparently the result of the lessons learned in many previous structures? It seems probable then that these are not 'relieving chambers' simply because they are not needed as such. The pyramid could stand firmly without them.

The answer to this puzzle and the likeliest is again vibration. These overhead chambers are formed by giant 70~ton slabs of red granite which are supported only by the walls of the King's Chamber and their five extensions above. In other words, they are designed to function like the "hanging stones'" of Stonehenge, - the dolmens (so-called "table rocks"), - and other such megalithic structures which support overhead stone slabs. ('Stonehenge', incidentally, literally means "stones hanging").

(Interestingly, it is known that these same 70 ton red granite blocks were brought from their quarry site situated 600 miles away. They were likely selected for their piezo-electric properties (and conductivity)).

While "hanging" as they do, these henge-stones are in mechanical tension, and as such - like a taut piano wire or drum membrane - they are especially sensitive to mechanical vibrations and will amplify such vibrations, - especially if they are in resonance, (the same frequency).

These vibrating granite slabs then while in tension, serve to cause electro-mechanical vibrations in the stones' crystal structure, which is then passed on to the constituent atoms by means of their interatomic bondings. These vibrating atoms in turn generate q. Energy as well as ultra-sonic emissions, and high frequency electric waves. (Audible sound vibrations are another factor; - to be explained). (All of this described activity is still occurring in the Pyramid, even now, as you read this!).

Granite is composed largely of quartz which is a piezo-electric crystal. This means that any mechanical vibration of the stone, with its attendant tension and compression, results in the generation of rapidly polarizing electric fields in the crystals whose frequency is precisely that of the granite's mechanical (ultra-sonic) vibration.

These oscillating electric fields in turn radiate electric field E-waves (of ultrasonic frequencies) which, impinging upon other matter, generates Energy in that matter - whether it be stone, flesh, bone, brain, or pineal gland, etc. To add even further to the generated q. Energy, the vibration of the stone pyramid was further enhanced during ceremonies taking place in the King's Chamber (A.K.A. "The Hall Of Initiation") and the Ascending (Grand) Gallery by the use of song and chant.

Indeed, *Madame Blavatsky*, the great Theosophist, occultist, and adventurer of the Victorian era, stated that the central chamber of this pyramid was used for such rituals. Blavatsky, founder of Theosophy and writer of 'The Secret Doctrine', strongly supported the long-held belief in the existence of a group of 'Masters', or a 'Brotherhood'.

According to her, these men or their ideological descendants, were the sacred guardians of secret knowledge and were associated with the Great Pyramid. This is in a manner similar to that in which the Pope and the organized papal infrastructure supports and maintains the Catholic religion, cathedrals, and church today).

Fourthly, the pyramid was a communication device. The so-called Grand Gallery of the pyramid was primarily designed as a hall of song and chant, (like megalithic sites and our present day cathedrals and churches, etc.). Here, during these ceremonies, a choir was assembled which perhaps sat on temporary wooden seats, (and allowed some water to ease their throbbing throats!). (The use of such seats would indeed explain the purpose of the twenty-eight socket-like recesses set sequentially into the two opposite walls of the ascending Grand Gallery). [See notes 5a & 5b].

The overhead lintel stones in the King's Chamber were and are designed to vibrate in resonance to the sounds generated by the choir assembled in the Grand Gallery. When they are set into vibration they generate ultrasonic frequencies, - and piezo-electric waves. It is likely that the red granite chosen for the King's Chamber was chosen for its high quartz content, - and attractiveness.

This phenomenon has been observed at many megalithic sites in England and Europe. These stones then generate electric waves of ultrasonic frequencies which interact with the field of the electrically charged apex. The result, combined with the created sound frequencies, is the generation of an electric carrier wave emitted and radiated out by the apex. This carrier wave generated by song and chant then is capable of transmitting lower frequency electric waves of audible sound frequency such as spoken words, etc. emanating from within the King's Chamber.

It is said that if one shouts in either of these two large cavities (either the Grand Gallery or the King's Chamber), the surrounding granite walls will ring like bells, i.e. they vibrate at these sound frequencies. (Anyone who has been inside a small tunnel in bedrock, such as a mine shaft and uttered a shout can confirm this "ringing" phenomenon).

One can now understand why the Grand Gallery was constructed to be high and vaulted. This type of construction in the Pyramid is ideally suited for the resonance and amplification of sounds generated within this cavity.

This same phenomenon can be experienced in cathedrals where structural stone is under great compression and tensional stress. Indeed the interior spaces of cathedrals and their stone columns were *designed* to "ring" in the same manner. [see note 6].

It is also likely that Greek, Egyptian, Babylonian, Phoenician, Israelite, etc., temples were designed to function in the same way. The tall stone columns, like the columns in cathedrals, under compression and supporting their massive lintels - in turn under tension - would all function like gigantic tuning forks or harp strings, vibrating in resonance to the raised voices of assembled worshippers. This is precisely why Mediterranean temples were constructed this way.

(This makes good sense since now we have a new discipline called "Archeaoacoustics". Apparently recent discoveries have been made by researchers linking ancient megalithic sites with the science of <u>sound</u> and its manipulation within these stone structures and their chambers).

The technique of using stone columns under compression goes back to Heliopolis in Egypt and probably beyond that. (Heliopolis was the first Egyptian religious center). From Egypt it was adopted by the Greeks and other Mediterranean civilizations such as the Phoenicians, and so on. From there it went to the Romans and then on to the middle eastern civilizations. The Templars, Crusaders, and perhaps merchants and pilgrims eventually brought this secret knowledge to medieval Europe.

The last remnants of extant knowledge were probably lost then with the demise of cathedral building and the advent of the *Renaissance* period. (This knowledge probably does still exist although now suppressed, locked away in cloistered custody in such places as the Vatican library and elsewhere, - available only to the initiated).

The King's Chamber is a large cavity which resonates to certain frequencies of sound. Sound travels at 743 miles per hour in air at sea level and at 32 degrees Fahrenheit (zero Celcius). The King's Chamber is reported to maintain a steady temperature of 68° F. so that in this room sound would travel at a velocity of 769 mph or 1128 feet per second. (The minor fact that the K.C. is not quite at sea level is of relative unimportance here) (Interestingly, 68° F. is precisely equal to 20° Celsius, or exactly one fifth the boiling point of water at sea level).

The King's Chamber (KC) therefore, which is 34 feet long and 17 feet wide, (and 19.2 feet high), would resonate primarily to sounds and harmonics of 33.2 cycles per second, 66.4 cps, 133 cps, 266 cps, and so on, – equivalent to the sounds produced by the deep voice of a male baritone, or a large drum, or harp, etc.. Such a voice can create sound frequencies in a range from 70 cps to 250 cps.

(It is also interesting to note that the length of the KC is exactly twice its width and that the length of the Grand Gallery (153 ft.) is exactly one fourth the length of the sloping edge of the pyramid's exterior (612 ft.), and is a precise multiple of 17, the width of the King's Chamber, (9 x 17)).

The KC is also connected by two channels proceeding from two opposite walls up into the masonry of the pyramid, so that there are two openings in these walls. When the King's Chamber was sealed shut by the three granite slabs (no longer in place) closing its entrance, it became a sealed room except for the two openings. (The granite slabs referred to once operated by sliding up and down, and it is said in ancient sources that they were operated simply by the use of <u>sound</u>).

The sealed chamber then together with these two small openings constitutes a large 'Helmholtz Resonator'. (A Helmholtz Resonator is a cavity with two small openings and is designed specifically to resonate at certain frequencies of sound. Such a cavity resonator has its length equal to a whole number of wavelengths of the contained energy).

It is easy to see that when the three granite slabs were lowered to seal the chamber that it would then become practically hermetically sealed and airtight except for the two small channels mentioned.

It is said by Bedouins living in the area that the Great Pyramid, from time to time - emits a deep sound. This may easily be explained by the action of the wind blowing across the faces of the pyramid at certain angles and interacting with the exposed ends of the air shafts leading to the King's Chamber. This weird effect is similar to that produced by someone blowing across the mouth of a bottle.

This action in turn causes an oscillation of air in the cavity within at low frequencies, which then becomes emitted sound waves, - the sound frequency being a function of the cavity size.

All in all, when the activities within the pyramid were in their heyday, the resulting effects must have been truly transcendental for those involved, - both literally and figuratively, objectively and subjectively. The pyramid must truly have been a thing of awe for every observer; - truly, a "house of the Gods".

Δ

SUMMARY; We can now summarize all of the technical mechanics involved in making the Pyramid a solid-state electronic device for the purposes of generating electronic flow and q. Energy.

They are;

1. the use of underground flowing water,
2. the use of naturally occurring electrical flow from the base to the apex,
3. the use of naturally occurring ultra-sound,
4. the use of sound vibration generated by song and chant,
5. the use of tension and compression in the stone elements of the structure, and
6. the electrically conducive impingement of electromagnetic

radiations which are constantly in the atmosphere at all times. We may also include

7. the generation of positive electrical charge at the apex by the action of wind and sun. (> *Electrodynamics*).

The Great Pyramid is without doubt, the grandest and most sophisticated pyramid and structure of the entire ancient world, embodying all the known science, physics, geometry, and architectural skills of the time. Since then the knowledge has seemingly slowly evaporated (or been suppressed), so that now virtually nothing is known about the structure, its purpose, its function, or construction. Probably the only remaining remnant and reminder of this glorious structure are the spires we see atop cathedrals.

Alternatively, it is even possible that the Pyramid was built to endure along with an organization of Adepts or 'Masters' with compelling ideology, - just as for the last 900 years, another compelling idea (Christianity) has existed, allied with the technique and technology of cathedral building.

The knowledge of stone cutting and dressing, geometry, and cathedral building, allied with mathematics and the technology of sound, was brought to the Western world chiefly by the Templars and Crusaders, and is largely responsible for the succeeding enlightenment of the *Renaissance* Age, beginning at the end of the thirteenth century.

Δ

< initiation > As part of the initiation ceremony, the initiate was sealed, by means of the three stone slabs, within the King's Chamber, and possibly also within the so-called 'coffer' or 'sarcophagus' by means of the lid (now missing). (This 'sarcophagus' might be more properly called "the Womb of Initiation", so that the initiate - upon arising from the initiation ceremony, would be "reborn" - having learned esoteric and secret knowledge). The chamber was then in effect, a *sensory deprivation* chamber. This, coupled with the 'bath' of 'life-energy' or quantum Energy in which the initiate was immersed, was to work upon the pineal, pituitary, other glands, and the endocrine system in general to bring about a powerfully transcendental state of mind or being.

It can be said then that the overall purpose of the Great Pyramid, (and other pyramids), was primarily twofold. It was to serve as a communication beacon, i.e. 'radio station' (to be explained), and also as a device for mental or spiritual improvement and enlightenment, i.e. a temple, (including perhaps another kind of "communication"). In this context, it is likely that the Chamber was also intended to be for sensory deprivation where out of body experiences may have been sought. The temple would no doubt also have been a powerful bio-healing agent.

(Diagrams have not been included for this article since many superb drawings and pictographs dealing with this subject may be found abundantly in many libraries).

Final Note: many books, and a great deal has been written in the past about so-called 'Pyramid Power'. Quantum Energy is evidently generated by the angular form and mass of the pyramid shape. This is a special case for quantum physics although it does not occupy an important place in this book. Many claim that experimental models of the pyramid must be of the precise proportions corresponding to the actual pyramid.

I personally believe this is whimsy and subjective perception, and that this proportion is unnecessary. The secret lies not in proportion but in the angular structure. There are two major reasons for this consideration.

Firstly, gigantic pyramids can be found all thoughout Egypt, and it is unlikely that such tremendous resources would be expended to construct things which don't work. This would be like building an elaborate clock then omitting the mainspring or connection for the battery. The second reason is that my own researches and those of others show that the Pyramid is an extremely sophisticated construction employing great mathematical genius, geometric principles, numbers, and proportions.

It is highly unlikely that such a structure would - out of stupendous fortuitousness and luck, also be the only proportion which generates Energy. Further is the consideration that if the geometric form of the G.P. is able to generate Energy, then why should not other geometric forms also be able to? In fact, _all_ matter generates and accumulates q. Energy. The pyramid form simply - but cleverly - utilizes this propensity of matter.

Δ

NOTE 1: That such monumental and time-defying constructs were left without a single word of explanation would render it almost obvious that these buildings were intended to be the engines of a mysterious and covert knowledge.

NOTE 2: cathedrals and the Gothic style made its appearance in Christian Europe in about 1130 AD. It appeared suddenly everywhere throughout Europe, always in association with Benedictine and Cistercian abbeys. They appeared immediately after the return in 1128 AD of the first nine Knights Templars.

The word "Gothic" derives from the Greek 'Goes' and the French word Goetie, meaning magic or fascination, i.e. 'goedic'. These in turn derive from the Cabbalistic secret alchemical language 'argotique'. (Possibly the hallucinogenic 'Ergot' (a rye mould) derives its name from this source).

NOTE 3: for those interested in pyramidology in general, the following is an article taken from the Vancouver Province newspaper (Canada) of Feb. 01, 1987. News Services, Nazlet el-Semman, Egypt.

> Japanese archaeologists using high tech equipment say they've discovered a concealed tunnel in the belly of the Great Pyramid.

"We discovered it by chance," said _Sakuji Yoshimura_, head of the team from Tokyo's Waseda University, as he worked yesterday inside the 134-Metre (436.6 feet) pyramid. (This may be the present height of the Pyramid, however it was originally 481 feet, including the capstone).

The Great Pyramid, tomb of the Pharaoh _Khufu_ who ruled Egypt around 2650 B.C., is the world's largest stone structure. (_Cheops_ in Greek).

The Japanese used a device that beams electromagnetic waves into solids to a depth of about ten metres, producing an image on a video screen. Yoshimura said his team confirmed the existence of unexplained cavities discovered last year by French experts.

NOTE 4; In 'The Holy Blood and The Holy Grail', by _Baigent, Leigh_, and _Lincoln_, Corgi books, 1982, It is given that "_San Grael_" meaning Holy Grail can also mean "_Sang Real_" meaning Royal Blood. This is a classical and very typical (and clever) _double entendre_ which could not have been accidental, due to the historical period, its importance and subject, and especially in a period when double entendres were commonly used. Indeed even if it were *unintentionally* used, it would imme-diately have been *recognized* as a double entendre in any event - and could not therefore have been used mistakenly.

NOTE 5a: researchers and investigators into the purposes and activities of large scale megalithic sites and the megalithic age are more or less agreed upon several facts. Firstly that such stone structures were tremendously important to the priest-hood and rulers of the community.

They served as gathering places, centers of healing, temples of worship, and community centers. Indeed, churches have served this purpose well into the twen-tieth century. Sadly this benefit to society has been largely lost. Such worship often took the form of song and dance, usually also accompanied by feasting and bonfires. Such a community center would also serve as a seat of government such as it was, population organization and control, exchange medium of news and gossip, net-working, project organization center, and so on.

Also generally unknown to the population were the secrets employed in con-struction of the sites, including the technology of sound manipulation and under-ground flowing water. These are two mandatory fundamentals of megalithic site engineering. It is more than passing likely therefore that the Great Pyramid, grand-father of all megalithic structures, would incorporate not only underground running water, but also the use of sound, song and chant.

NOTE 5b: twenty eight seats would mean at least twenty eight members of a choir and very likely more; - the more, the better. Each seat would be capable of seating four, so that the choir would number 112.

NOTE 6: Now the vaulted 'Grand Gallery' - so-called - is a mystery of magni-tude and its purpose is unknown. It is difficult to conceive of any practical use for it unless one allows that it was used as a chamber of song and chant. With this in mind, the hollow cavity makes good and practical sense. Furthermore, if one were to design such a cavity intended for use by a choir of initiates inside a pyramid of stone, this hollow geometry would appear to be an excellent purpose-built design.

Further, in consideration of the twenty eight sockets, they are set into the stone at precise and regular intervals, they are identical in measure and geometry. Since each pair is set opposingly in the stone, it appears obvious that each socket of the pair worked in conjunction with each other. All the sockets would of course serve identical purposes. The simplest and most logical explanation is that they were an-chor points supporting a cross-member which was portable in nature, i.e. made of wood. These cross members were simply used as seats.

160 B ~ THE GOLD APEX & COMMUNICATION

Chemists will tell you that the element Cesium (a metal) is the most reactive solid. This is due to its electron shell configuration, (2,8,18,8,1) with a single elec-tron in the outermost or 'valence' shell. The most reactive of all the elements is fluorine. Actually however this is not precisely true. There is a solid element which is more "reactive" than Cesium and yet paradoxically is extremely stable, - that element is <u>gold</u>.

(All these elements have one feature in common which is that they possess but one electron in their valence (outermost) shells).

In the special case of gold, the single outer electron is held so weakly by the gold atom that it will give it up almost as if it had no use for it. Therefore any piece of pure gold left standing in an atmosphere containing - say oxygen (as in air) - will gain a slight positive charge.

If you were to pick up such a piece of gold there would actually be an exchange of electricity or re-distribution of charges between your body and the gold - (if the gold has not already been grounded). (Because of this potential for electrical activity in the air, gold is sometimes recommended to be placed in contact with a surface abrasion or bruise to facilitate healing. (as described in 161).

Gold then is certainly a "noble" metal in that it behaves in an "aloof" or non-associative manner. However gold can chemically react with, and be dissolved by Hydrochloric acid – the same acid in your stomach's digestive juices. It is therefore possible to eat gold (in limited amounts), and this is done in parts of Asia where it is believed that doing so is good for the health and psyche: [see note 1].

> If a gold 'capstone' was ever placed upon the Apex of the Great Pyramid, it would very quickly become highly electrically charged under the action of wind and Sun. (Moving oxygen gas and ultra-violet light being the key active agents, i.e. as in electro-chemistry and photo-electricity). This would make the capstone a positively charged terminal surrounded by its positive electric field, which would then extend down through the pyramid rock structure, thus providing a boost for the negative electron flow upwards from the base. This electron flow would percolate through the granite structure as previously explained.

In other words, the pyramid base would be a negative electrical terminal and source of electrons (i.e. a cathode) which would then percolate up through the granite and limestone toward the positively charged apex (anode). Hence would be established an overall powerful electric current through the bulk of the pyramid. (See previous 160 A, *The Great Pyramid*).

In the 1930's, *Sir Wm. Siemens* - a British engineer - demonstrated this to his Arab guide. He wrapped a wet paper around his glass water bottle thereby converting it into an electrical capacitor. This capacitor then quickly accumulated a high voltage charge. Then by using his own hands, he completed a circuit between the water in the bottle and the wet paper - causing a loud spark to issue forth and frighten his friend, ('Guide to Pyramid Energy'…).

After reaching the gold apex, these excess electrons and Energy would be wafted away by the moving air currents so that the entire volume of air above and around the pyramid would be constantly supplied and saturated with electrons and negative ions, - and quantum Energy. This in turn would have a resulting modifying effect on the local weather, causing clouds to form in the pyramid's vicinity. (*Reich's Cloud Buster*).

The described fortuitous electronic quality of gold is precisely why gold was chosen for the apex. It has been reported by pilots flying directly overhead of the pyramid that their compasses and electronic equipment go haywire and do not function properly. On occasion, it has also been reported by Bedouins camping in the area that mysterious lights are sometimes seen hovering above the apex.

A light beam passing through a source or field of Energy is electrically modified and thus becomes a carrier of Energy in much the same way as an FM radio carrier wave carries a modifying frequency. The apex of a pyramid is a source of electrons and q. Energy, and a beam of light as from the Sun or Moon passing the apex will "pick up" or " absorb" and become a carrier of healing q. Energy.

Therefore if one were to stand at the point on the ground where the shadow of the pyramid's apex passes, one would receive a few minutes bathing in Pyramid generated healing radiation. (This conjures up the poetic and intuitive image of the pyramid at night with the full Moon hovering magically above its apex. And who knows, perhaps by tracing a path on the ground while following the Moon or Sun in this way, some interesting discovery might be made regarding the pyramid!).

(Similarly, a tree is a good source of Energy, and sunlight passing through the leaves will gather and transmit a large dose of Energy being emitted by these leaves to any organism happening to stand or sit in its shade). (I have called this phenomenon "the *Jesus Effect*", and if you have read the Bible's *New Testament* you will understand why).

It is also possible that Egyptian obelisks and megalithic standing stones ('Menhirs') were similarly used in this way. In ancient rituals where people would dance in a circle around these tall stones, they would pass momentarily a number of times through the stone's shadow.

It is entirely conceivable, even likely - given the importance and relevance of the structure (though not recorded) - that such processions were a ritual of worship conducted in the pyramid's shadow from time to time. It is also conceivable that such celebrations were timed to coincide with the full Moon since during such times, the atmosphere would be favorably replete with Energy and calmness.

Δ

< the pyramid and communication >

The gold apex of the Great Pyramid was/is charged to a high voltage with static electricity and has a positive charge. This positive charge is maintained by the effects of wind and Sun. The voltage of this charge can be roughly estimated. The foundation bedrock of the pyramid is emitting electrons into the structure's base. The voltage potential in such a structure, or in a tree for that matter, increases with altitude at roughly 100~300 volts per meter. If you are standing on the ground long enough with your bare feet, an electric charge will accumulate in your body, and if you are about six feet tall the voltage potential will eventually reach 200~600 volts or more at the top of your head.

This accumulation would take considerable time however, due to the small number of free electrons available at the ground level, the small area of your foot-soles, the natural electrical resistance of the body, (which is about twelve thousand Ohms - (12,000 Ω)), and the fact that there is no positive charge at your head inducing the electrons to flow upward. The Great Pyramid by contrast has been standing for thousands of years, and this accumulation is of course being constantly maintained.

Since this pyramid would originally be 481 feet tall (with an apex), the voltage accumulated at this point (!) would be roughly from 50,000 to 150,000 volts! This is an estimate based on a structure which is not even "funnel shaped" - as is the pyramid. This upward electronic flow is condensed, flowing as it is from a wide base (roughly thirteen acres, i.e. 756 x 756 feet) towards a virtual point, so that a very much larger potential voltage accumulates at the apex. The voltage here may be then as great as a million volts or more: [see note 2].

The pyramid - as explained, is in constant mechanical vibration at many frequencies, particularly ultra-sonic. This means the apex is also vibrating, and since it would be highly electrically charged, this mechanical vibration would naturally be translated into electrical waves of ultra-sonic frequencies, i.e. above 20,000 cps, so that the gold apex would be constantly emitting and radiating electric field wave frequencies.

During ceremonies within the pyramid as explained, a choir giving song and chant would serve to enhance this vibration - thus creating a powerful electric carrier wave of sonic and ultra-sonic frequencies beamed out in all directions by the apex.

Hence any sound or words spoken within the King's Chamber cavity of the pyramid would add to this vibration a coherent message which in turn would be carried along by the carrier wave being beamed out. The "carrier wave" vibrations would be provided by the priestly choir in the 'Grand Gallery' while the real communication would be exercised in the KC.

The lintels, or red granite blocks forming the ceiling of the King's Chamber are arranged in a fashion so that there are in fact five such 'ceilings', each one above the other and each separated by an air space - with a total of five such air spaces. The need for so many spaces may be explained by the physics of sound propagation. ('Sound Waves And Light Waves', Winston E. Kock).

If sound is generated in the King's Chamber, the sound waves are transmitted to the surrounding stone and then travel outward through this material. Sound travels faster through stone than through air. This means that the sound vibrations travelling upward through the rock above the chamber interact with these air spaces.

These alternating slabs of rock and air in turn serve to act as a focussing 'lens' system for the sound. (Parallel to the way in which multiple lenses in a camera focus the light to a small area on the photographic film). [note 3].

This means that the sound travelling upward through the stone is now focussed into a beam by this arrangement of lintels and continues upward until it is further focussed by the conical shape of the upper part of the pyramid. It then impinges upon the electrically broadcasting gold apex.

In other words, the pyramid is a kind of radio transmitter using sound, ultra sound, and electric fields, - with the obvious intention of the broadcast message being received! Anyone having a *suitably designed receiver* would of course be able to duplicate the intended message. (> *Ark of the Covenant*).

Δ

< the gold apex >

Actually the apex or capstone was probably not entirely pure gold, although its outer, visible structure may have been, and from an aesthetic and functional point of view certainly should have been, especially with so important a structure as the Great Pyramid. (It would have contrasted beautifully with the polished white Tura limestone which originally covered the pyramid).

According to ancient Arab traditions in Egypt, the apex was constructed of three metals, i.e. gold, silver, and brass. The brass would undoubtedly have given structural strength and perhaps formed a framework, while silver is certainly the best electrical conductor. Then again it is possible that the metal was in fact an alloy of all three. This would certainly have been an interesting alloy, - not likely to be duplicated in modern times, (and perhaps having unusual electrical properties!).

Edgar Cayce in his readings has given that the apex was of gold, silver, and bronze. (Bronze is an alloy of copper and tin and is quite strong as ancient metals go, while brass is an alloy of copper and zinc, (also quite strong). (Copper, silver, and gold are adjacent metals in the periodic table of elements). Either way, the evidence is persuasive enough to support the notion that an apex did once adorn the pyramid's peak and that it contained primarily gold as well as silver and copper.

There is also an inscription found by the explorer *Jequier* at the pyramid of *Queen Udjebten* referring to a gilded capstone. If her small pyramid were in fact capped by a gold apex, there can be no doubt then that the grandest of all the Pyramids was also crowned. (It is conceivable that the concept of a gold crown worn by kings actually originated with the pyramid!). (Furthermore the Great Pyramid at present has a small platform at its apex. This apex also has a stone fixture, a part of which contains an original socket hole which appears to be an obvious receptacle for some other construction to be lowered onto it).

Now if we entertain the likelihood that two ancient pyramids were thusly adorned, it certainly does open the forum to speculation that indeed this was a common practice in the ancient world of pyramid building. Intuitively this certainly feels correct, especially when we consider that pyramids in general were regarded as *major* cultural, religious, social, and technological investments; - indeed the foundations of culture - much like the Catholic Church.

Since we can surmise that pyramids must have occupied a very important place in the culture and religious belief systems, and since we know that gold was commonly and widely used for religious decoration, we must allow that it therefore is very likely that gold apices were *de rigeur* for all such monumental structures. We also know that even small obelisks were topped by a gilded pyramidion.

Pure gold, as already delineated, very readily becomes electrically charged under the actions of wind and sun. It is also very beautiful in appearance and - importantly in this application, does not tarnish - always presenting a *pure* metallic surface. It can then be appreciated why gold was considered to be the "royal metal", the "*substance extraordinaire*".

NOTE 1: In fact this may be quite true for it has recently come to light that a certain form of gold was used, (i.e. ingested), by the ancient nobilities - notably in Egypt, to achieve an enlightened state of mind.   The form of gold referred to is presently called "Ormus" by modern researchers in the West, and is prepared by passing through the gold an electric arc, if you will - until it turns to a fine white powder. This powder is said to be monatomic gold consisting of isolated and unbonded gold atoms. (Very likely it is some combination of gold and oxygen, e.g. AuO_2).

This "Ormus" then resembled a very fine flour - and was in fact mixed with flour and included in special baked bread for the nobility. Obviously, the question arises concerning the connection between ancient Egypt and electric arcs. This intriguing subject has been broached in several books, including ones by *Zachariah Sitchin*. One is obliged to wonder here at the many marvelous properties of gold, almost as if it were a metal invented by the gods*!*

NOTE 2: the voltage field may be this intense, however there is no mechanism by which this tremendous electrical potential could be translated into a single discharge, due to the electrical resistance of granite. In order for this to occur, the pyramid itself would have to be of a superior conductor such as metal. Further, a suitable discharge terminal would have to exist also constructed of metal leading into a conductive ground base.

NOTE 3: Another interesting way in which sound can be focussed is by a straight row of trees or stone columns as found at ancient temples such as the Acropolis or the Parthenon, or those in Heliopolis, etc..

If you stand at one end of this row and speak, these sound waves will travel parallel to this row and will be focussed by the columns to a listener standing at the other end of the row. Hence a priest standing at one end of the temple would be able to communicate with a seeker of wisdom standing at the other end of the row of columns.

The seeker would ask a question and would receive an answer from an invisible source. An ancient Man would naturally conclude that this reply was from the god of this temple. If you ever travel to Greece, visit the Parthenon and try it out. (These Ancients were very crafty!).

161 ~ THE SECRETS OF GOLD

Gold - as already suggested, has a few peculiar properties which makes it rather unique among the metals. For example, it is reported that gold veins in the ground have the magical ability to grow in situ in their native rock! Indeed it is an interesting question as to how the gold got there in the first place because gold does not, under natural conditions, combine chemically with any other element or mineral, etc..

The reason for this is the configuration of the gold atom. Elements such as gold, cesium, copper, and silver - with one electron in the valence shell - tend to lose this electron more easily than those having more, which is why these four elements are superior electrical conductors.

According to *E. Rabinowicz* in 'Exo-electrons', (Scientific American; January 1977), electrons are emitted by a freshly exposed metal surface. A gold surface certainly qualifies as a freshly exposed surface as its chemical and physical characteristics are always as if it is freshly exposed or cut. (That is, the surface, unlike almost all other metals - never tarnishes, oxidizes, or combines steadfastly with other chemicals or elements (except mercury) under normal conditions. Therefore any gold mass always presents a surface of pure metal! (All references to gold in this book imply that the gold in question is pure twenty-four carat gold).

Next to Gold, the other metal having this characteristic is Silver. However silver is unfortunately subject to tarnishing by corrosive industrial gases in the atmosphere such as sulfur dioxide. In ancient times however silver would have retained its pristine metallic surface. [note].

If an atom of oxygen should contact an atom of gold, i.e. a gold surface, - instead of combining with the gold atom, the atom simply takes the electron from the gold atom and wafts away with it, - willy-nilly. This leaves the gold atom electro-positively charged. This would naturally happen on a large scale if a large surface area of gold were exposed to air, (which is normally composed about one fifth of oxygen).

Dr. Naomi Kanof, a Washington dermatologist who in 1980 taught at Georgetown University Medical School, has often reported success in treating skin ulcers by the application of gold leaf to the affected site while keeping the area clean with alcohol and a swab. The gold leaf is the only dressing used and it may be changed after seven days or so.

NOTE: At some points in history, silver culinary and eating utensils were used by high officials and royalty. This is due to the fact that many poisons contained corrosive elements or compounds (often with sulfur) which would be detected by the silver implements, - they turn black.

162 ~ ELECTRO-STATIC FIELD WAVE RADIATION

This type of radiation is different from Electro-Magnetic radiation in that Em radiation has both an electric and a <u>magnetic</u> field component. The electric field is provided by the field surrounding the individual electrons which are flowing through a conductor as part of an electric current. The magnetic field component is generated by the *motion* of the charged particles and their attendant electric fields through a conductor.

If an electric flow is pulsed or oscillated at a reasonably high frequency, it will cause this combined electric~magnetic field to become an electro-magnetic wave which will then separate from the conductor. Thereafter it expands and radiates away into the surrounding space, passing through non-conductive and non-reflecting matter such as air, ceramic, or wood, etc.. (Diagram 25). (Incidentally, water - especially salt water is a good reflector of Em radiations hence any electronic wave emitting technology including radar may not be used for underwater research. Hence, sound wave techniques (Sonar) are used).

(In radio technology, the desired frequencies carrying information are always lower than those of the carrier-wave. the higher the frequency of the emitted carrier wave, the more effective and efficient is the radiated signal. This is due to the fact that the <u>power</u> of the emitted and radiated waves is proportional to their frequencies. The "polarization" of the magnetic field surrounding the conductor is perpendicular to the direction of motion of the electrons (i.e. electric current), and perpendicular also to the direction of the electric field. (Refer to the "Right-hand Rule" in electronics technology). (What is not known is that the carrier wave can carry energy having a higher frequency than itself. It is simply a matter of modifying the electric field of the carrier wave: (more elsewhere)).

In other words, imagine that the expanding wave assumes the form of a spheroid or an orasphere surrounding the emitting conductor which is a straight wire. (See diagram 25). The expanding oraspherical wave pulse then is represented by the surface of the orasphere and the magnetic polarization is represented by the lines of '<u>latitude</u>'. (> 207 AD).

In the case of electrostatic field vibration and radiation from a charged sphere for example, the oscillating electric charges, i.e. the electro-statically charged <u>surfaces</u>, are physically moving to and from the observer. Hence there is no magnetic component to the oscillation of the electrostatic field - nor to the resulting field vibration compressional wave which radiates away from the object. The object may be an electro-statically charged bell, or metal sphere, etc..

An electro-magnetic pulse wave is called a longitudinal wave because it is generated by the back and forth oscillatory motion of electrons in a long conductor (or aerial). Latitudinal or compressional waves are those produced by a vibrating flat, two-dimensional surface and are similar in form and geometry to sound waves emitted by the membrane of a drum or the metal of a struck bell, etc..

163 ~ THE ARK AS A CAPACITOR

The *Ark Of The Covenant* as described in the Bible is constructed of Acacia wood and is covered both inside and outside with gold sheet. Many researchers have interpreted this to mean that the Ark technically is an electrical capacitor. In classical electronic technology, such a capacitor would consist of two electrically conductive plates separated by an electrically insulating material, i.e. a dielectric, such as for example a resinous wood.

These two plates must not be in contact with each other and must remain separated, i.e. electrically isolated, since each plate must carry an opposite charge. The gold covering the inside surfaces of the box or ark would necessarily be electrically isolated from the exterior gold cover if the Ark was to represent such a capacitor. This separation and isolation feature would for practical reasons have been at the rim or upper edge of the open box. At the rim therefore there would have been exposed wood separating the inner layer of gold from the outer one.

For several reasons however, the Ark as described could not have functioned as this "two-plate" type of capacitor. To begin with the Ark as a box - we are told - was covered by a lid of solid gold. Gold of course is an electrical conductor and as such would have provided an electrical circuit between the outer and inner layers of gold of the Ark's construction - at the rim of the box opening.

It is actually likely - as will become clear, that this rim would also have been covered with gold so that not a tiny bit of wood remained exposed, and this is intuitively correct. If the ark were to function as a two-plate type capacitor there would have to have been an exposed portion of wood, thereby providing a separation between the two gold plates. Furthermore because of the voltages involved, this gap would have to have been - for practical considerations - larger in width than a hand span, or five inches.   Besides all this, if there had been any exposed wood separating the sheets of gold, the relatively narrow gap between the gold plates would have presented a means for shorting the capacitor, - especially if the weather or air had been just a little damp.

Therefore in view of the fact that a better and simpler method exists, it is likely that the Ark would not have functioned as a two-plate capacitor. Secondly, the Ark and its construction was carefully and precisely described, but there are no specific instructions given in the Bible for the inner and outer layers of gold to be electrically insulated. If this were a necessary part of the construction, it is very unlikely that such an important and fundamental feature would not have been noted. Thirdly, there was *no need* for this type of construction, since the Ark would have been adequately charged even without this elaborate structure.

Fourthly, when *Uzza* touched the (charged) Ark - as recorded in the Exodus of the Old Testament, he was struck dead as if by lightning. He had of course grounded the Ark with his conductive body and been electrocuted. Now if the Ark were a capacitor of the two-plate variety, this grounding effect and consequent electrocution could not have occurred.

Grounding the single outer plate alone in such a two-plate box capacitor would not cause a discharge of the stored electrical energy. Therefore the mechanism for charging the Ark with a static electric charge must have been a different one, - as described previously: (*Secrets Of Gold*, 161). Since therefore there could be no practical separation between the inner and outer plates of gold, - the gold covering, both inside and outside the Ark, was of one continuous layer - including the gold lid.

In 1971 students at the Cambridge University in Massachusetts built a replica of the Ark following instructions given in the Bible. When left alone, it was found that the Ark - while locked safely in its own room, would develop by itself, a high voltage electrostatic charge in its gold covering. As high as 750 volts was recorded. The device was deemed to be so dangerous to personnel that it was quickly dismantled.

There is also the very strong argument that a two-plate capacitor has virtually no exterior electric field - this field being contained internally between the two plates and in the dielectric material, i.e. the Ark's Acacia wood construction. A single plate type however would have a fairly strong exterior field extending some distance from the Ark, thereby establishing its usefulness in electric-wave communication.

Furthermore, unusual incidents are recorded for the behaviour of the Ark at certain times. These characteristics wouold be compatible with an object supporting a high value electric field.

Instructions in the Bible were given that the Ark was to be carried using gold covered wooden poles inserted into gold rings attached to the Ark. In this case, the Ark would be grounded through the poles and the bearers handling them while being carried - and thus could not accumulate a charge under these conditions. (Even if these men were shod and thereby not directly grounded, - the fact that they were moving outdoors, exposed to the breezes and grasses, the sweat evaporating off their bare skin and hair, would all have ensured that any accumulation of static charge on their skins would have been neutralized by the surrounding air). Alternatively, while the Ark was being transported by wooden oxcart however, it would have been insulated from ground and therefore could accumulate said high voltage charge.

It therefore seems likely that it was the intention of the Ark's designer that it should - at predetermined times or due to necessity, be carried by either poles or by wheeled cart. This is evident since strict instructions were given that the Ark should not be touched while being transported by cart. It seems likely therefore that provisos were made in order that the Ark could be transported efficiently over almost any type of terrain.

It is not explained in the Bible, but it is obvious that special techniques would have been employed in removing the (charged) Ark from the cart to a safe haven. The carriers of the Ark would have been specially clothed in some way so as to protect them from the high voltage charge and consequent electrocution. Alternatively the Ark - while on the cart, could have been draped with cloth or silk ropes woven with gold thread. This cloth or rope touching the ground would ensure the continual grounding of the accumulated charge. In any event, the priesthood in charge of the Ark would no doubt have been trained in its function and operation. Of course all these techniques would have been conducted under the guise of religious ritual; - indeed, it *was* religious ritual!

164 ~ THE ARK OF THE COVENANT, ITS TRUE PURPOSE

The Ark was primarily a technical electronic device and it was in fact nothing less than a two way radio communicator - albeit based on a simpler, and perhaps a more ingenious and elegant technology. The secret of its function lies with the design of its basic structure and a particular and generally unknown property of gold, as already explained. [see note].

The Ark's structure is basically that of a large wooden box. The wood used in its construction was Acacia ('Shittim') wood which is a very strong, resinous wood, and hence very durable (and incidentally a good electrical insulator). Acacia, being very resinous, is also very resistant to dampness - a very important feature in electrical applications. The most important feature however would have been its durability and resistance to rot.

The Ark was constructed having large areas of exposed gold metal surfaces. It was normally left standing exposed to air for some length of time and insulated from grounding. This would allow oxygen in the air to constantly "absorb" (i.e. take up) electrons belonging to the gold.

Over a period of time, it can be seen that through this described mechanism, a very large electro-positive charge would accumulate on the surface and in the mass of the gold structure of the Ark. This electrical process would be enhanced if the gold surfaces were exposed to an actinic light such as ordinary sunlight.

The Ark of the Covenant is otherwise known as the "Ark of the Lord", "Ark of God", and "Ark of the Testimony". Its purpose was ostensibly to be moved as a container for precious relics such as the tablets inscribed with the Ten Commandments. The Ark was then a "*double entendre*". The tablets within would represent "the word of God" which would thereafter speak audibly from the Ark itself.

The Ark was covered with a layer of sheeted gold both inside and outside and provided with two gold covered wooden poles for carrying. On the lid were mounted two ornamental representations of "cherubim", i.e. what we would call demons, with outstretched wings. According to the Bible, God spoke to the priesthood from between the two cherubim. In other words from the box or the Ark itself.

> If you have ever carried an empty wooden or cardboard box, suitcase, or a large biscuit tin in a place where loud noises, vehicular traffic, or music is heard, you may have noticed that the tin or box structure vibrates in resonance to these noises. This phenomenon can also be experienced if you are carrying an open umbrella, as described. The taut cloth will pick up the vibrations in a similar manner. These objects - and similar ones having large areas in mechanical tension capable of resonating, will pick up and resonate to sounds - including music and spoken words.

Now imagine two or more tuning forks (or harps, etc.) vibrating in sympathy. If one fork is caused to vibrate, this vibration is transmitted to the others by means of the sound vibrations set up in the air. However these forks could be placed further apart and then <u>electro-statically charged</u>. Again, if one is struck, the vibration will be transmitted to the others - not by means of air waves, but by the vibrations or waves in the electric field communicating between them. When these two tuning forks are in a vibrating mode, spoken words will be transmitted between the two forks by means of the carrier wave established.

Now imagine that you have such a box or device which is electro-statically charged. If the box is caused to resonate and vibrate in response to a sound, the vibrating and electro-statically charged panels of the box will generate electric waves in the surrounding electric field. These waves will be of the same frequency as the resonating sound waves.

Imagine now another such electro-statically charged box some distance away. The two electrostatic fields are able to interact with each other if they are vibrating via these oscillating electric fields. The second electrostatic field will vibrate in resonance to the first. If the second box is of roughly similar dimensions to the first, it too will resonate to similar electrically induced sound frequencies. (The Reader may recall the electrically charged harps and bells discussed earlier).

The formula is as follows: - the box (Ark) is made to vibrate in resonance to sound, i.e. speech. This vibration causes the gold-covered electro-statically charged panel to vibrate. Electric waves are generated and emitted by the vibrating gold panels. These electric waves radiate and are detected by another similarly charged box (Ark).

By means of a reverse procedure, the electric waves then cause the charged panels of the second box to vibrate mechanically. This mechanical vibration is then translated into sound waves which are identical to those emitted from the first box, - e.g. speech. Hence the voice and words of a speaker at one box will be transmitted and picked up and delivered to a listener stationed at the second box.

This method of communication between the boxes is more efficient if there is already a carrier wave emitted by one box. In ancient technology this carrier wave would have been established by means of setting the box in a vibratory mode with the use of chanting and song and the use of musical instruments. The final result of all this would have been a set of devices (Arks) which were capable of electric and sound communication.

It may be seen then that any number of such identical devices could be used to communicate with each other. They would be 'radio' receivers and transmitters in a true sense of the words, without requiring any electrical power input and requiring virtually no maintenance. Musical instruments using components which are in mechanical tension - such as drums, harps, bells, and so on could also be used. It is even likely that this technology allowed the priesthood to install such communication devices in all their major temples.

NOTE: the last mention of the Ark in the Bible is in Kings viii, 12; wherein it is inferred that Solomon built his temple to be a permanent resting place for the "Holy of Holies". It is still a mystery what became of this powerfully important relic of the religious Israelites' God, *Yahweh*. My own personal belief is that this technology has been suppressed and hidden by those who understand it, such as the priesthoods and others. The general population would not have access to this arcane knowledge and would be ignorant of the technology involved.

165 ~ ELECTRIC FIELD WAVES

In our everyday experience there are three kinds of field originating from essentially electrical sources. They are the magnetic, the electric, and the electromagnetic. The first is that field which surrounds or emanates from a simple bar magnet or an electro-magnet. (Diagram 03). An electric field is that which surrounds an electron, an ion, or an electro-static charge as may be found on the sphere of a Van De Graaf Generator, - or again, surrounding a conductor which is carrying a current flow of any given voltage.

Other powerful electric fields are found in nature surrounding clouds, as in an electric storm. An electro-magnetic field is the commonest to our experience, - being found surrounding an electric current, such as a lightning strike, or an electron flow through a conductor.

The conductor may take the form of a copper wire or perhaps a liquid, or an ionized gas, i.e. a plasma. (If a plasma (an ionized gas] is caused to flow, that flow is an <u>electronic</u> flow and will consist of particles of both signs, i.e. negative and positive).

An energized conductor is surrounded by both a magnetic and an electric field which are mutually inter-dependant. If a steady DC electrical flow through a conductor is pulsed or changed to an AC flow, the resulting rapidly changing field does something amazing, (although we take it for granted). (> *Separation Of Radio Waves*). (It may be noted here that electromagnetic fields are generated by charged particles *in motion* while electric fields are generated by charged particles not necessarily so).

These pulses or rapid changes in the field become electromagnetic <u>waves</u>, which then separate from the original field and radiate away from the source. They become independent entities which travel outward as spheroid (oraspherical) layers of perturbations in the space-time continuum. (Diagram 25). [See note 1].

The ultimate cause of this expansion is the mutual repulsion of the electric field waves. The magnetic field oscillations passively "hitch a ride" so to speak on the electric waves, like passengers in a car, - which is actually quite an amazing phenomenon in itself.

As will become clearer however, it is the magnetic field which is the 'formative' or binding force maintaining the oraspherical forms of the wave (see Glossary), while it is essentially the electric field which causes its expansion. (It is important to remember that it is the magnetic field which is responsible for <u>form</u> while the electric field is the dynamic force of expansion, i.e. <u>function</u>). These two fields then give "Form and Function" to the Em wave. (The importance of this will be discussed later. The nature of pure electric wave expansion has been explored in 111 AB).

Essentially, free *magnetic* waves do not and cannot exist. They cannot separate from the main field, but can exist however within the confines of the field itself. (this can be imagined when a bar magnet for example vibrates mechanically, i.e. is struck like a bell). Even if such waves existed they could not be detected. [please see note 2].

Thirdly and importantly, electro-static (i.e. electric) fields can be a source of electric waves. (Electromagnetic and electric waves are older than Man since these wave types are generated by electrical storms around the planet. Lightning generates electro-*magnetic* waves while the cloud itself, being electro-statically charged, generates and emits compressional *electric* waves when impinged upon by sound waves, such as are generated by thunder, etc..

The electric field waves emitted by the cloud have a frequency corresponding to that of the sound waves. In other words, when you hear thunder you are also receiving compressional electric waves of the same frequency from the charged clouds (actually of course, you will receive the electric waves long before you hear the thunder). If you had a suitable, electrically charged receiver which could resonate to these sonic frequency electric waves, you would hear them emanating from your device as thunder before the actual sound waves reach your ear.

If an electro-statically charged surface, (e.g. the metal sphere of a Van De Graaf generator), is caused to vibrate mechanically like a bell, it will naturally emit audible sound waves or ultra-sonic vibrations in the air. It will also emit electric waves of the same frequency. Indeed, if a charged sphere is impinged upon by speech waves, these too will be converted to electric waves.

Just as a smaller bell emits a higher note, i.e. a higher frequency sound, so it is in the case of a charged sphere. The smaller the sphere, the higher the frequency of the emitted electric wave would be. It follows logically then that if the sphere were as small as an atom, this frequency would be correspondingly extremely high and of quantum Energy frequencies.

(In fact frequencies approaching those of electrons themselves are ultimately possible, i.e. 3×10^{-8}). (Please see - *What is E-Energy ?*).

NOTE 1: an interesting comparison may be made between an AC pulse wave and a DC pulse wave. Using a given amount of input energy, an AC pulse wave will deliver a greater volume and hence strength of oscillation, whereas a DC pulse wave mode is capable of delivering a higher frequency. Hence, AC field technology is used for radio and television broadcasting over great distances.

Think of a wood-saw having teeth. A hand-saw is designed to be pulled back and forth so that this motion may be thought of as an AC motion. It can be seen that the frequency of teeth passing over one centimeter of wood is low. This frequency is limited by the design of the saw and momentum. Another kind of saw blade is the circular saw where the motion of the sawteeth can be related to a pulsed DC mode. It can be seen that in this case, the saw teeth can pass over the wood at a much higher velocity and therefore with a much higher frequency. The same is true in electronic wave emission.

> *A pulsed DC wave can be emitted with a higher frequency than can an AC pulse. This is likewise true of an electric wave having a 'square-wave' pattern.*

NOTE 2: this is very interesting for it parallels the phenomenon of perception. If something is not perceived, does it exist? We may argue that it does or should exist, but then how does one support his argument? If it is not perceived, we cannot say it exists - but we can say it does not exist. Therefore, the deciding factor really is perception. This is the only thing we can base our agreement and reality upon*!*

166 ~ ATOMIC VIBRATION & ENERGY EMISSION

When two or more atoms or ions, such as in the surfaces and interfaces of solids, liquids, gases, etc., - collide, they may form bonds of an electronic nature. As the atoms are pulled apart, their respective surfaces, i.e. valence shell bonds are "stretched". An analogy would be as if two inflated balloons were glued together at one point and then pulled apart.

The result is a sudden release of the bonding forces with the resulting vibration of the balloons' surfaces. In the case of an atom, the surface of the balloon represents the valence shell, (i.e. outer electron shell). A similar vibration occurs when an electron (or other charged particle) passes an atom in an energized conductor, either a solid, a liquid, or a plasma. It is this vibration of the electrified atomic surface which generates and emits the compressional electric quantum E-Energy in the same way that a bell generates and emits sound waves.

As explained elsewhere, it is not only the valence shell which is involved in the vibration. Structurally this shell forms the atom's surface and hence is the more important influence. However all the shells are in vibration and are therefore sources of q. Energy. Since these inner shells are "stronger" or "tighter" in tension they - like more rigid bells emitting a higher note of sound, - emit higher frequencies of q. Energy. In effect, a large atom such as gold which has six shells is like a carillon of six bells, - each one ringing with its own frequency.

167 ~ E – ENERGY FREQUENCIES

Just as in a lightning cloud or a Van de Graaf generator's sphere, there are intermediate energy levels between the state of no charge, (i.e. "ground state"), and the highest permitted energy level where the spontaneous release of electromagnetic energy occurs.

In the thunder cloud also, the static charge slowly builds up until a critical or threshold charge is reached where the cloud can no longer hold the charge, and then it is spontaneously released as a giant spark or lightning bolt. It is interesting to note that electrically charged clouds can generate and radiate electric energy fields even before a lightning discharge. (This "threshold crisis" limit and spontaneity is typical of quantum phenomena, i.e. slow increase followed by very rapid 'spontaneous' decrease).

If such clouds, while being charged - are impinged upon by sound waves such as thunder, explosions, or aircraft noises etc., they will vibrate in resonance, - being virtually a gas (i.e. water vapor, mist, etc..). The clouds then being surrounded by an electric field, their charged particles (ions and water droplets) will in turn emit and radiate electric waves at the same audio frequency. (See Diagram 06). [see note].

In the old days of black and white television, the screen would sometimes show a 'ghost' or double image of a picture. This was due to the set receiving the original signal and in addition the reflected or re-broadcast signal from another source. In certain weather conditions, this other source would be the electrically charged clouds overhead.

(It is also conceivable therefore that electrically charged storm clouds separated by miles of sky could be used to communicate by means of such sound ~ electric waves, and indeed could be used to relay such electric wave signals in the same manner that a satellite relays radio communication). In other words, an electric storm would function as a relay station for the electric waves from the ground to another storm, and then again to a local ground receiver many miles away.

A similar mechanism occurs in the atom's shells and especially the valence shell. There are various energy levels in the shell(s) during the accumulation of energy between the first moment of discharge of a photon (when the shell drops to the ground state) and the next spontaneous release of a light quantum where the shell has again attained the fully energized quantum state.

It is these intermediate levels of valence shell "electro-tension" energy which exist and therefore determine the frequency of the emitted E-wave, so that the higher the energy level of the emitting atomic shell (like the tightness of a drum's membrane), the higher is the emitted 'note', i.e. frequency of quantum q. Energy. Higher frequencies are also emitted (and absorbed) by the inner shells.

Another analogy is that of an inflated balloon. If the balloon is half inflated and then subjected to ambient noise in the environment, it will resonate and vibrate to the lower sound frequencies and will emit sound of the same frequencies when struck. If the balloon is now inflated fully so that the membrane is tauter, it will now resonate to and vibrate with higher audible frequencies. The frequencies will also depend upon the type of atom, i.e. gold, copper, aluminium, quartz, etc..

For example one may have two identical balloons fully inflated, yet one will have a higher frequency of vibration than the other. This is because one may be a smaller balloon to begin with yet which has been over-inflated to match the size of the second one, and so on.

Electric waves of sufficient intensity will ionize a gas so that it then becomes an electronic plasma. Generally the higher the frequency of the emitted wave energy, the more efficiently is the gas ionized. The frequency is important due to the electron resonances in the gas atoms (and free electrons) to the incident electric waves or particles.

Ultra-violet light and X-rays are good ionizing radiations. In fact, any radiation with frequencies higher than X-rays will be very efficient ionizers, and quantum Energy of course falls into this category. Not only does q. Energy represent an excellent ionizer of gas/plasma but conversely - a plasma is also an efficient producer and generator of quantum Energy.

SUMMARY: the various factors which contribute to the range of generated q. Energy frequencies are as follows. The smaller the electron shell in an atom, the higher the emitted frequency of q. Energy. Likewise, frequency also depends upon the energized state of the particular shell in question. Hence we can see that a complete range of Energy may be produced by a small number of atoms in a mass. This is useful to know of course for it means that we can select those frequencies which we require.

NOTE: in fact it would be very easy to devise your own instrument for detecting such sounds. If you have a harp or guitar, etc., and are able to electro-statically charge it, some of the taut strings will resonate to such electric waves in the atmosphere and thereby emit a sound corresponding to these radiations. A cardboard box covered with aluminium foil and likewise charged will also work. If you have another such box, they may be used to communicate with each other. (Caution: Electricity can be dangerous, - take the advice of an expert!).

168 ~ QUANTUM ENERGY WAVE CHARACTERISTICS

Another major difference between electro-magnetic energy and q. Energy is the respective emitted wave-forms. Em. energy as we know it exists primarily and naturally as a dipole wave-form. This is to say that regardless of the emitted wave geometry (i.e. sine-wave, square-wave, saw-tooth wave, etc.) the complete wave cycle itself has two magnetic polarities. An emitted radio wave is generated by the oscillating (i.e. AC) movement of electrons in a conductor. This two-way movement creates two opposite magnetic polarities in the wave-form cycle. [see note].

A visible photon is created by the movement of an electron from one energy band in an electron shell to a lower one. Evidently, the electron behaves in a manner which is equivalent to its oscillating back and forth between the two bands - in order to generate the photon. An analogy of this is the back and forth motion of electrons (and protons) in a lightning strike. We find this dipole phenomenon at all levels of Em activity.

Quantum Energy on the other hand is peculiar in that this dipole characteristic does not exist since the Energy is 'uni-polar'. This unipolar geometry can easily be duplicated with Em expanding wave radiation such as radio waves. If the full bipolar cycle is rectified in the conductor, the result is a DC pulsed flow. In this event, the successive radio-waves are all magnetically polarized in the same direction. As a result, the radiation is aided in expansion due to the simple fact that similarly polarized magnetic fields repel, they are not mutually attractive. [see note].

In practical terms, this means that less input energy is required to create a broadcasted wave-train. The generated and emitted wave-form then becomes – not a bipolar wave form (e.g. sine-wave) - but a unipolar "garland" wave type whose energy ranges from zero to its energy peak and back to zero again.

NOTE: the dipole characteristic that we are all familiar with in electronic technology is due to the fact that the electronic wave emissions produced are done so with alternating vectors of electronic flow. In a pulsed direct current flow however the situation is different. Radio-waves can still be generated, but the magnetic polarity of the wave cycle does not reverse. This means that instead of a North-South magnetic field during one cycle, we would have a North-North field, for example. This is actually less efficient than AC, but what the heck - if you have power to waste, a pulsed DC mode is a lot simpler.

Δ

BIOGRAPHY

Stanley North presently lives in London, Ontario, Canada. At an early age, in the nineteen-seventies, he became interested in the 'New Age' subjects which were being promoted in the popular media such as so-called 'Earth energies', 'pyramidology', UFO's, life on other planets, the hollow earth/moon theories, Poltergeist phenomena, and so on. This initiated a process of research for unusual energy forms including a possible solution to the enigma of gravity. He became convinced early in his career that indeed there must be a levitating or "anti-gravity" force, especially after perusing the many corroborated reports of unusual *Poltergeist* phenomena. Armed with this conviction and with the conviction also that a solution and a means to generating this energy could be eventually found, he dedicated himself to its discovery.

He has studied philosophy, psychology, and sociology, as well as quantum physics, electronics, astrophysics, high energy electro-dynamics, para-physics (including Poltergeist and ball-lightning activities) travelled extensively around the world, and has taught English as a second language in the Far East. He was especially impressed with the culture of Japan with its highly structured and organized social system, and the good manners with which every Japanese is inculcated.

It also became clear to him that Asians generally were convinced of the reality of a form of energy with which we in the West are completely unfamiliar. This energy is known as *Ch'i* in China and '*Ki*' in Japan. So real is this general belief there that the entire social structure is partly based upon this reality, and this is even truer in China.

His research and studies of almost thirty years in varied subjects has culminated in the book which you now hold in your hands.

You will note that the book cover displays two other names as co-authors. In fact, these names are word plays and were placed there to amuse the reader. The second name can be reversed to yield *Nicola Tesla* which the author believes to be one of the greatest scientist/inventors who ever lived. The third name - 'Peter M. South', is an anagram for *Prometheus*, the Greek Hero who brought fire to Mankind.

32